How does behaviour develop in humans and animals? What are the causal mechanisms governing this behaviour? These important questions are addressed in this book. All the significant conceptual and empirical advances in the study of behavioural development are discussed in this volume by a wide range of scientists from different disciplines. A special feature of the approach taken here is that learning, as studied by experimental psychologists, is considered to be one process contributing to the development of the individual from conception to death. The development of the brain is also discussed in relation to behavioural processes.

The book is dedicated to Jaap Kruijt, whose pioneering approach to the study of causal mechanisms of behavioural development has inspired many workers in this field. It will be an invaluable resource for advanced undergraduates, graduate students, and researchers in animal behaviour, psychology, and neuroscience.

CAUSAL MECHANISMS OF BEHAVIOURAL DEVELOPMENT

CAUSAL MECHANISMS
OF BEHAVIOURAL
DEVELOPMENT

Edited by
JERRY A. HOGAN
Department of Psychology, University of Toronto

and

JOHAN J. BOLHUIS
Department of Zoology, University of Cambridge

CAMBRIDGE
UNIVERSITY PRESS

Published by the Press Syndicate of the University of Cambridge
The Pitt Building, Trumpington Street, Cambridge, CB2 1RP
40 West 20th Street, New York, NY 10011-4211, USA
10 Stamford Road, Oakleigh, Melbourne 3166, Australia

© Cambridge University Press 1994

First published 1994

Printed in Great Britain at the University Press, Cambridge

A catalogue record for this book is available from the British Library

Library of Congress cataloguing in publication data

Causal mechanisms of behavioural development/edited by Jerry A.
Hogan and Johan J. Bolhuis.
p. cm.
Festschrift for Jaap Kruijt.
Includes bibliographical references and index.
ISBN 0-521-43241-3 (hardback)
1. Genetic psychology. 2. Developmental psychology.
3. Psychology, Comparative. I. Hogan, Jerry A. II. Bolhuis, Johan J.
III. Kruijt, J. P.
[DNLM: 1. Behaviorism. 2. Behavior, Animal. 3. Human
Development. 4. Motor Skills. 5. Mental Processes. BF 199 C355
1994]
BF701.C38 1994
155.7–dc20 93-23580 CIP

ISBN 0 521 43241 3 hardback

TAG

Contents

Contributors

Gerard P. Baerends
University of Groningen, Zoological Laboratory, PO Box 14, 9750 AA Haren, The Netherlands

Peter D. Balsam
Department of Psychology, Barnard College, Columbia University, 606 West 120 Street, New York, NY 10027, USA

Kent C. Berridge
University of Michigan, Department of Psychology, Neuroscience Laboratory Building, 1103 East Huron Street, Ann Arbor, MI. 48104–1687, USA

Hans-Joachim Bischof
Lehrstuhl für Verhaltensphysiologie, Universität Bielefeld, Postfach 8640, 4800 Bielefeld 1, Germany

Elliott M. Blass
Department of Psychology, Cornell University, Ithaca, NY 14853–7601, USA

Johan J. Bolhuis
University of Cambridge, Department of Zoology, Downing Street, Cambridge CB2 3EJ, England

Nicky S. Clayton
University of Oxford, Edward Grey Institute, Department of Zoology, South Parks Road, Oxford OX1 3PS, England

Timothy J. DeVoogd
Department of Psychology, Cornell University, Ithaca, NY 14853, USA

Alison S. Fleming
Department of Psychology, Erindale College, University of Toronto, Mississauga, Ontario, Canada L5L 1C6

Ton G. G. Groothuis
University of Groningen, Zoological Laboratory, PO Box 14, 9750 AA Haren, The Netherlands

Jerry A. Hogan
University of Toronto, Department of Psychology, Toronto, Canada M5S 1A1

Mark H. Johnson
Carnegie Mellon University, Department of Psychology, Pittsburgh, Pennsylvania 15213, USA

John L. Locke
Neurolinguistics Laboratory, Massachusetts General Hospital, 101 Merrimac Street, Boston, MA 02114, USA

Ian P. L. McLaren
Department of Psychology, University of Warwick, Coventry CV4 7AL, England

David F. Sherry
Department of Psychology, University of Western Ontario, London, Canada N6A 5C2

Sara J. Shettleworth
University of Toronto, Department of Psychology, Toronto, Canada M5S 1A1

Rae Silver
Department of Psychology, Barnard College, Columbia University, 606 West 120 Street, New York, NY 10027, USA

Carel ten Cate
University of Leiden, Zoological Laboratory, Kaiserstraat 63, 2311 GP Leiden, The Netherlands

Foreword
Introducing Jaap Kruijt

GERARD P. BAERENDS

Unravelling the causal processes through which species-characteristic behaviour develops in the individual is one of the most important issues of ethology. This analysis is not only of purely theoretical interest, but has an applied value as well in the prevention and cure of behavioural disturbances. Jaap Kruijt's retirement at the end of the academic year 1992/93 concludes a period of almost 40 years in which he devoted himself to developing this branch of ethology within the framework of the Zoological Laboratory of the University at Groningen (The Netherlands). By focusing his research on relatively complex social behaviour and by strongly opposing the dichotomous classification of behaviour into innate and learned, Kruijt has rendered an internationally unique contribution to the study of behavioural development. This makes it worthwhile to dwell briefly upon some phases of his own ontogeny.

Jakob Pieter Kruijt was born in Amsterdam in 1928. In his boyhood he developed a strong interest in wildlife, influenced – like many of us – by the writings of two Dutch naturalists, E. Heimans and Jac. P. Thijsse, and also by participating in the activities of the Dutch Youth Association for Nature Study. This interest made him ripe for the study of biology, which he started in 1946 at the University of Utrecht. At that time zoological research in Utrecht was mainly concerned with laboratory work; comparative and developmental physiology were strongly represented. In both fields Jaap did experimental work for his master's degree, which he obtained in 1953. However, during these years in Utrecht his earlier acquired appetite for observing intact animals in their normal surroundings was not extinguished. It led him – as the final requirement for his degree – to join a field project supervised by me in Groningen. The project was meant to explore the information-processing mechanism enabling a herring gull to distinguish its eggs as objects for incubation. It was a continuation of pilot studies started in the mid-thirties by Niko Tinbergen at the University of Leiden and in which I had had the privilege of assisting him (Kruijt & Hogan-Warburg, 1982). Having Jaap in our group was such a great pleasure that means to continue our co-operation were sought.

Another item on our programme dealt with the phylogenetic origin of display activities. We were interested in using the 'conflict hypothesis' to explain the ways in which the radiating differentiation of the forms of homologous displays in different closely related species comes about. This hypothesis had been developed in Niko Tinbergen's school to explain how the forms of visual displays may have evolved. It suggests that displays have originated from the short-lasting and apparently irrelevant acts that occur when different, largely incompatible, behavioural systems (such as those subserving attack, escape or sexual behaviour) are simultaneously activated in the animal. Type and form of the activity induced in this way would be determined by the degree to which each of the interacting systems is involved, i.e. the motivational balance of the animal during performance. We hoped to find a relationship between differences in the form of homologous displays in related species and the species-characteristic variance of their motivational balances. We supposed that the variance arose because of the different ecological conditions in the species' habitat. For this enterprise we needed species-rich taxonomic units, and consequently chose the *Cichlidae* among fishes and the *Galliformes* among birds. When a grant for this research was obtained in 1954, I was happy to find Jaap willing to start with the birds. After the expiration of this grant, 3 years later, Jaap was appointed a permanent member of staff. Eventually, he became my successor.

In accordance with the plans, Jaap began with observing, describing and comparing the displays of as many different species as he could lay hands on. However, when he tried to relate the occurrence of these displays to specific equilibria of interacting tendencies, he felt dissatisfied with the methods being used for this purpose, and looked for better ones. In particular he was sceptical about arriving at an insight into the motivational background of displays if only the adult stage were studied. He reasoned that the complexity of social behaviour and its complete expression were likely to be built up gradually in the individual under the influence of (at least partly changing) internal and external stimulation, according to developmental programmes which are ultimately based on information in the genome. Consequently he elected to give priority to the ontogenetic aspect of displays, which was only feasible if all efforts would be concentrated on a single species. The choice fell on the Burmese red junglefowl. By shifting the emphasis of his research towards ontogeny, Jaap founded his own niche in the laboratory. Rapidly, the study of behavioural development became an essential aspect of our institute. We need not regret that this necessarily delayed the development of the original plans.

When constructing his theory on the nature of 'instinct', Lorenz (1937, 1965) had strongly stressed – in opposition to extreme Watsonian behaviourism – the importance of species-specificity in animal behaviour and its ultimate basis in the genome. This had made him feel a need for distinguishing between 'innate' and 'learned' behaviour, a practice generally accepted in ethology until the early fifties. However, shortly before Jaap started the junglefowl study, Lehrman (1953) published his sharp critique of Lorenzian ethology, and in particular of Lorenz's endeavour to label behaviour as either innate or learned. Jaap's background in developmental physiology facilitated his receptivity to this criticism. He had learned to see ontogenetic development as a process of close *interaction* of genetic and environmental factors, and strongly disagreed with the Lorenzian view that the ontogeny of behaviour would proceed through superposition of learned *components* on innate ones. Actually, he totally rejected Lorenz's dichotomous classification of behaviour as 'innate' or 'learned'. He considered it a serious impediment to all efforts to unravel the developmental process because it sidesteps the issue of how genetic information becomes expressed in the course of development. In particular, Kruijt (1968, 1978) warned against labelling behavioural *elements* as 'innate' or 'learned' on the basis of results from experiments in which young animals were reared in the absence of environmental factors which the human observer considered to have functional relevance in natural circumstances ('Kaspar Hauser' experiments).

Kruijt (1964) studied form and occurrence of social behaviour of the junglefowl from hatching up to adulthood, using small groups of artificially incubated chicks. He found escape behaviour to be present from the time of hatching. At first it was a reaction to relatively strong stimulation, but it was not a response to other chicks until after the first 2 weeks. Aggressive pecking at other chicks appeared in the third week. From that time on, attack and escape often occurred in alternation. This behaviour was usually combined with the performance of interruptive activities which could be interpreted as resulting from simultaneous arousal of the incompatible tendencies to attack and to flee. After the sixth week, the agonistic contacts became more severe, and the display repertoire became extended with motor patterns which could again be attributed to interactions of the agonistic tendencies. At that time also, a tendency to copulate became expressed in the males; it exerted an inhibiting influence on the tendencies to attack and to flee and thus affected display behaviour. These observations supported the conflict hypothesis: they made it likely that the interactions of motivational factors, postulated to have given rise to the evolution of social displays, are proximately involved in the course of ontogeny (Kruijt, 1962a, 1964). They further inspired penetrating research by Jaap and his associates into the causation of some particular displays and interruptive activities (e.g. Feekes, 1972; Groothuis, this volume).

By manipulating the social environment in which chicks were raised and then testing them as adults under normal conditions, Jaap tried to obtain information about the effect of contact with conspecifics on the course of development of social behaviour. Raising male chicks in visual isolation from conspecifics had no effect on their ability to perform the motor patterns of the displays (Kruijt, 1962b, 1964, 1971). However, depending on the length of this treatment and the age of the chick, visual isolation caused disturbances in the capacity of a bird to establish the social relations normal for the species. The organisation among the components of behaviour was often disturbed or the behaviour was oriented inadequately. Irregular fits of attack and escape behaviour were seen frequently. Comparable experiments with rhesus monkeys, independently carried out by Harlow and his co-workers in the same period, yielded similar effects. These findings made Jaap recognise that interaction of the developing animal with various elements of the environmental situation plays an essential role in the developmental programmes. Feedback from responses of a bird to specific stimulation appeared to influence the impact of such stimuli on the developmental processes involved (Kruijt, 1985).

Having recognised this, Jaap looked at studies tracing the origin of an animal's responsiveness to the appearance of its parents, prospective sex partners or conspecifics in general, and wondered whether sufficient account had been taken of a possible relationship between interacting with an object and learning about its features. Especially in experiments on filial and sexual imprinting, inanimate and therefore unresponsive target objects have often been used in a standardised static environment. Such unnatural conditions might have mistakenly led to suggestions about the developmental programme, in particular with respect to the existence and importance of predisposed preferences of the subject for specific features of the target, or absence of responsiveness to others. This notion made Jaap scrutinise the results of the experiments on sexual imprinting carried out by Schutz (1965) in mallards and by Immelmann (1972) in zebra finches. Both studies had not so much been aimed at tracing the developmental process of partner recognition as at establishing whether the information present about the appearance of the partner should be called learned or innate.

Schutz had watched the partner choice of male and female mallards that had been raised with foster mothers of alien duck species and then released into a mixed population. He concluded that in the male mallard, partner recognition is predominantly based on imprinting, but in the female, it is based on a predisposed bias for species-specific features of the male. The fact that not all females chose a conspecific male made Jaap suspicious about Schutz's conclusion. He realised that in the test situation, females were likely to have received most courtship from drake mallards. Using a wild and a white mallard strain, he and some co-workers (Bossema & Kruijt, 1982; Kruijt, Bossema & Lammers, 1982) repeated the experiments under more stringently controlled conditions. They found that females chose between a wild or a white male on the basis of who was most actively courting her. If both types of male were equally active, females tended to opt for the rearing strain. Therefore, the apparent sexual difference in imprintability as reported by Schutz, must have been an artefact due to the testing situation.

Immelmann had also, by means of a cross-fostering technique, demonstrated imprinting in zebra finches: male zebra finches raised by Bengalese finches strongly preferred to court females of the foster species. On the other hand, such a male occasionally was found to prefer females of its own species. Moreover, males raised by a mixed pair, consisting of a Bengalese and a zebra finch, predominantly courted zebra finch females. From these latter facts, Immelmann concluded that zebra finch males also possess a predisposed bias for the appearance of their own species. However, Jaap's pupil, Carel ten Cate, found that zebra finches care for their young more intensively than do Bengalese finches: thus the young would have had more interactions with a zebra finch foster parent than with a Bengalese finch foster parent. Ten Cate manipulated the amount of interaction with the parents, suppressing it in the zebra finch parent and stimulating it in the Bengalese one, and found that this experience, and not a predisposed bias for zebra finch features, made the adult zebra finch males preferably court their own kind. The occasional courting of zebra finch females by males reared by Bengalese pairs was found to be due to experience with the other young in the nest; it did not occur in males raised alone or with only one sibling (ten Cate, 1982; Kruijt, ten Cate & Meeuwissen, 1983).

Early in his career, Jaap combined his laboratory studies on development with field studies of the behaviour of black grouse on their leks. He and his students published one of the first modern accounts of the organisation and function of the complex social behaviour of a species in its natural environment (Kruijt & Hogan, 1964; Kruijt, De Vos & Bossema, 1972; De Vos, 1979). However, the need for experimentation under well-controlled conditions forced Jaap to spend most of his time in the laboratory. It also made him contribute considerably to the design of the animal observatory in Groningen, in which all sorts of animals could be kept and reared under semi-natural conditions and observed closely. In all his work, Jaap Kruijt remained keenly aware of the importance of insight into the functioning of the animal in its biotope. He also continually emphasised the importance of apparently trivial or subtle details for understanding behavioural development. These values, in combination with his critical attitude, he has passed on to a new generation of behavioural scientists.

References

Bossema, I. & Kruijt, J. P. (1982). Male activity and female mate acceptance in the mallard (*Anas platyrhynchos*). *Behaviour*, 79, 313–324.

De Vos, G. J. (1979). Adaptedness of arena behaviour in black grouse (*Tetrao tetrix*) and other grouse species (*Tetraoninae*). *Behaviour*, 68, 277–314.

Feekes, F. (1972). Irrelevant ground pecking in agonistic situations in Burmese red jungle fowl (*Gallus gallus spadiceus*). *Behaviour*, 43, 186–326.

Immelmann, K. (1972). Sexual and other long-term aspects of imprinting in birds and other species. *Advances in the Study of Behavior*, 4, 147–174.

Kruijt, J. P. (1962a). On the evolutionary derivation of wing display in Burmese Red Junglefowl and other gallinaceous birds. *Symposia of the Zoological Society of London*, 8, 25–35.

Kruijt, J. P. (1962b). Imprinting in relation to drive interaction in Burmese Red Junglefowl. *Symposia of the Zoological Society of London*, 8, 219–226.

Kruijt, J. P. (1964). Ontogeny of social behaviour in Burmese red junglefowl (*Gallus gallus spadiceus*). *Behaviour supplement* 12, 1–201.

Kruijt, J. P. (1968). *Instinct in de hoenderhof*. Openbare Les (Inaugural Lecture). University of Groningen. Groningen: Wolters-Noordhoff.

Kruijt, J. P. (1971). Early experience and the development of social behavior in junglefowl. *Psychiatria, Neurologia, Neurochirurgia* 74, 7–20.

Kruijt, J. P. (1978). Die Jugendentwicklung von Verhalten. In *Psychologie des 20. Jahrhunderts*, Bd. 6., ed. R. A. Stamm & H. Zeier, pp. 231–247. Zürich: Kindler Verlag.

Kruijt, J. P. (1985). On the development of social attachments in birds. *Netherlands Journal of Zoology*, 35, 45–62.

Kruijt, J. P., Bossema, I. & Lammers, G. J. (1982). Effect of early experience and male activity on mate choice in mallard females (*Anas platyrhynchos*). *Behaviour*, 80, 32–43.

Kruijt, J. P., De Vos, G. J. & Bossema, I. (1972). The arena system of Black Grouse. *Proceedings of the 15th International Ornithological Congress*, 399–423.

Kruijt, J. P. & Hogan, J. A. (1964). Social behaviour on the lek in black grouse, *Lyrurus tetrix tetrix* (*L*). *Ardea*, 55, 203–240.

Kruijt, J. P. & Hogan-Warburg, A. J. (1982). The effect of different egg features for the egg-retrieval response: Ch. 5: speckling. In *The herring gull and its egg*, ed. G. P. Baerends & R. H. Drent., *Behaviour*, 82, 117–136.

Kruijt, J. P., ten Cate, C. J. & Meeuwissen, G. B. (1983). The influence of siblings on the development of sexual preferences in zebra finches. *Developmental Psychobiology*, 16, 233–239.

Lehrman, D. S. (1953). A critique of Konrad Lorenz' theory of instinctive behavior. *Quarterly Review of Biology*, 28, 337–363.

Lorenz, K. (1937). Über die Bildung des Instinktbegriffes. *Naturwissenschaften*, 25, 289–300, 307–318.

Lorenz, K. (1965). *Evolution and Modification of Behaviour*. Chicago: University of Chicago Press.

Schutz, F. (1965). Sexuelle Prägung bei Anatiden. *Zeitschrift für Tierpsychologie*, 22, 50–103.

ten Cate, C. (1982). Behavioural differences between zebra finch and Bengalese finch (foster) parents raising zebra finch offspring. *Behaviour*, 81, 152–172.

Preface

This book is dedicated to Jaap Kruijt. His pioneering approach to the study of the causal mechanisms of behavioural development has inspired many, as is obvious in many of the contributions to this volume. We have both had the pleasure of working with him, as postdoctoral fellow and as doctoral student, respectively. Chapters in this book were commissioned from a wide range of scientists interested in behavioural development. Some of these have been former students or colleagues of Jaap Kruijt, but many are workers in the field with interests and approaches that are compatible with the approach Kruijt has taken to development. Each author was asked to review a specific area of the field in such a way that we think almost all the important conceptual and empirical advances in the study of development have been covered in one chapter or another. A special feature of the approach taken here is that learning, as studied by experimental psychologists, is considered to be one process contributing to the development of the individual from conception to death. Learning is seen to play an important part in development in several chapters in the book, and the relation of learning to other developmental processes is discussed specifically in the last three.

We are grateful to Bob Lockhart and Rob Honey who have read several of the chapters in manuscript form and commented on them. Thanks are due to Alan Crowden and Sara Trevitt of Cambridge University Press, who have been extremely helpful throughout the development of this book.

Toronto,   J. A. H.
Cambridge,   J. J. B.
February, 1993

Part one

Introduction

1

The concept of cause in the study of behavior

JERRY A. HOGAN

The title of this book, *Causal Mechanisms of Behavioural Development*, comprises four substantive words, all of which are relatively common in English. They are words that most speakers of the language, and certainly most scientists, would probably say they know the meaning of. Yet all four can be misleading for that very reason; all of these words are used in many different ways, and few people are actually aware of these differences. At the risk of seeming pedantic, I shall devote the first few pages of this book to a discussion of some distinctions that should make clear how we are using these words. Some of this discussion will tend to the philosophical, but I trust that most readers will find it useful for putting the subsequent chapters into a general perspective. I shall take each word in turn.

Cause

In his well-known article 'On aims and methods of ethology', Tinbergen (1963) stated that there are four kinds of questions one can ask about biological phenomena: causation, survival value, ontogeny, and evolution. These distinctions are important, and have had a great influence on students of animal behavior. However, as I have pointed out elsewhere (Hogan, 1984), these questions do not cover all the important aspects of behavior. A more general classification of questions can be derived from some distinctions made originally by Aristotle.

In discussing physics and metaphysics, Aristotle pointed out that one and the same thing could be described or explained in four different ways. These types of explanation have been called 'causes' and are usually listed as: 1) material, 2) efficient, 3) formal, and 4) final cause. A standard example is the description of a chair: it may be made of wood (cause 1),

3

have been manufactured in such and such a way by Mr Smith (cause 2), have four legs, a seat, and a back arranged in a particular way (cause 3), and be used to sit in (cause 4). Just as Tinbergen argued that our understanding of any biological phenomenon is increased by asking all four questions, so Aristotle argued that our understanding of any phenomenon is increased by considering all four of its causes.

If we apply Aristotle's classification to behavior, we discover that it can subsume Tinbergen's four questions, and that additional questions arise as well. It will be helpful first to translate Aristotelian terminology into terms that are more suitable for the analysis of behavior. I shall talk about 1) the matter, 2) the causation, 3) the structure, and 4) the consequences of behavior.

For most purposes, the matter of behavior is the same regardless of the kind of behavior being considered, so this question is primarily relevant to the definition of behavior itself. I define behavior as the expression of the activity of the nervous system. This definition raises a number of issues that I discuss in a later section. I shall here consider some of the implications of Aristotle's other three distinctions for the analysis of behavior, and also indicate some of the relationships among the questions that can be asked. At the end of this section I return briefly to the Aristotelian notion of 'cause'.

Causation of behavior (efficient 'cause')

Aristotle defines efficient 'cause' as 'the primary source of the change or coming to rest; e.g. the man who gave advice is a cause, the father is cause of the child, and generally what makes of what is made and what causes change of what is changed' (1947, p. 122). In the remainder of this chapter I shall use the terms cause and causal, without quotation marks, to refer to Aristotle's efficient 'cause', but retain quotation marks for Aristotelian usage. The causes of behavior include stimuli, the internal state of the animal, various types of experience the animal has had during its development, as well as the genes with which it is endowed. Understanding the mode of action of these causal factors, including their interaction with each other, is the primary goal of a causal analysis of behavior. I find it useful to distinguish among three types of causation in terms of the changes that are of interest.

The first type of causation is *motivational*, in which the immediate effects that causal factors have on behavior are of primary interest. This use of the word corresponds to Tinbergen's causal question; it is also similar to the phrase 'proximate causation', as used by many biologists (Baker, 1938).

An electric shock causes the rat to jump, hunger causes the chicken to eat, advice causes the man to spend his money in a certain way. Note that the term motivation is used here in a much broader way than is often the case in that it refers to external as well as internal causes of behavior. In fact this use of the term corresponds to its original sense: that which causes motion.

A second type of causation is *developmental* (Tinbergen's ontogenetic question), in which changes in behavior during the course of an individual's lifetime are of primary interest. What effects does the presentation of a small red cube to a young chick have on its behavior when it grows up, and how do those effects come about? Will the adult chicken prefer to mate with the red cube rather than with a conspecific? The causal factors in development are the same stimuli, internal states, and so on that are important for motivation, but one is primarily interested in their longer term effects. The red cube certainly affects the behavior of the young chick when it is presented, as does the father of the child affect the behavior of the mother prior to conception, but it is the origins and course of development of adult sexual behavior or of the child that one wishes to explain.

It should be noted that distinguishing between motivation and development on the basis of time scale (e.g. Tinbergen, 1972) is not fundamental because a more basic distinction is whether the changes brought about by the causal factors are reversible or permanent (i.e., result in a change in behavioral structure – see below). For example, the increase in gonad size in some birds and fish as a result of increasing day length is generally considered to be a problem in motivation, whereas one-trial learning to avoid electric shock in a rat is a problem in development. The former changes occur over days or weeks, whereas learning occurs in a few seconds. On the other hand, reversible and permanent are themselves relative concepts and in any particular example the two can fade into each other. Thus, it is probably best to consider that motivational and developmental changes often reflect a different time scale, but more basically reflect underlying causal mechanisms that are more or less reversible.

A third type of causation is *phylogenetic* (Tinbergen's evolutionary question), in which changes in behavior over generations are of primary interest. Here the concern is what effect causal factors have on the genes – primarily on genetic recombination and mutation – because in most species only genes can be passed on from one generation to the next. However, in species in which specialized modes of conspecific communication have developed, as in humans, there is no reason to exclude culturally transmitted information as a causal factor in the phylogeny of

　　　　　　　　　　Jerry A. Hogan

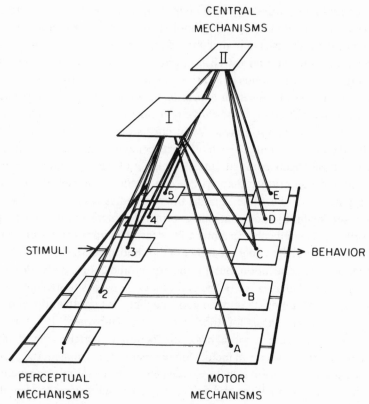

Figure 1.1 Conception of behavior systems. Stimuli from the external world are analyzed by perceptual mechanisms. Output from the perceptual mechanisms can be integrated by central mechanisms and/or channeled directly to motor mechanisms. The output of the motor mechanisms results in behavior. In this diagram, Central mechanism I, Perceptual mechanisms 1, 2, and 3, and Motor mechanisms A, B, and C form one behavior system; Central mechanism II, Perceptual mechanisms 3, 4, and 5, and Motor mechanisms C, D, and E form a second behavior system. 1-A, 2-B, and so on can also be considered less complex behavior systems. From Hogan (1988). Reproduced with permission.

behavior. Finally, it may be surprising to some readers that I consider phylogenetic changes to be a causal problem. I discuss this issue later in the section on the relationship between cause and function.

Structure of behavior (formal 'cause')

Aristotle's formal 'cause' refers to the parts and the relations among them. A chair consists of four legs, a seat, and a back arranged in a particular way with respect to each other. By analogy, one can also describe the form or *structure* of behavior. What are the parts or units of behavior? I have proposed motor, central, and perceptual mechanisms as the basic units of behavior (Hogan, 1988). All of these parts are viewed as corresponding to structures within the central nervous system. They are conceived of as consisting of some arrangement of neurons (not necessarily localized) that acts independently of other such mechanisms. They are called behavior mechanisms because their activation results in an event of behavioral interest: a specific motor pattern, an identifiable internal state, or a particular perception. This conception can also include entities such as ideas, thoughts, and memories, which are the cognitive structures proposed by many psychologists.

Behavior mechanisms can be connected with one another to form larger units called behavior systems, which correspond to the level of complexity indicated by the terms feeding, aggressive, incubation, and play behavior (Baerends, 1970; 1976; Hogan, 1988). The organization of the connections among the behavior mechanisms determines the nature of the behavior system. A behavior system can be defined as any organization of perceptual, central, and motor mechanisms that acts as a unit in some situations (Hogan, 1971; Hogan & Roper, 1978). A pictorial representation of this definition is shown in Figure 1.1. The first part of this definition is structural and is basically similar to Tinbergen's definition of an instinct (1951, p. 112); the second part is causal. Ideally, it might be desirable to have a purely structural definition, but at present we can only infer structure from causal analysis.

Many of the chapters in this book deal specifically with the development of behavior mechanisms and systems. As we shall see, there are various levels of perceptual and motor mechanisms, and the connections among them can become highly complex. A diagram such as Figure 1.1, if expanded to encompass all the facts that are known, would soon become unmanageable. In the extreme it would become congruent with a wiring diagram of the brain. The main use of such a diagram is to give direction to our thinking.

Consequences of behavior

The consequences, or effects, of a behavior such as pecking, for example, may include closing a microswitch in a Skinner box, enjoying the performance of pecking itself, reducing hunger, providing nutrients necessary for egg production, or increasing fitness. Closing a microswitch and enjoying the activity are immediate consequences of performing the behavior, hunger reduction takes many minutes, producing eggs takes hours or days, and increasing fitness can only be measured over generations.

These various consequences introduce the problem of time scale (cf. Zeiler, 1992), which we have already met when discussing the causes of behavior. Time itself is no more satisfactory a criterion for distinguishing among consequences of behavior than it was for distinguishing among causes of behavior. A better criterion might be the type of changes being considered. Thus, one could distinguish motivational, developmental, and phylogenetic consequences, depending on the question of interest. Eating reduces hunger, reinforces a stimulus–response association, and allows an animal to pass its genes to future generations.

Alternative terms could also be used to describe the different effects of behavior. For example, feedback is a concept that includes many motivational consequences of behavior, and reinforcement would be a primary developmental consequence of behavior. Most biologists would use the word function for the phylogenetic consequences of behavior (e.g. Hinde, 1975). In this context, function is synonymous with survival value, which is Tinbergen's remaining question, and (inclusive) fitness, that is, consequences that have been acted upon by natural selection. It is also similar to the phrase *ultimate causation* (Baker, 1938), but I return to this phrase later.

I have previously used the word function to refer to all these various consequences of behavior (Hogan, 1984; see also Zeiler, 1992), but I now think it is better to follow biological usage and restrict the word function to the phylogenetic consequences of behavior and use consequences as the general term.

Aristotle's final 'cause', and the problem of teleology

Aristotle defines final 'cause' as 'that for the sake of which', and uses the following example to illustrate his meaning: 'health is the cause of walking about. ("Why is he walking about?" we say. "To be healthy ...")' (1947, p. 123). Another example is provided by various answers to the question: Why did the chicken cross the road? To get to the other side; to avoid being run over; to get food; to increase the likelihood of passing on its genes to subsequent generations. These answers all specify the goal or purpose of the behavior, and all imply that some outcome that has not yet occurred is the cause of the behavior. That this approach is fallacious can be seen by considering, for example, that the chicken might be hit by a car when crossing the road, or that there might be no food on the other side when it gets there; the man might drop dead of a heart attack because of walking about too strenuously. *The outcome of behavior can never determine its occurrence.* At best, the outcome of behavior can be one of the determining (causal) factors of future occurrences of similar behavior.

All the examples given above are examples of teleology, 'the belief that natural processes are not determined by mechanism but rather by their utility in an overall natural design' (*American Heritage Dictionary*, 1969). In this sense, teleology does not lend itself to scientific investigation because it implies that the goal of the behavior (what it is designed for) has become its efficient 'cause'. Unfortunately, Aristotle's final 'cause' is basically a teleological concept, and the biologists' concept of *ultimate causation* (Baker, 1938) is often used teleologically as well. It is possible to rephrase both the questions and answers so that teleological implications are avoided, and one can then talk quite scientifically about the outcomes or consequences of behavior (cf. Pittendrigh, 1958). I return to this issue in the section on the definition of behavior.

Finally, it should be noted that some authors use the word teleology in a somewhat different sense from the definition given above (e.g. Mayr, 1988; Hopkins & Butterworth, 1990). Such authors consider that teleological concepts can be used to explain the occurrence of some behavior. It is my opinion that it is better not to use the word teleology when discussing causes of behavior, but further discussion of this issue is beyond the scope of this chapter.

Mechanism

The *American Heritage Dictionary* (1969) defines a mechanism as 'the arrangement of connected parts in a machine' or 'any system of parts that operate or interact like those of a machine'. It defines a machine as 'any system ... formed and connected to alter, transmit, and direct applied forces in a predetermined manner to accomplish a specific objective'. Clearly, a behavior system, as defined above, is a machine that directs internal and external causal factors that produce a specific outcome. (It is also clear that the phrase *causal mechanism* is redundant: mechanism implies cause.)

The problem of level of analysis

The concept *behavior mechanism* has been introduced above as the unit of analysis in the causal study of behavior. For example, in fowl, activation of the perceptual mechanism responsible for recognizing dust plus activation of the central mechanism responsible for accumulating motivational factors for dustbathing causes the motor mechanisms for dustbathing to be activated which in turn cause the observed motor patterns of dustbathing. This example is a causal account of the occurrence of dustbathing. Further causal analysis could investigate the properties of the stimuli that activate the dust-recognition mechanism, and the stimuli and chemicals that activate the central mechanism.

Some investigators, however, would argue that the word mechanism, with respect to behavior, implies analysis at a neural level. It can be seen that this is certainly not true in terms of the dictionary definition of mechanism. This confusion is one related to the level of analysis. For example, the ethological concept of releasing mechanism (a kind of perceptual mechanism) is one defined in terms of its consequences at the neural level: it is postulated that somewhere in the nervous system there must be an organization of neurons that is responsible for the properties attributed to releasing mechanisms. These properties could come about as a result of any of several different neural organizations. The actual neural organization and its location in the brain are not of special interest when analyzing behavior; what is of interest are the consequences of the activation of that mechanism. It is in this sense that von Holst & von St. Paul (1960) and Baerends (1976) refer to the functional organization of behavior. At the behavioral level, however, the concept of the releasing mechanism is a causal concept: it is postulated that activation of a particular releasing mechanism is one of the controlling (causal) factors for a certain behavior.

A further problem with assuming that mechanism implies analysis at the neural level is that the correspondence between the neural and behavioral levels is seldom, if ever, one to one (see Hogan, 1980, p. 6; Berridge, this volume, p. 174). Nonetheless, as several chapters in this book attest, neural and chemical analyses of behavior can provide important insights into the causal mechanisms of behavior. I return to this issue in the section on the definition of behavior.

The relation of cause and function

I have argued on several occasions that cause and function are logically independent concepts (Hogan, 1984, 1988). Nonetheless, the biological definition of function as 'consequences that have been selected' and current biological usage of the phrase 'ultimate cause' both imply some sort of causal relationship between cause and function, a relationship that I maintain is teleological (and un-Darwinian). Several authors have argued specifically, however, that such usage is not teleological (e.g. Daly & Wilson, 1983, p. 15; Alcock, 1989, Chapter 1). Their argument is that natural selection is a mechanism, and therefore a cause, of evolution. But here is where the confusion of levels of analysis arises. From the point of view of the individual (or the gene itself), the cause of evolution is the mutations or recombination of genes that bring about variability among individuals. Natural selection is the consequence or outcome of the fact that some individuals are more successful than others. Natural selection is only a mechanism of evolution at the level of the group or species, where it is a mechanism for changing the frequency of genes in the population pool. That is, in some particular environment, individuals with a certain genotype will be more successful than individuals with a different genotype. However, that environment does not cause the successful genotype to appear; it only maintains that genotype once it has been caused by other (causal) factors. The situation is similar to that in operant psychology. Reinforcement is an outcome when a response occurs; the response must occur first 'for other reasons' (Skinner, 1938).

I should make explicit here that although I believe many (reputable) scientists confuse the concepts of cause and function, it has not been my intention to denigrate any group of individuals, but rather to point out some conceptual fuzziness that arises from the way people use language. I also believe the study of function and the adaptive significance of behavior is an important and necessary activity of behavioral scientists. In my opinion, Tinbergen was correct to emphasize that a complete understanding of behavior requires answers to all his questions.

Behavior

As already mentioned, the definition of behavior is an issue that requires some comment. One definition that is used by some behavioral scientists, and corresponds in many ways to common sense, is that which an animal does, and what it does consists of muscular contractions and glandular secretions. This definition, however, does not include many phenomena, such as perceptions and feelings, that intuitively belong in the concept. Another definition brings in the concept of mind, because mind does include all the phenomena one expects. Although I am sympathetic to this solution, I prefer a more corporeal concept. I define behavior as the activity of the nervous system, which may be manifested as activity in muscles and glands (Hogan, 1984). This definition is closely related to Tinbergen's definition of an instinct (1951, p. 112) as 'a hierarchically organized nervous mechanism .... '. Tinbergen's definition is essentially the same as our definition of a behavior system. The activation of an instinct produces instinctive behavior and the activation of a behavior system produces behavior. In both cases, behavior is the expression of activity of the nervous system.

An important point to be made about this definition is that it does not imply that the study of behavior involves neurophysiology. The study of behavior is the study of the functioning of the nervous system and must be carried out at the behavioral level, using behavioral concepts, as I have argued above. Physiology in general and neurophysiology in particular may provide useful insights into the functioning of the nervous system, but the major concern of behavioral science is the *output* of the nervous system, manifested as perceptions, thoughts, and actions.

The content of behavior then becomes the behavioral structures discussed above: perceptual, central, and motor mechanisms. The activation of these mechanisms produces perceptions, ideas, memories, intentions, and the like, as well as feelings, emotions, and actions. Using this framework, it is possible to rephrase the answers to many 'why' questions so as to avoid teleology and the notion of purpose. For example, we can say the man is walking about because he believes that walking about promotes health, and he wants to be healthy. This belief and desire reflect the activation of specific cognitive mechanisms that exist before the behavior occurs, and can be considered its cause. In people, such beliefs and desires reflect cognitive structures that have been built up by the specific experience of the individual. The causal explanation for why the man is walking about would require an investigation into the developmental causes of those specific cognitive structures and the immediate causes of their activation.

This conception of cognitive structures allows a way to resolve some of the current controversies in the field of animal cognition (e.g. see Ristau, 1991; Beer, 1992; Sherry, this volume). We can begin with the notion that cognition only implies knowledge, and that knowledge is another way of saying that cognitive structures exist. I have already argued that activation of a cognitive structure is the cause of behavior. Insofar as ideas, beliefs, intentions, purposes, and the like are considered to be cognitive structures, the activation of the structures representing these entities can be a cause of behavior. In this framework the concept of consciousness becomes an epiphenomenon, an outcome of the activation of a cognitive mechanism, and definitely not a cause of behavior. Interestingly, this approach was already espoused many years ago by Huxley (1893) who wrote: 'The consciousness of brutes would appear to be related to the mechanism of their body simply as a collateral product of its working, and to be as completely without power of modifying that working as the steamwhistle which accompanies the work of a locomotive engine is without influence upon its machinery.' (p. 240).

It should be noted, however, that as with any cognitive structure, a causal analysis requires investigation of the developmental and phylogenetic causes of the structure, as well as the motivational causes of its activation. If we say only that the chicken crossed the road because it intended to get to the other side, we have merely restated the fact that it was seen to cross the road. A causal analysis would inquire into the origins of the intention (was it a product of phylogenetic experience or individual experience; and what were the selective forces or the particular experiences presumed to be necessary) and the causal factors activating the intention (what specific stimuli or internal state lead to activation of that specific intention). It should be clear that without such an analysis, the use of cognitive language is no advance on the behavioristic language that has been used in the past.

Development

Development is the study of changes in the structure of behavior. Kruijt (1964) proposed that, in young animals, the motor components of behavior often function as independent units, and that only later, often after specific experience, do these motor components become integrated into more complex systems such as hunger, aggression, and sex. Hogan (1988) generalized this proposal by considering that the units of structure are the various behavior mechanisms–perceptual, central, and motor–defined above. The study of development then includes changes in the organization of specific behavior mechanisms, as well as changes in the connections among these components. It is this general framework that has been used to organize this book.

Acknowledgements

Lidy Hogan-Warburg, Martin Daly, Robert Hinde, and Robert Lockhart read a draft of this chapter and made many useful comments on various issues discussed. They all disagree, sometimes quite strongly, with some of my opinions. Nonetheless, their input has helped me define my own position more clearly, and I thank them for their efforts. Robert Hinde drew my attention to the book by Butterworth & Bryant (1990). Perusing that book made me realize that there is still a long way to go before there is general agreement on many of the issues raised in this chapter.

References

Alcock, J. (1989). *Animal Behavior: An Evolutionary Approach.* Sunderland, Massachusetts: Sinauer.

Aristotle. (1947). *Introduction to Aristotle,* ed. R. McKeon. New York: Random House.

Baerends, G. P. (1970). A model of the functional organisation of incubation behaviour in the Herring Gull. In *The Herring Gull and Its Egg,* ed. G. P. Baerends & R. H. Drent. *Behaviour Supplement,* 17, 261–312.

Baerends, G. P. (1976). The functional organisation of behaviour. *Animal Behaviour,* 24, 726–738.

Baker, J. R. (1938). The evolution of breeding systems. In *Evolution: Essays on Aspects of Evolutionary Biology,* ed. G. R. de Beer, pp. 161–177. Oxford: Oxford University Press.

Beer, C. (1992). Conceptual issues in cognitive ethology. *Advances in the Study of Behavior,* 21, 69–109.

Butterworth, G. & Bryant, P., eds. (1990). *Causes of Development.* London: Harvester Wheatsheaf.

Daly, M. & Wilson, M. (1983). *Sex, Evolution & Behavior,* 2nd edn. Boston: Willard Grant Press.

Hinde, R. A. (1975). The concept of function. In *Function and Evolution in Behaviour,* ed. G. P. Baerends, C. Beer & A. Manning, pp. 3–15. Oxford: Oxford University Press.

Hogan, J. A. (1971). The development of a hunger system in young chicks. *Behaviour,* 39, 128–201.

Hogan, J. A. (1980). Homeostasis and behaviour. In *Analysis of Motivational Processes,* ed. F. M. Toates & T. R. Halliday, pp. 3–21. London: Academic Press.

Hogan, J. A. (1984). Cause, function, and the analysis of behavior. *Mexican Journal of Behavior Analysis,* 10, 65–71.

Hogan, J. A. (1988). Cause and function in the development of behavior systems. In *Handbook of Behavioral Neurobiology,* Vol. 9, ed. E. M. Blass, pp. 63–106. New York: Plenum Press.

Hogan, J. A. & Roper, T. J. (1978). A comparison of the properties of different reinforcers. *Advances in the Study of Behavior,* 8, 155–255.

Hopkins, B. & Butterworth, G. (1990). Concepts of causality in explanations of development. In *Causes of Development,* ed. G. Butterworth & P. Bryant, pp. 3–32. London: Harvester Wheatsheaf.

Huxley, T. H. (1893). *Collected Essays,* Vol. 1. *Method and Results.* London: Macmillan.

Kruijt, J. P. (1964). Ontogeny of social behaviour in Burmese red junglefowl (*Gallus gallus spadiceus*). *Behaviour, supplement* 12.

Mayr, E. (1988). The multiple meanings of teleological. In *Toward a New Philosophy of Biology,* pp. 38–66. Cambridge, Mass.: Harvard University Press.

Pittendrigh, C. S. (1958). Adaptation, natural selection, and behavior. In *Behavior and Evolution,* ed. A. Roe & G. G. Simpson, pp. 390–416. New Haven: Yale University Press.

Ristau, C. A. (1991). *Cognitive Ethology*. Hillsdale, NJ: Lawrence Erlbaum .

Skinner, B. F. (1938). *The Behavior of Organisms*. New York: Appleton-Century-Crofts.

Tinbergen, N. (1951). *The Study of Instinct*. London: Oxford University Press.

Tinbergen, N. (1963). On aims and methods of ethology. *Zeitschrift für Tierpsychologie*, 20, 410–433.

Tinbergen, N. (1972). Ethology. (1969) In *The Animal in Its World*, Vol. 2, pp. 130–160. London: Allen & Unwin.

von Holst, E. & von St. Paul, U. (1960). Vom Wirkungsgefüge der Triebe. *Naturwissenschaften*, 47, 409–422. (Trans.: On the functional organisation of drives. *Animal Behaviour*, 1963, 11, 1–20.)

Zeiler, M. D. (1992). On immediate function. *Journal of the Experimental Analysis of Behavior*, 57, 417–427.

2

Neurobiological analyses of behavioural mechanisms in development

JOHAN J. BOLHUIS

In the previous chapter, Hogan discussed the concept of behaviour systems (cf. Hogan, 1988). In a development of earlier suggestions by Kruijt (1964), Hogan has proposed that behavioural ontogeny involves the development of various kinds of perceptual, motor, and central mechanisms and the formation of connections among them. In this chapter, I evaluate some of the ways in which the neural substrates of developing behavioural mechanisms have been investigated. The chapter is divided into two main parts. First, research aimed at discovering neural localisation of function is discussed, and I provide a brief review of some of the plastic changes at the neuronal level that have been found to occur during the development of specific perceptual mechanisms. In the second half of the chapter, I show how the analysis of neural mechanisms of behaviour can be important for understanding the causal organisation of behaviour during development. That is, neurobiological interventions may enable us to distinguish between, and independently manipulate, different behavioural mechanisms. I discuss a number of different techniques and evaluate what conclusions can be drawn from them as to the organisation of behaviour. We shall see that the results of this research often lead to a further behavioural analysis of perceptual mechanisms. Specific examples are taken mainly from research into the perceptual mechanisms involved in filial imprinting and song learning in birds. As Hogan observes in Chapter 10, in contrast to motor mechanisms, the development of perceptual mechanisms is often dependent on functional experience. For this reason, most neurobiological analysis of the development of perceptual mechanisms related to imprinting and song learning is concerned with the formation of representations on the basis of specific experience. However, in the same species, the development of some perceptual mechanisms may proceed partly in the absence of functional experience, as in the case of 'predispositions', and this will also be discussed.

Neural localisation of function

The ultimate aim of much of the neurobiological research in developing animals is to characterise the neuronal mechanisms of the substrates underlying behaviour. To this end, some form of localisation, whether it is topographical or in terms of a particular mechanism (e.g. a particular receptor mechanism or neurotransmitter system), is important. Ever since Lashley's (1950) unsuccessful attempts at localising 'the engram' using brain lesions, researchers have expressed scepticism at the possibility of identifying a particular region of the brain as the locus of the storage of information. With the advance of neuroscience, many other techniques have become available that can be used to study localisation, and in contemporary research often these methods will be used first.

Broadly speaking, two partially overlapping strategies can be distinguished in the neural analysis of developing perceptual mechanisms. In one strategy the expression and distribution of certain markers for neuronal plasticity are studied, in relation to the behaviour in question. An example of this approach is the work by Horn and his collaborators (e.g. Horn, 1985, 1990, 1992) on filial imprinting in the chick. Filial imprinting is the process through which social preferences become restricted to a particular stimulus as a result of exposure to that stimulus (see Bolhuis, 1991, for a review). In the early work on the neural substrates of imprinting, the chick brain was simply divided into three parts (forebrain roof, forebrain base and midbrain), and the incorporation of radioactive precursors into RNA was measured after imprinting. The only theoretical presupposition was one that is shared by most researchers in the field of learning and memory, namely that memory storage involves plastic changes at the neuronal/synaptic level (e.g. Hebb, 1949) and that these plastic changes presumably involve protein synthesis (Horn, 1962, p. 276; Horn, 1985; Dudai, 1989).

The second research strategy has been used for instance in the neural analysis of song learning in birds (see DeVoogd, this volume). Here, the neural pathways between auditory input and vocal output were charted first, after which the function of the different forebrain nuclei involved was studied by means of lesions, electrophysiology, or morphometric analyses (see below and DeVoogd, this volume, for details).

I will now briefly describe a few examples of the first research strategy. These examples will serve to illustrate to what degree localisation of

(a)   **(b)**

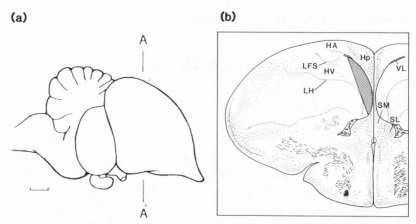

Figure 2.1 Schematic drawings of the brain of the domestic chick. **(a)** Lateral aspect of the brain. The vertical lines AA' indicate the approximate plane of the coronal section outline of the brain in (b). Scale bar: 2 mm. Adapted from Horn (1991b). According to Horn (1991a), the rostro-caudal extent of the IMHV is roughly from A 6.4 to A 9.6 in the stereotaxic atlas by Kuenzel & Masson (1988). **(b)** Simplified diagram of a coronal section of the chick brain at the level of the IMHV (approximately A 7.6 in the atlas by Kuenzel & Masson (1988)). The extent in the coronal plane of the left IMHV, as removed in biochemical studies (see text), is indicated by the hatched area. Adapted from Ambalavanar *et al.* (1993). *Abbreviations*: HA, hyperstriatum accessorium; HV, hyperstriatum ventrale; Hp, hippocampus; LFS, lamina frontalis superior; SL, nucleus septalis lateralis; SM, nucleus septalis medialis; LH, lamina hyperstriatica; VL, ventriculus lateralis.

function could be demonstrated, and how a lack of control experiments complicates the interpretation of the results. The technique of measuring the incorporation of radioactive aminoacids into proteins or uracil into RNA during the development of perceptual mechanisms has been used extensively by Horn and co-workers in filial imprinting (see Horn, 1985, 1990, 1992). Use of these techniques led to the identification of the IMHV (the intermediate and medial hyperstriatum ventrale; see Figure 2.1) as a region that is crucially involved in the learning process, probably as a site of memory storage.

Another much-used method is that of the tracing of the uptake of radioactive 2-deoxy-D-glucose (2DG) in neurones (Sokoloff *et al.*, 1977). 2DG is very similar to glucose, the major source of energy of cells, but once taken up by the cell and phosphorylated, 2DG is not metabolised further. The amount of 2DG that is trapped inside the cell in this way is used as a measure for metabolic rate, which in turn is supposed to be related to neuronal activity.

There are several differences between chicks that are trained and those that are not, apart from the fact that the former will have learned the characteristics of the imprinting stimulus and the latter will not. For instance, compared to chicks in an untrained control group, chicks in the experimental (imprinted) group may have had more visual experience, they may have been more stressed, or they may have shown more motor activity. In a number of ingenious control studies, Horn *et al.* were able to eliminate these non-specific factors as possible causes for the changes at the neural level.

Recently, Kabai, Kovach & Vadasz (1992), using 2DG autoradiography, provided some evidence for the involvement of a forebrain area that may be IMHV in visual imprinting in Japanese quail chicks. In this study, as in an experiment by Kohsaka *et al.* (1979) involving domestic chicks, the chicks did not receive preference tests with the imprinting object and a novel object. Thus, it is not clear whether the chicks were actually imprinted. Furthermore, neither of these studies used control experiments to investigate whether the changes were related to learning or to visual or other experience.

Kossut & Rose (1984) used the 2DG method in a one-trial passive avoidance task in day-old chicks (see Andrew, 1991a, Chapter 1, for details). These authors reported a significant increase of 2DG uptake not only in IMHV, but also in *Paleostriatum augmentatum* and *Lobus parolfactorius*. Chicks in the experimental group were presented with and allowed to peck a coloured bead coated with the aversive substance methylanthranilate. Most chicks learned to avoid pecking a bead of the same colour in one trial. In a control group the bead was coated with water. Thus, the difference between the two groups of chicks was not only that the majority of the chicks in the experimental group had learned to avoid and the chicks in the water group had not, but also that the former had experienced the disagreeable event of smelling and tasting methylanthranilate and the latter had not. This difference between the two groups may have contributed to the changes at the neural level and complicates the interpretation of these changes (see Bourne *et al.* (1991) for some evidence to support this possibility).

Recent techniques involving the expression of immediate early genes (IEGs) are more specific yet, since the products of their expression can be identified in the nuclei of single cells. These techniques involve the visualisation of the expression of certain genes, or their protein products, which are involved early on in responses to neuronal stimulation of some kind (Dragunow & Faull, 1989). Thus, immediate early gene expression has been used as a marker for cellular activation and it is likely to be involved whenever there are plastic changes in the neurone. The advantage of this method is that nuclei of activated cells are stained, in contrast to the 2DG method where metabolic activity is also measured in neuropil. In the latter case it is not known whether the changes in glucose metabolism are in neurones in the region involved or in terminals of afferents to that region (Sagar, Sharp & Curran, 1988). Using the IEG technique, the involvement of the IMHV in filial imprinting (McCabe & Horn, 1993) and passive avoidance learning in the chick (Anokhin *et al.* 1991) was confirmed.

A recent study using the IEG technique in the song-learning paradigm produced unexpected results. Mello, Vicario & Clayton (1992) studied expression of the immediate early gene ZENK (also known as *NGFI*-A, Egr-1 or *zif*/268) in adult canaries and zebra finches after exposure to taped conspecific song, heterospecific song, or tone bursts. Exposure to conspecific song led to a significant increase in the expression of ZENK in the medial caudal neostriatum and in the medial part of the hyperstriatum ventrale, compared to the two control groups. The result was surprising because there was no significant effect of exposure to song on ZENK expression in the traditional 'song control nuclei', such as the high vocal centre (HVC), area X, LMAN or RA (see DeVoogd, this volume), nor was there an effect in field L, a primary auditory projection area. These stimulus-specific and area-specific effects make it unlikely that the IEG result reflects increased levels of arousal. As Mello *et al.* note, the regions showing significant IEG expression have not traditionally been implicated in the production or perception of song. It is interesting, however, that from the figures supplied, the location of the 'active' region in the medial hyperstriatum ventrale in the canary and zebra finch appears to be similar to the IMHV in the chick (see above, and Horn, 1985). It is entirely possible that exposure to song would lead to the expression of different immediate early genes in other brain regions (cf. Wisden *et al.*, 1990). Furthermore, it may be that during song *acquisition* there is expression of immediate early genes in brain areas that have previously been associated with song learning (DeVoogd, this volume). Nevertheless, there is an intriguing suggestion that the same brain region may be involved in the neural substrate for the perceptual mechanisms involved in visual imprinting and in song learning. Furthermore, Brown & Horn (1992; see below) found that there were neurones in the left IMHV of unanaesthetised chicks, trained by simultaneous exposure to an auditory and a visual stimulus, that responded to auditory stimuli during testing. Responses to auditory (and to visual) stimuli were also found in dark-reared chicks. It would be interesting to analyse the effects of auditory learning on the responsiveness of neurones in this brain region in the chick, as well as in songbirds. Furthermore, it is important to establish whether lesions to the medial hyperstriatum (and the medial caudal neostriatum) in songbirds affect the learning and/or production of song.

Changes in the neural substrate as a result of specific experience

When there is strong evidence for localisation of function, the next phase of research is a neuronal characterisation of the brain substrates that have been identified. In this section I will briefly review some of the plastic changes that have been associated with particular developing perceptual mechanisms (see also DeVoogd, this volume). I will discuss structural, neurochemical and electrophysiological changes.

Structural changes: synapses, spines and cells

An important reason for investigating changes at the cellular level is that it has been suggested that in early neural development, experience may lead to weakening or eliminating of synaptic connections, while later experience causes the growth of new connections (e.g. Greenough, Black & Wallace, 1987; see also DeVoogd, Johnson, Hogan, this volume).

Bradley, Horn & Bateson (1981) and Horn, Bradley & McCabe (1985) demonstrated that imprinting led to a significant increase in the length of the postsynaptic density, a thickening of the area of the postsynaptic membrane where there is a high density of receptors. This increase was limited to synapses on dendritic spines; there was no such increase in synapses directly on the shafts of the dendrites. Furthermore, the increase only occurred in synapses in the left IMHV, not in the right, and was correlated with approach activity during training (Horn *et al.*, 1985). Similar changes in synaptic morphology within the IMHV were reported by Stewart *et al.* (1984) after passive avoidance training in the chick (cf. Stewart, 1991). The changes in postsynaptic density in the latter study are likely to have occurred at a different class of synapses than in the imprinting studies (see Horn, 1991b, for further discussion).

Using the Golgi method of staining, some authors have reported changes in the number of dendritic spines in some brain areas after early learning. For instance, Patel, Rose & Stewart (1988) reported, *inter alia*, a small, non-significant increase in the number of dendritic spines per unit length of dendrite in the left IMHV in chicks that had learned a passive avoidance task, compared to chicks that had not. Comparison of these findings with those of a similar study (Patel & Stewart, 1988), in which a near doubling of spine numbers was found, suggests that non-specific aspects of training may be responsible for much of the reported changes. Also, the absence in the former study of significant changes in spine shape, that had been reported in the Patel & Stewart (1988) study, prompted Patel *et al.* (1988) to conclude that 'changes in spine shape may not be connected with engram formation *per se*'.

In contrast to the findings after passive avoidance training, Wallhäusser & Scheich (1987) reported a *reduction* in the number of dendritic spines on a particular class of neurones, after an auditory imprinting session in guinea fowl chicks (see also DeVoogd, this volume). The changes were found in a region in the medial hyperstriatum (MNH) that may have some overlap with the anterior part of IMHV. This is one of the few studies to report a reduction in spines in relation to learning. However, this study had a very small sample size, and methodological inadequacies (such as a selection of the experimental animals) prevent an unambiguous interpretation of the results (see also Horn, 1985, for a discussion of this work).

Nottebohm *et al.* (1990) discuss the significance of neurogenesis seen in RA-projecting neurones in HVC (and perhaps in area X) for some time after hatching and into adulthood in canaries and zebra finches. At present, it is not clear whether these new neurones, which become functionally integrated in existing circuits, are necessary for the central representation of song. For instance, there is no relationship between neurogenesis and seasonality or whether or not the birds modify their songs in adulthood (see Nottebohm, 1991; DeVoogd, this volume, for further discussion of these issues).

Receptors and intracellular mechanisms

It has been suggested that neural plasticity involving neuronal N-methyl-D-aspartate (NMDA) receptors might play an important role in information storage in the brain (e.g. McCabe & Horn, 1988; Morris, 1989; see also Horn, 1962, for an implication of changes in receptor numbers in learning). McCabe & Horn (1988) found that imprinting led to a 59% increase in the number of NMDA receptors in the left IMHV, but not in the right IMHV. Furthermore, it was found that in the left IMHV, the number of NMDA receptors was positively correlated with preference for the training stimulus. This result supports the suggestion that the changes are indeed related to learning. This class of receptors also plays a crucial role in the induction of an artificial form of synaptic plasticity known as long-term potentiation (LTP; Bliss & Lømo, 1973). Similar synaptic mechanisms to those underlying LTP have been suggested to be involved in the neural mechanisms of memory storage (e.g. Morris, 1989). From an imprinting perspective, it is interesting that recent findings by Bradley, Delisle Burns & Webb (1991) suggest that an LTP-like phenomenon may also be induced in the brain of the chick.

The synaptic changes found as a result of filial imprinting could increase the effectiveness of neurotransmission and form the neural basis of memory (Horn, 1985). A significant increase in the number of NMDA receptors in the left IMHV was found at 8 hours after the end of training, but not at 3 or 6 hours after training (McCabe & Horn, 1991). This raises the issue of what mechanism would be involved in memory storage for the first few hours after training. Horn (1990) speculated that this mechanism could be presynaptic. That is, in analogy with mechanisms underlying LTP, there could be an early phase of presynaptic activity such as increased release of neurotransmitter, followed by postsynaptic changes such as increased postsynaptic densities and receptor numbers. Stewart *et al.* (1984) reported increased numbers of vesicles per synaptic bouton in the left IMHV after passsive avoidance training, which gives some support to the possible involvement of a presynaptic mechanism.

Recent research into the neural mechanisms of imprinting has concentrated on subcellular events. The results suggest the involvement of protein kinase C-dependent protein phosphorylation in the learning process (e.g. Sheu *et al.*, 1993; Ambalavanar *et al.*, 1993).

Electrophysiological recording of neuronal activity

When there is strong evidence for the function of a particular region of the brain in behaviour, electrophysiological recording of the activity of neurones or groups of neurones is extremely valuable. Ideally, one should be able to study the characteristics of perceptual mechanisms, and their development, at the neuronal level. Although some authors have stressed the importance of recordings from single neurones, especially in the context of elementary cortical mechanisms of vision (see Hogan, this volume), electrophysiological data are extraordinarily difficult to interpret in relation to complex behaviour. For example, although much is known about the properties of neurones in the rat hippocampus (e.g. O'Keefe & Nadel, 1978), this knowledge does not tell us what the role of this structure in memory is. Of course, recording from the hippocampus has been inspired by an approach in which the presumed importance of this structure for information storage resulted from neuropsychological findings. Recent electrophysiological findings by Brown and collaborators (Brown, Wilson & Riches, 1987) suggest that the majority of neurones in the monkey temporal lobe with possible memory-related properties are located in the perirhinal cortex rather than in the hippocampus proper (cf. Zola-Morgan *et al.*, 1989).

Even when there is reason to assume that the site of electrophysiological recording is involved in information storage, there are still logical problems involved in the interpretation of the results. Suppose, for instance, that a particular region of the brain is likely to be a neural substrate for information storage. Within the generally accepted theoretical framework for neural mechanisms of memory storage (e.g. Horn, 1985; Dudai, 1989), it might be supposed that a group (a network) of n neurones would be involved in the storage of information about a particular event. However, it is also presumed that the actual neural representation is implemented by a particular distribution of synaptic strengths, which may be altered (strengthened or weakened) by experience. A single neurone may have up to 60000 afferent synaptic connections (estimates for cat motor cortex, Cragg, 1967). Thus, in principle there would be a pool of $n \times 60000$ 'connections' available for the neural representation of an event. Especially in brain structures that are not layered as clearly as, for instance, the mammalian hippocampus, the principal measure derived from extracellular recording with a microelectrode is the number of spikes (action potentials) per unit of time, from a single neurone or from a group of neurones. It could be argued that when changes in behavioural mechanisms lead to a redistribution of synaptic weights (i.e. changes in the efficacy of individual synapses) in the neural substrate underlying these mechanisms, the net result of this redistribution will be a change in the probability of neurones firing. Thus, changes in firing rate of neurones would reflect this change in synaptic organisation. In other words, even if it is assumed that individual neurones give us little information as to the neural representation of behavioural mechanisms, their activity may reflect the properties of the network as a whole. This is especially important when the analysis of electrophysiological findings is conducted at the level of neuronal populations (e.g. McLennan & Horn, 1992; see below).

A number of studies have used electrophysiological recording in the investigation of developing perceptual and motor mechanisms. The neurobiological analysis of bird song learning has employed an approach in which the presumed importance of a brain region for the behaviour is proposed on the basis of its neuroanatomical connections, and on the effects of lesions of these nuclei. As Nottebohm *et al.* (1990, p. 121) put it, 'We do not know yet how learned sounds are processed...' and 'We do not know how and where learned sounds are stored.' This lack of knowledge of the function of the various 'song control nuclei' complicates the interpretation of electrophysiological recordings from them. Margoliash & Konishi (1985) recorded from units in the HVC, one of the forebrain nuclei presumed to be involved in song learning, in adult white-crowned sparrows. They found that most of the responsive units responded preferentially to the bird's own song, even when compared to the song of its tutor. In order to evaluate the significance of these findings, it is important to record from the HVC of juvenile birds before they start singing themselves. In that case, it is possible to enquire whether neurones in the HVC respond differently to the tutor song, compared to unfamiliar songs (see also DeVoogd, this volume).

Bradford & McCabe (1992) recorded spontaneous activity from neurones in the IMHV of anaesthetised chicks after imprinting. They reported a significant positive correlation between preference score and spontaneous firing rate in the left IMHV. McLennan & Horn (1992) conducted electrophysiological recordings in the forebrain of anaesthetised chicks that had been trained previously by exposing them to a rotating red box or a rotating blue box, or that had been reared in darkness. The trained chicks all strongly preferred the training object to the novel object in a preference test. McLennan & Horn recorded from groups of neurones in the left IMHV. For their analysis, they averaged for each chick the neuronal activity at *all* recording sites in the left IMHV. It was found that the response to a blue box was different from the response to a red box in dark-reared chicks. This difference was changed significantly by training. In blue-trained chicks the responses to both boxes resembled the response to the blue box in the dark-reared chicks; in red-trained chicks the responses to both boxes resembled the response to the red box in the dark-reared birds. The two boxes rotated at the same speed and were of the same shape and size; they differed in both colour and brightness. The electrophysiological findings suggested that one of the transformations brought about by training is to place the left IMHV neuronal responses to the two stimuli into the same category. Bolhuis & Horn (1992) inquired whether the implied neuronal generalisation had a counterpart at the behavioural level. Using stimuli similar to those used by McLennan & Horn (1992), they found that chicks significantly preferred a stimulus that differed from the training object in one aspect (colour) to an object that differed from the training object in more aspects (colour, shape, and pattern). In control tests, the chicks strongly preferred the training object to the object that differed from it only in colour. Thus, in this instance electrophysiological evidence prompted behavioural research into the nature of the perceptual mechanism involved in imprinting and yielded analogous results.

In recent studies, Brown & Horn (1993) recorded from the left IMHV of unanaesthetised chicks that had been trained by exposure to an imprinting object. In general, neuronal responsiveness to presentation of an imprinting object was much greater in these chicks than in the anaesthetised chicks of McLennan & Horn (1992), suggesting that anaesthesia profoundly depresses neuronal responding in this brain region. Brown & Horn found that in imprinted chicks, some neurones in the left IMHV responded preferentially to the training stimulus. Thus, taken together, the electrophysiological recording studies suggest that some neurones in left IMHV may be involved in 'generalisation' between imprinting objects, whilst other neurones respond in a highly specific way to the stimulus on which the chicks are imprinted (see also Horn, 1992).

Dissociations and the logic of lesions: general considerations

The remainder of this chapter is concerned with an evaluation of neurobiological manipulations that may provide insights into the organisation of perceptual mechanisms. That is, by means of restricted lesions, electrical stimulation or by analysing hemispheric asymmetries, different perceptual mechanisms may be affected differently. The function of these mechanisms may thus be influenced, and/or they may be dissociated.

The most widely used tool to achieve this objective is that of restricted electrolytical or neurochemical lesions to the brain. There are general conceptual problems with the lesion technique (see Olton, 1986; Irle, 1990, for recent accounts). That is, the logic of inferring the function of a brain region by studying the effects of ablating it has been questioned (cf. Gregory, 1961). Further, electrolytic and aspiration lesions to a particular brain locus have been shown to have diverse effects beyond that locus (Schoenfeld & Hamilton, 1977). The advent of more sophisticated neurotoxic lesion techniques has led to improvements. For instance, infusion of ibotenic acid (e.g. Jarrard, 1989) is thought to destroy cells, leaving fibres of passage intact. These novel techniques can profoundly change our views of the function of the brain regions involved (e.g. Jarrard, 1985).

As described above, Horn *et al.* demonstrated that a restricted region of the chick forebrain, the IMHV, was involved in the learning process (Horn, McCabe & Bateson, 1979; see also Horn, 1985). The early studies were correlative. To establish that the region was necessary for learning, it was essential to impair its function. Thus, if the IMHV is necessary for the recognition memory of imprinting, then the destruction of the IMHV should impair both the acquisition and the retention of filial preference in imprinting. In a series of studies, it was demonstrated that bilateral lesions to IMHV indeed had this effect (Horn, 1985; McCabe, 1991).

DeVoogd (this volume) reviews the use of restricted lesions in the bird song paradigm (see also Konishi, 1985; Nottebohm *et al.*, 1990 for reviews). There is an interesting parallel between results of lesion experiments by Brenowitz (1991) for song perception and Bolhuis *et al.* (1989b) for imprinting. Brenowitz (1991) reported that female canaries show copulation solicitations to conspecific song, but not to songs of the white-crowned sparrow. However, females with bilateral lesions to HVC, one of the nuclei in the song-related pathways, showed copulation solicitations to heterospecific as well as to conspecific song. Bolhuis *et al.* (1989b) reared female domestic chicks in small groups with a male. When the animals were approximately 3 months old, simultaneous preference tests were conducted involving the familiar male, an unfamiliar male from the rearing strain, and an unfamiliar male from a novel strain. It was found that the females spent significantly more time with the novel male of the rearing strain than with either of the other two males. Females that had received bilateral lesions to the IMHV spent approximately equal time with all three males. These results suggest a role for the IMHV in sexual imprinting, possibly by impairing the ability of the birds to recognise individuals and hence to discriminate between them. Similarly, lesions to the HVC may impair the ability of female canaries to recognise particular songs.

It should be pointed out, however, that the HVC in songbirds is not in the same relative neuroanatomical location as the IMHV in the chick. The HVC (high(er) vocal centre) used to be called HVc (hyperstriatum ventrale, pars caudale), when it was assumed that it comprised the caudal part of the hyperstriatum ventrale (e.g. Konishi, 1985). Since then, it has become clear that 'HVC' is in fact a part of the neostriatum and that in the songbird brain it is caudal to the intermediate and medial hyperstriatum ventrale (e.g. Alvarez-Buylla, Theelen & Nottebohm, 1988).

Single dissociations: can a lesion dissociate different perceptual mechanisms?

A single dissociation refers to a differential effect of a neural manipulation (such as a lesion) on behaviour. Single dissociations are widely used in experimental psychology, often in an attempt at analysing the function of a particular brain structure. Prominent examples are the neuropsychology of human amnesia and the animal models that resulted from it. Weiskrantz (1990) and Tulving (1992) review a number of different types of memory that have been suggested as a result of neuropsychological findings. For example, damage to the temporal lobe in humans leads to severe anterograde impairments of memory for specific events, but does not affect memory for skills. Cohen & Squire (1980) have called these two types of memory declarative and procedural memory, respectively. Others have used different terms for similar distinctions (e.g. Squire, 1987; see Tulving, 1992, for a critical discussion). A danger with these distinctions is that they may become attempts at characterising the behavioural impairment, without acquiring psychological meaning. An example is the much-used distinction between working memory and reference memory, that was introduced by Olton, Becker & Handelmann (1979). Olton *et al.* proposed these terms purely as operational definitions of the behavioural tasks involved. That is, in a working memory task the information to be remembered changes from trial to trial, whilst in a reference memory task the to-be-remembered information is constant throughout the experiment. Damage to the hippocampus was thought to affect memory in working memory tasks but not in reference memory tasks. Subsequently, authors have used these two terms as if they actually referred to two different types of memory. There is little evidence, however, to show that that is the case. Thus, a simple dissociation as a result of a neural manipulation may be difficult to interpret (see also Horn, 1991b, pp. 248–250), and in the next section I will discuss a few examples involving early learning during development.

Is the IMHV involved exclusively in imprinting?

An example of a lesion-induced dissociation which is similar to the distinctions made in primate amnesia research, is the distinction between performance in an instrumental learning task and in imprinting after bilateral lesions to the IMHV (Johnson & Horn, 1986). Johnson & Horn (1986) trained chicks on an operant task in which the birds had to depress one of two pedals in order to be exposed to a rotating imprinting object. Subsequently the birds were given simultaneous choice tests involving the presentation of this object and a novel object. Normal chicks learn to press the correct pedal and will imprint on the reinforcing stimulus, showing a significant preference for it in the subsequent simultaneous choice test. Chicks with bilateral lesions to the IMHV learned the operant task at the same speed as control chicks, but they were impaired significantly in the imprinting test. Johnson & Horn (1986) argued that the lesion dissociated associative learning and recognition memory.

Johnson & Horn's interpretation implies that successful performance in the operant task does not require stimulus recognition. In learning terminology, the rotating object is a conditioned stimulus (CS) in the case of imprinting and an unconditioned stimulus (US) or reinforcer in the case of operant learning (cf. Shettleworth, this volume). Chicks do not need to be familiar with an imprinting object for it to act as a US in operant learning (Bateson & Reese, 1968). In operant conditioning, pre-exposure to the US impairs subsequent conditioning with that US (Mackintosh, 1983). This may not be the case in imprinting where, for instance, pre-exposure to a stimulus may impair subsequent conditioning with a novel stimulus (Bateson & Reese, 1969). Thus, it is conceivable that imprinting on a particular stimulus increases the reinforcing capability of that stimulus. If that were the case, lesions to the IMHV might be expected to impair operant conditioning with a familiar imprinting object as US, compared to control chicks. On the other hand, IMHV-lesioned chicks might be expected to be less affected by a change in reinforcing stimulus than control birds.

In filial imprinting as well as in conventional examples of associative learning, the animal needs to form a representation of the stimulus involved (the CS in the case of conditioning). In the chick, the IMHV may be involved in both. Suppose, for example, that in the operant task used by Johnson & Horn (1986) the two pedals would have a distinctive colour and pattern and their spatial position was exchanged regularly, with the 'active' pedal always having the same colour and pattern. In that case, the task would acquire Pavlovian properties requiring recognition of the characteristics of the 'active' pedal. If the IMHV is indeed involved in recognition memory, it is predicted that in this case chicks with bilateral lesions to this structure would be impaired in the acquisition of the task. At any rate, it is important to realise that the lesion experiment by Johnson & Horn does not show that imprinting is not a form of associative learning (cf. Horn, 1985; Bolhuis, de Vos & Kruijt, 1990). That is, lesions to the

IMHV may impair recognition of the imprinting stimulus, but the learning process of imprinting might still involve the association of a representation of the stimulus with a representation of reinforcing aspects of that stimulus (Hoffman & Ratner, 1973; Bolhuis *et al.*, 1990; see Horn, 1985, pp. 118–126, for a similar view, and for a detailed discussion of these issues).

Given the significance of the IMHV in imprinting, it is important to know whether lesions to this structure would also affect other forms of early learning. In a study of one-trial passive avoidance learning (see above and Andrew, 1991a), Davies, Taylor & Johnson (1988) placed bilateral lesions in the IMHV, or in a lateral cerebral area, or performed a sham-operation in newly hatched chicks. The next day the chicks were allowed to peck a coloured bead that was coated with aversive methylanthranilate. Sham-operated control chicks tended to avoid a bead of that colour (but not beads of a novel colour) on subsequent exposure. IMHV-lesioned chicks did not avoid the training bead, whereas the other lesioned chicks behaved like the sham-operated controls. The authors speculate that the IMHV lesions may either have impaired the chicks' ability to recognise individual stimuli, or have interfered with the inhibition of what they termed 'inborn' responses or 'spontaneous tendencies' (cf. Gaffan *et al.*, 1984). Because the IMHV-lesioned chicks approached both beads during the test, the latter alternative seems more likely (cf. Horn, 1985, pp. 123–125; Johnson, 1991; Davies, 1991, for detailed discussions).

More recent results, however, suggest that the the IMHV does play a role in stimulus recognition (Patterson & Rose, 1992). Patterson, Gilbert & Rose (1990) had found that when bilateral lesions encroaching on the IMHV were placed *before* passive avoidance training, the lesioned chicks did not avoid the training bead at test. When the lesions were placed *after* training, the lesioned birds avoided the training bead. Patterson *et al.* (1990) concluded that the IMHV is not important for memory retention. Importantly, Patterson *et al.* (1990) did not give their chicks a dis-crimination test after passive avoidance training, but presented the birds with the training bead only. In a more recent study, Patterson & Rose (1992) gave the chicks a discrimination test involving the training bead and a bead of a different colour. Using similar lesions, placed *after* passive avoidance training, they found that lesioned chicks avoided both beads during the test. Sham-operated control chicks avoided the training bead more than the novel bead. Thus, the lesioned chicks appeared to have learned to avoid beads, but they could not discriminate between individual beads.

In conclusion, lesions to the IMHV not only affect the development of the perceptual mechanism involved in filial imprinting. It is likely that the IMHV also plays a role in sexual imprinting, and it is involved in passive avoidance learning. It may be that the IMHV is the neural substrate for the storage of representations of specific stimuli.

Temporal dissociations of lesion effects in song learning

An interesting dissociation after brain lesions comes from work on song learning (Bottjer, Miesner & Arnold, 1984; Scharff & Nottebohm, 1991; see also DeVoogd, this volume), where it was found that lesions to the forebrain nuclei LMAN or area X disrupted song development in juvenile zebra finch males, but that similar lesions placed in these nuclei in adult birds did not affect song output (see Nottebohm *et al.*, 1990; DeVoogd, this volume, for examples). In this case, effects of the lesion on the birds' sensory or motor abilities can be ruled out, unless it is assumed that lesions to LMAN or area X affect these abilities in young birds but not in adults, which seems unlikely. Thus, these results suggest that LMAN and area X may be involved in the acquisition of song. This suggestion is supported by findings in adult canaries. In contrast to adult zebra finches, adult canaries continue to update their song repertoire (Nottebohm *et al.*, 1990). Under certain conditions, bilateral lesions to LMAN lead to a deterioration of song in adult canaries (Nottebohm *et al.*, 1990; see DeVoogd, this volume, for a further discussion of this issue).

Double dissociations

Teuber (1955) was the first to introduce the term 'double dissociation', which is used to designate situations where a neurological patient who is better at task A than at task B is complemented by a different patient with the reverse dissociation. Shallice (1979) considered the phenomenon to be of crucial importance in that 'strong neuropsychological evidence for the existence of neurologically distinct functional systems depends on double dissociation of function'. Shallice (1988, p. 234ff) provides a detailed analysis of the concept of double dissociation, and the different ways in which it has been used in neuropsychology.

It is not known whether lesions to HVC or RA affect the acquisition of song in zebra finches before crystallisation has occurred. If they did not, such findings, together with the age-dependent effects of lesions to LMAN and area X discussed above, would indicate an interesting double dissociation (see Nottebohm *et al.*, 1990, for further discussion). The closest equivalent of a double dissociation in the development of perceptual mechanisms is that of imprinting. Horn & McCabe (1984) evaluated a series of experiments involving bilateral lesions to IMHV. Lesions to this structure significantly impaired filial imprinting. However, when the results of four lesion studies were analysed together, it emerged that the effect of the lesions was dependent upon the type of stimulus used. The two imprinting objects used were a rotating red box and a rotating stuffed jungle fowl, about the same size and rotating at the same speed. There was a significant impairment of filial preference after bilateral IMHV lesions with both stimuli. However, the effect of the lesion was much greater when the training stimulus used was the red box. IMHV lesions led to random performance at testing in box-trained chicks. In contrast, fowl-trained chicks with IMHV lesions, although impaired significantly, still showed a significant preference for the training object during testing. Similar differential effects were obtained in a study in which forebrain noradrena-line levels were depleted by means of injection of the noradrenergic neurotoxin DSP4 (Davies, Horn & McCabe, 1985). Using the same two imprinting objects, Bolhuis, McCabe & Horn (1986) found that there was a significant positive correlation between preferences for the training stimulus and plasma testosterone levels in fowl-trained chicks, whilst there was no such relationship in box-trained chicks. Furthermore, injections of testosterone-enanthate significantly enhanced preferences in fowl-trained chicks but were without effect in chicks trained with the red box. The results of these three studies taken together suggest a double dissociation of function. In this case, the double dissociation was demonstrated by using different techniques.

Predispositions

On the basis of their evaluation of the lesion experiments, Horn & McCabe (1984) proposed that filial preferences in the chick may be partly under the influence of a predisposition for particular stimuli (see also ten Cate, this volume). In Hogan's (1988) terminology, such a predisposition is a 'prefunctionally developed perceptual mechanism', specifically an object recognition mechanism. As described in the previous section, Horn & McCabe's single dissociation was followed by an endocrinological investigation (Bolhuis *et al.*, 1986) that, together with the lesion results and those of Davies *et al.* (1985), suggested a double dissociation. Thus, the distinction between perceptual mechanisms involved was initially a result of neurobiological intervention experiments. Following these neurobiological findings, a series of behavioural experiments confirmed the existence of at least two perceptual mechanisms involved in the development of filial preferences and suggested that the two mechanisms are behaviourally separable. The predisposition is for stimuli in the head and neck region and develops over time, after a certain amount of non-specific experience (Johnson, Bolhuis & Horn, 1985, 1992; Johnson & Horn, 1988; Bolhuis, Johnson & Horn, 1985, 1989a; see Johnson & Bolhuis, 1991, for a review). Bilateral lesions to IMHV impaired filial imprinting, but did not affect the development of the predisposition (Johnson & Horn, 1986). Thus, the neural substrate for the predisposition appears to be outside the IMHV, but has not been identified as yet.

It should be pointed out here that the IMHV is involved in imprinting with stimuli such as a stuffed fowl. That is, when chicks are exposed to such stimuli, not only will the predisposition (involving structures other than the IMHV) influence their behaviour, but, furthermore, the animals will form a representation of the characteristics of the stimulus. For this latter learning process the integrity of the IMHV is required. For instance, Johnson & Horn (1987) demonstrated that after imprinting exposure to a stuffed jungle-fowl hen, chicks preferred the training model to a novel stuffed jungle-fowl hen. Bilateral lesions to the IMHV prevented the acquisition of such a preference. Further evidence that the IMHV is involved in the recognition of individual stimulus animals comes from the study by Bolhuis *et al.* (1989b; see above).

A predisposition for species-specific sounds has been demonstrated in song learning in certain species (Marler, 1991). When fledgling male song sparrows and swamp sparrows were exposed to taped songs that consisted of equal numbers of songs of both species, they preferentially learnt the songs of their own species. Males of both species are able to sing the songs of the other species, thus it appears that perceptual mechanisms are involved in what Marler (1991) called 'the sensitization of young sparrows to conspecific song' (p. 200). As in the case of filial imprinting, 'The physiological basis of this ability remains to be discovered' (ibid; see also Nottebohm *et al.*, 1990; DeVoogd, this volume).

In filial imprinting, the interaction between the two perceptual mechanisms is likely to be at the behavioural level. That is, the predisposition does not prevent or constrain learning about stimuli that are not relevant for the predisposition (Bolhuis *et al.*, 1989a). Rather, it is likely that the predisposition to approach and follow certain stimuli directs the animal's attention to such a stimulus and that it then learns the individual characteristics of this stimulus. Similarly, Marler (1991) states 'Thus, once sparrows have focused attention on songs that satisfy certain innately specified criteria, they then proceed to learn them' (p. 200).

Cerebral asymmetry and behavioural lateralisation

Cerebral asymmetry of function and behavioural lateralisation suggest that the two cerebral hemispheres play different roles in the causation of behaviour. Rogers (1991) reviewed the evidence for asymmetry of the fibres connecting the retina to the forebrain in the young chick. For instance, there is a transient asymmetry in the strength of the connections from the thalamus to the contralateral hyperstriatum, which is related to the sex of the animal and to visual experience (Boxer & Stanford, 1985). Behavioural experiments, some involving pharmacological or endocrinological manipulations, confirmed the neuroanatomical findings and demonstrated that the asymmetries influence visually guided behaviour.

Neural lateralisation at the level of perceptual mechanisms was demonstrated by Horn, McCabe and colleagues in filial imprinting (cf. Horn, 1985, 1990; McCabe, 1991). In a series of studies it was found that the left IMHV plays a different role in the storage of information during imprinting than does the right IMHV. On the one hand, the left IMHV is likely to be a site of long-term information storage. On the other hand, over a period of about 6 hours after the end of imprinting exposure, the right IMHV gradually ceases to be crucial for information storage and another part of the brain takes over (see Horn, 1985, and McCabe, 1991, for detailed discussions).

Although many of these findings were a result of experiments involving electrolytic lesions, it is important to point out that the original suggestion of lateralisation was arrived at through the discovery of cerebral asymmetry in synapse morphology that accompanies imprinting (Bradley *et al.*, 1981; Horn *et al.*, 1985; cf. Stewart *et al.*, 1984). Subsequent investigations of glutamate receptor binding (McCabe & Horn, 1988, 1991) and protein phosphorylation (Sheu *et al.*, 1993) revealed a similar asymmetry between the left and the right IMHV. Experiments involving lesions or electrophysiology confirmed the previous findings and suggested the time course for the dynamic changes in memory storage capacities of the left and right IMHV (See Horn, 1985, 1991b; and McCabe, 1991, for reviews). There is some evidence to suggest cerebral asymmetry related to birdsong. Nottebohm (1991) recently summarised these findings as follows: 'The origin and significance of functional asymmetry in brain pathways for song control – particularly as it relates to learning – remains a mystery' (p. 208).

Horn, McCabe & Cipolla-Neto (1983) provided intriguing evidence for effects of electrolytic lesions to the left IMHV influencing the function of the right IMHV. Lesioning of only the right IMHV, either before or after training with an imprinting object, does not significantly affect learning. These and other results led to a model suggesting that the left IMHV is a possible permanent store, whilst the right IMHV may act as a temporary store (Horn, 1985; McCabe, 1991). However, when the left IMHV was lesioned before training, post-training lesions to the right IMHV *did* produce amnesia. Thus, it appears that the right IMHV may take on another function as a result of damage to the left IMHV.

Behavioural lateralisation is analysed primarily by studying behaviour while restricting visual input to one eye. Andrew (1991b) reviews the evidence for behavioural lateralisation in the young chick. In most bird species there is almost complete crossing over of the optic fibres (cf. Rogers, 1991). That is, virtually all fibres from the left eye project to the right hemisphere and *vice versa*. However, the two hemispheres are not isolated anatomically (e.g. Horn, 1985; Rogers, 1991) and information may pass through the supraoptic commissure. Only when the supraoptic commissure is severed surgically (Horn, Rose & Bateson, 1973) is it possible to make more specific statements about the hemispheres involved in a particular behaviour. The absence of anatomical isolation of the two hemispheres in unoperated animals led Andrew and his collaborators to use the terms left eye system (LES) and right eye system (RES) when describing the results of their experiments. Andrew (1991b) catalogued the different functions that the two eye systems are presumed to have in the young chick. He suggests that the left eye system is involved in detailed analysis of stimuli and detects changes in their properties and spatial position. The right eye system is thought to play a role in assigning stimuli to categories and in the 'choice of the response appropriate to a stimulus' (p. 536). Recent evidence suggests that the LES is important for the recognition of conspecifics in the chick (Vallortigara, 1992).

Electrical stimulation of the brain

Most of the arguments I have presented with regard to the use of lesions in the analysis of behaviour apply, *mutatis mutandis*, to electrical stimulation of the brain. The problems with the interpretation of brain stimulation are illustrated by a debate as to its implications for the causation of behaviour, especially with regard to Tinbergen's (1952) conflict hypothesis (see, for example, Brown & Hunsperger, 1963; cf. Baerends, Groothuis, Hogan, this volume; see Baerends, 1975, for a review). Brown & Hunsperger suggested that if display were the result of conflict between the influence of two different behaviour systems, then 'threat' behaviour would occur only as a result of simultaneous electrical stimulation of the brain areas involved with attack and escape, respectively. The results of their experiments with cats did not support this suggestion, as they found that threat could in fact be elicited by stimulation of a single area. Baerends (1975), in his evaluation of the conflict hypothesis, has provided a detailed critique of these studies. He suggested a number of alternative explanations on neurophysiological grounds, for example that the stimulation may affect areas outside the immediate vicinity of the electrode. On the other hand, he argued that in the course of evolution, displays may come under the influence of a new motivational system, a process called 'emancipation' (see Groothuis, this volume). If it is the case that emancipation also occurs at the neural level, then it would be predicted that stimulation of the neural substrate would activate the display. These different interpretations combined make it very difficult to attain testable hypotheses with regard to the effects of brain stimulation on behaviour, as alternative explanations could not be ruled out, whatever the result. Baerends concluded that the results of the stimulation experiments do not argue against the conflict hypothesis, and that these kinds of findings are in fact not suitable for the analysis of behaviour. For a recent discussion of the meaning of electrical stimulation in the analysis of behaviour, see Kruk (1991) and Berridge & Valenstein (1991).

Electrical stimulation of the brain was used in filial imprinting by McCabe, Horn & Bateson (1979). In this case the stimulation did not evoke a particular behaviour, but was shown to mimic sensory input, thus 'introducing information directly into the IMHV by artificial means.' (Horn, 1985, p. 175). Briefly, it was found that chicks that had received bilateral electrical stimulation of a particular frequency into the IMHV, later preferentially approached a light that flickered with the same frequency to a light flickering with a different frequency. Importantly, this effect was not found after electrical stimulation of two different regions of the forebrain, the ectostriatum and the hyperstriatum accessorium. These two brain structures are visual projection areas and project either directly or via a multi-synaptic pathway to the IMHV. Horn (1985) argued that this indicates that input to the IMHV from a visual pathway was not sufficient for the chicks to 'store' this information, and that direct stimulation of the IMHV presumably activated two different inputs (or their postsynaptic elements) that are necessary for such information storage. Horn speculated that these two inputs were from sensory systems and from an 'incentive system', respectively (see Horn, 1985, for further discussion). These findings confirmed a role for the IMHV in information storage.

Conclusions

Neuronal plasticity underlying the development of perceptual mechanisms

Research into the cellular plasticity underlying the development of perceptual mechanisms has made important advances in recent years. The analysis of the characteristics of the neurones and networks involved may elucidate the organisation of perceptual mechanisms. For example, the way in which, through training, an imprinting stimulus is represented in the neuronal networks in IMHV is becoming clear both at the electro-physiological and at the cellular levels of analysis. Such characterisation will be crucial for the development of realistic neural network models of the perceptual mechanisms involved.

Of the perceptual mechanisms studied, only those that develop under the influence of specific experience have yielded evidence as to the underlying neural plasticity. This evidence supports the widely held hypothesis that information storage in the brain involves changes in neuronal connectivity (e.g. Hebb, 1949; Horn, 1962; see Horn, 1985, Squire, 1987, Dudai, 1989, for recent reviews). Both the filial imprinting and passive avoidance learning paradigms in the chick have shown that learning is accompanied by increased protein synthesis, protein kinase C-mediated protein phos-phorylation, neuronal activation, changes in the postsynaptic density and density of postsynaptic receptors in a restricted region of the forebrain (Horn, 1985, 1992; Rose, 1991). The suggestion that a possible mechanism for early learning is synapse selection through loss of spines (Wallhäusser & Scheich, 1987; DeVoogd, this volume) has only received weak support so far. Also, the relationship between increased neurogenesis and bird song learning (e.g. Nottebohm *et al.*, 1990) is not clear.

Experience-expectant and experience-dependent information storage?

Several contributors to the present volume (DeVoogd, Johnson, Hogan) discuss the interesting distinction that Greenough *et al.* (1987) have made between 'experience-expectant' and 'experience-dependent' information storage. The distinction is based on the type of information that is stored, which is ubiquitous and the same for all members of the species in the former, and unique to the individual in the latter. However, Greenough *et al.* suggest that their categorisation is not only based on the type of information stored, but also on the neuronal mechanisms involved. Experience-expectant information storage is thought to involve synaptic selection or 'pruning', whilst experience-dependent storage is suggested to involve the formation of new synapses. The definition of these terms at two different levels of explanation limits their usefulness as explanatory principles. For instance, does the increase in postsynaptic densities in the IMHV related to filial imprinting and to passive avoidance learning in the chick mean that these forms of learning are 'experience-dependent'? On the other hand, if Wallhäusser & Scheich (1987) were to demonstrate conclusively that auditory imprinting leads to loss of spines, would this kind of learning then be termed 'experience-expectant? Hogan and DeVoogd (this volume) suggest this, and the latter author mentions sensitive periods, constraints and resistance to change as indications of the 'experience-expectant' character of song learning and imprinting. As we have seen above, however, the neural changes accompanying visual imprinting are of the 'experience-dependent' kind (Greenough *et al.* (1987, p. 551) acknowledge this). Furthermore, at the ethological/psychological level it it obvious that during song learning and imprinting, the young animals form representations of specific stimuli and events in their environment that are unique to individuals (e.g. particular aspects of the tutor song or visual characteristics of individual imprinting stimuli). Is there perhaps a major distinction between auditory and visual imprinting on the basis of the information stored? That seems very unlikely, as in auditory imprinting, too, the young birds can store very specific information (e.g. Gottlieb, 1988; Van Kampen & Bolhuis, 1991; cf. Bolhuis & Van Kampen, 1992).

   This brief discussion illustrates the difficulties with the interpretation of Greenough *et al.*'s classification. That is not to say that it may not be a useful heuristic tool. As we have seen above, the development of perceptual mechanisms often involves the influence of both functional and prefunctional experience (Hogan, 1988, this volume). This classification does not map on to Greenough *et al.*'s hypothesis, as functional experience may well be the same for all members of the species (see also Hogan, Chapter 10). Nevertheless, in terms of Greenough *et al.*'s classification, for instance during the development of filial preferences, both 'experience-dependent' and 'experience-expectant' information may be stored (see Kraemer, 1992, for a similar view). It would seem extraordinarily difficult to tease apart the mechanisms involved in these two putative processes, let alone relate them to specific neural changes.

Neurobiological paradigms and the organisation of behavioural mechanisms

The examples involving single or double dissociations after localised lesions, or of asymmetry of function that were discussed illustrate that some form of localisation of function preceding these analyses facilitates their interpretation. Single dissociations are particularly useful when there is firm evidence as to the function of the manipulated brain region. When this is not the case, it is difficult to interpret the behavioural results of neural manipulations such as lesions, as it is not clear at which point in the neural chain of events this manipulation is effective. The differential effects in adult and in juvenile zebra finch males of lesions to LMAN or area X led to the suggestion of two forebrain circuits that may have a different function in the development of bird song (see DeVoogd, this volume). An example that is closest to a double dissociation is the demonstration of the existence of different perceptual mechanisms in the development of filial preferences, that are separable both behaviourally and neurobiologically. Current evidence suggests that these mechanisms interact at the behavioural level. These results show the importance of a continual interplay between behavioural and neurobiological research.

Acknowledgements

This chapter is dedicated to Jaap Kruijt, on the occasion of his retirement. I am grateful to him for his guidance and support, and for the many constructive discussions that we enjoyed throughout the years of our collaboration. Thanks to Gabriel Horn, Jerry Hogan, Mark Johnson and Tim DeVoogd for their comments on the manuscript, and to Rob Honey and William Greenough for discussion. Some of the research discussed in this chapter was supported by the Agricultural and Food Research Council (UK).

References

Alvarez-Buylla, A., Theelen, M. & Nottebohm, F. (1988). Birth of projection neurons in the higher vocal center of the canary forebrain before, during, and after song learning. *Proceedings of the National Academy of Sciences, USA*, 85, 8722–8726.

Ambalavanar, R., Van der Zee, E. A., Bolhuis, J. J., McCabe B. J. & Horn, G. (1993). Co-expression of *Fos* immunoreactivity in protein kinase C-gamma (PKCγ)-positive neurons: quantitative analysis of a brain region involved in learning. *Brain Research*, 606, 315–318.

Andrew, R. J. (1991a). *Neural and Behavioural Plasticity: The Use of the Domestic Chick as a Model.* Oxford: Oxford University Press.

Andrew, R. J. (1991b). The nature of behavioural lateralization in the chick. In *Neural and Behavioural Plasticity: The Use of the Domestic Chick as a Model*, ed. R. J. Andrew, pp. 536–554. Oxford: Oxford University Press.

Anokhin, K. V., Mileusnic, R., Shamakina, I. Y. & Rose, S. P. R. (1991). Effects of early experience on c-fos gene expression in the chick forebrain. *Brain Research*, 544, 101–107.

Baerends, G. P. (1975). An evaluation of the conflict hypothesis as an explanatory principle of display. In *Function and Evolution of Behaviour*, ed. G. P. Baerends, C. Beer & A. Manning, pp. 236–239. Oxford: Oxford University Press.

Bateson, P. P. G. & Reese, E. P. (1968). Reinforcing properties of conspicuous objects before imprinting has occurred. *Psychonomic Science*, 10, 379–380.

Bateson, P. P. G. & Reese, E. P. (1969). The reinforcing properties of conspicuous stimuli in the imprinting situation. *Animal Behaviour*, 17, 692–699.

Berridge, K. C. & Valenstein, E. (1991). What psychological process mediates feeding evoked by electrical stimulation of the lateral hypothalamus? *Behavioral Neuroscience*, 104, 778–795.

Bliss, T. V. P. & Lømo, T. (1973). Long-lasting potentiation of synaptic transmission in the dentate area of the anaesthetised rabbit following stimulation of the perforant path. *Journal of Physiology*, 232, 331–356.

Bolhuis, J. J. (1991). Mechanisms of avian imprinting: A review. *Biological Reviews*, 66, 303–345.

Bolhuis, J. J. De Vos, G. J. & Kruijt, J. P. (1990). Filial imprinting and associative learning. *Quarterly Journal of Experimental Psychology*, 42B, 313–329.

Bolhuis, J. J. & Horn, G. (1992). Generalization of learned preferences in filial imprinting. *Animal Behaviour*, 44, 185–187.

Bolhuis, J. J., Johnson, M. H. & Horn, G. (1985). Effects of early experience on the development of filial preferences in the domestic chick. *Developmental Psychobiology*, 18, 299–308.

Bolhuis, J. J., Johnson, M. H. & Horn, G. (1989a). Interacting mechanisms during the formation of filial preferences: The development of a predisposition does not prevent learning. *Journal of Experimental Psychology: Animal Behavior Processes*, 15, 376–382.

Bolhuis, J. J., Johnson, M. H., Horn, G. & Bateson, P. (1989b). Long-lasting effects of IMHV lesions on social preferences in domestic fowl. *Behavioral Neuroscience*, 103, 438–441.

Bolhuis, J. J., McCabe, B. J. & Horn, G. (1986). Androgens and imprinting: Differential effects of testosterone on filial preference in the domestic chick. *Behavioral Neuroscience*, 100, 51–56.

Bolhuis, J. J. & Van Kampen, H. S. (1992). An evaluation of auditory learning in filial imprinting. *Behaviour*, 122, 195–230.

Bottjer, S. W., Miesner, E. A. & Arnold, A. P. (1984). Forebrain lesions disrupt development but not maintenance of song in passerine birds. *Science*, 224, 901–903.

Bourne, R. C., Davies, D. C., Stewart, M. G., Csillag, A. & Cooper, M. (1991). Cerebral glycoprotein synthesis and long-term memory formation in the chick (*Gallus domesticus*) following passive avoidance training depends on the nature of the aversive stimulus. *European Journal of Neuroscience*, 3, 243–248.

Boxer, M. I. & Stanford, D. (1985). Projections to the posterior visual hyperstriatal region of the chick: an HRP study. *Experimental Brain Research*, 57, 494–498.

Bradford, C. M. & McCabe, B. J. (1992). An association between imprinting and spontaneous neuronal activity in the hyperstriatum ventrale of the domestic chick. *Journal of Physiology*, 452, 238P.

Bradley, P., Delisle Burns, B. & Webb, A. C. (1991). Potentiation of synaptic responses in slices from the chick forebrain. *Proceedings of the Royal Society of London, Series B*, 243, 19–24.

Bradley, P., Horn, G. & Bateson, P. P. G. (1981). Imprinting: an electron microscopic study of chick hyperstriatum ventrale. *Experimental Brain Research*, 41, 115–120.

Brenowitz, E. A. (1991). Altered perception of species-specific song by female birds after lesions of a forebrain nucleus. *Science*, 251, 303–305.

Brown, J. L. & Hunsperger, R. W. (1963). Neuroethology and the motivation of agonistic behaviour. *Animal Behaviour*, 11, 439–448.

Brown, M. W. & Horn, G. (1992). Neurones in the intermediate and medial part of the hyperstriatum ventrale (IMHV) of freely moving chicks respond to visual and/or auditory stimuli. *Journal of Physiology*, 452, 102P.

Brown, M. W. & Horn, G. (1993). The influence of learning (imprinting) on the visual responsiveness of neurones of the intermediate and medial part of the hyperstriatum ventrale (IMHV) of freely moving chicks. *Journal of Physiology*, 459, 161P.

Brown, M. W., Wilson, F. A. W. & Riches, I. P. (1987). Neuronal evidence that inferomedial temporal cortex is more important than hippocampus in certain processes underlying recognition memory. *Brain Research*, 409, 158–162.

Cohen, N. J. & Squire, L. R. (1980). Preserved learning and retention of pattern analyzing skill in amnesia: dissociation of knowing how and knowing that. *Science*, 210, 207–209.

Cragg, B. G. (1967). The density of synapses and neurones in the motor and visual areas of the cerebral cortex. *Journal of Anatomy*, 101, 639–654.

Davies, D. C. (1991). Lesion studies and the role of IMHV in early learning. In *Neural and Behavioural Plasticity: The Use of the Domestic Chick as a Model*, ed. R. J. Andrew, pp. 329–343. Oxford: Oxford University Press.

Davies, D. C., Horn, G. & McCabe, B. J. (1985). Noradrenaline and learning: the effects of the noradrenergic neurotoxin DSP4 on imprinting in the domestic chick. *Behavioral Neuroscience*, 99, 652–660.

Davies, D. C., Taylor, D. A. & Johnson, M. H. (1988). The effects of hyperstriatal lesions on one-trial passive avoidance learning in the chick. *Journal of Neuroscience*, 8, 4662–4666.

Dragunow, M. & Faull, R. (1989). The use of *c-fos* as a metabolic marker in neuronal pathway tracing. *Journal of Neuroscience Methods*, 29, 261–265.

Dudai, Y. (1989). *The Neurobiology of Memory*. Oxford: Oxford University Press.

Gaffan, D., Saunders, R. C., Gaffan, E. A., Harrison, S., Shields, C. & Owen, M. J. (1984). Effects of fornix transection upon associative memory in monkeys: role of hippocampus in learned action. *Quarterly Journal of Experimental Psychology*, 36, 173–221.

Gottlieb, G. (1988). Development of species identification in ducklings: XV. Individual auditory recognition. *Developmental Psychobiology*, 21, 509–522.

Greenough, W. T., Black, J. E. & Wallace, C. S. (1987). Experience and brain development. *Child Development*, 58, 539–559.

Gregory, R. L. (1961). The brain as an engineering problem. In *Current Problems in Animal Behaviour*, ed. E. H. Thorpe & O. L. Zangwill, pp. 307–330. Cambridge: Cambridge University Press.

Hebb, D. O. (1949). *The Organization of Behavior*. New York: John Wiley & Sons.

Hoffman, H. S. & Ratner, A. M. (1973). A reinforcement model of imprinting: Implications for socialization in monkeys and men. *Psychological Review*, 80, 527–544.

Hogan, J. A. (1988). Cause and function in the development of behavior systems. In *Handbook of Behavioral Neurobiology*, Vol. 9, ed. E. M. Blass, pp. 63–106. New York: Plenum Press.

Horn, G. (1962). Some neural correlates of perception. In *Viewpoints in Biology*, Vol. 1, ed. J. D. Carthy & C. L. Duddington, pp. 242–285. London: Butterworth.

Horn, G. (1985). *Memory, Imprinting, and the Brain*. Oxford: Clarendon Press.

Horn, G. (1990). Neural bases of recognition memory investigated through an analysis of imprinting. *Philosophical Transactions of the Royal Society of London B*, 329, 133–142.

Horn, G. (1991a). Technique for removing IMHV from the chick brain. In *Neural and Behavioural Plasticity*, ed. R. J. Andrew, pp. 44–48. Oxford: Oxford University Press.

Horn, G. (1991b). Imprinting and recognition memory; a review of neural mechanisms. In *Neural and Behavioural Plasticity: The Use of the Domestic Chick as a Model*, ed. R. J. Andrew, pp. 219–261. Oxford: Oxford University Press.

Horn, G. (1992). Brain mechanisms of memory and predispositions: Interactive studies of cerebral function and behavior. In *Brain Development and Cognition: A Reader*, ed. M. H. Johnson, pp. 481–509. Oxford: Blackwell.

Horn, G., Bradley, P. & McCabe, B. J. (1985). Changes in the structure of synapses associated with learning. *Journal of Neuroscience*, 5, 3161–3168.

Horn, G. & McCabe, B. J. (1984). Predispositions and preferences. Effects on imprinting of lesions to the chick brain. *Animal Behaviour*, 32, 288–292.

Horn, G., McCabe, B. J. & Bateson, P. P. G. (1979). An autoradiographic study of the chick brain after imprinting. *Brain Research*, 168, 361–373.

Horn, G., McCabe, B. J. & Cipolla-Neto, J. (1983). Imprinting in the domestic chick: The role of each side of the hyperstriatum ventrale in acquisition and retention. *Experimental Brain Research*, 53, 91–98.

Horn, G., Rose, S. P. R. & Bateson, P. P. G. (1973). Monocular imprinting and regional incorporation of tritiated uracil into the brains of intact and 'split -brain' chicks. *Brain Research*, 56, 227–237.

Irle, E. (1990). An analysis of the correlation of lesion size, localization and behavioral effects in 283 published studies of cortical and subcortical lesions in old-world monkeys. *Brain Research Reviews*, 15, 181–213.

Jarrard, L. E. (1985). Is the hippocampus really involved in memory? In *Brain Plasticity, Learning, and Memory*, ed. B. Will, P. Schmitt & J. C. Dalrymple-Alford, pp. 363–372. New York: Plenum Press.

Jarrard, L. E. (1989). On the use of ibotenic acid to lesion selectively different components of the hippocampal formation. *Journal of Neuroscience Methods*, 29, 251–259.

Johnson, M. H. (1991). Information processing and storage during filial imprinting. In *Kin Recognition*, ed. P. G. Hepper, pp. 335–357. Cambridge: Cambridge University Press.

Johnson, M. H. & Bolhuis, J. J. (1991). Imprinting, predispositions and filial preference in chicks. In *Neural and Behavioural Plasticity: The Use of the Domestic Chick as a Model*, ed. R. J. Andrew, pp. 133–156. Oxford: Oxford University Press.

Johnson, M. H., Bolhuis J. J. & Horn, G. (1985). Interaction between acquired preferences and developing predispositions during imprinting. *Animal Behaviour*, 33, 1000–1006.

Johnson, M. H., Bolhuis. J. J. & Horn, G. (1992). Predispositions and learning: Behavioural dissociations in the chick. *Animal Behaviour*, 44, 943–948.

Johnson, M. H. & Horn, G. (1986). Dissociation between recognition memory and associative learning by a restricted lesion to the chick forebrain. *Neuropsychologia*, 24, 329–340.

Johnson, M. H. & Horn, G. (1987). The role of a restricted region of the chick forebrain in the recognition of conspecifics. *Behavioural Brain Research*, 23, 269–275.

Johnson, M. H. & Horn, G. (1988). Development of filial preferences in dark-reared chicks. *Animal Behaviour*, 36, 675–683.

Kabai, P., Kovach, J. & Vadasz, C. (1992). Neural correlates of genetically determined and acquired color preferences in quail chicks. *Brain Research*, 573, 260–266.

Kohsaka, S., Takamatsu, K., Aoki, E. & Tsukada, Y. (1979). Metabolic mapping of chick brain after imprinting using [14C]2-deoxy-glucose technique. *Brain Research*, 172, 539–544.

Konishi, M. (1985). Bird song: from behavior to neuron. *Annual Review of Neuroscience*, 8, 125–170.

Kossut, M. & Rose, S. P. R. (1984). Differential 2-deoxyglucose uptake into chick brain structures during passive avoidance training. *Neuroscience*, 12, 971–977.

Kraemer, G. W. (1992). A psychobiological theory of attachment. *Behavioral and Brain Sciences*, 15, 493–511.

Kruijt, J. P. (1964). Ontogeny of social behaviour in Burmese red junglefowl (*Gallus gallus spadiceus, Bonaterre*). *Behaviour*, suppl. 12.

Kruk, M. R. (1991). Ethology and pharmacology of hypothalamic aggression in the rat. *Neuroscience and Biobehavioural Reviews*, 15, 527–538.

Kuenzel, W. J. & Masson, M. (1988). *A Stereotaxic Atlas of the Brain of the Chick (Gallus domesticus)*. Baltimore: Johns Hopkins University Press.

Lashley, K. S. (1950). In search of the engram. *Symposia of the Society for Experimental Biology*, 4, 454–482.

Mackintosh, N. J. (1983). *Conditioning and Associative Learning*. Oxford: Clarendon Press.

Margoliash, D. & Konishi, M. (1985). Auditory representation of autogenous song in the song system of white-crowned sparrows. *Proceedings of the National Academy of Sciences of the USA*, 82, 5997–6000.

Marler, P. (1991). Song-learning behavior: the interface with neuroethology. *Trends in Neurosciences*, 14, 199–206.

McCabe, B. J. (1991). Hemispheric asymmetry of learning-induced changes. In

Neural and Behavioural Plasticity, ed. R. J. Andrew, pp. 262–276. Oxford: Oxford University Press.

McCabe, B. J. & Horn, G. (1988). Learning and memory: Regional changes in N-methyl-D-aspartate receptors in the chick brain after imprinting. *Proceedings of the National Academy of Sciences of the USA*, 85, 2849–2853.

McCabe, B. J. & Horn, G. (1991). Synaptic transmission and recognition memory: time course of changes in N-methyl-D-aspartate receptors after imprinting. *Behavioral Neuroscience*, 105, 289–294.

McCabe, B. J. & Horn, G. (1993). Imprinting leads to elevated *Fos*-like immunoreactivity in the intermediate and medial hyperstriatum ventrale (IMHV) of the domestic chick. *Journal of Physiology*, 459, 160P.

McCabe, B. J., Horn, G. & Bateson, P. P. G. (1979). Effects of rhythmic hyperstriatal stimulation on chicks' preferences for visual flicker. *Physiology & Behavior*, 23, 137–140.

McLennan, J. G. & Horn, G. (1992). Learning-dependent responses to visual stimuli of units in a recognition memory system. *European Journal of Neuroscience*, 4, 1112–1122.

Mello, C. V., Vicario, D. S. & Clayton, D. F. (1992). Song presentation induces gene-expression in the songbird forebrain. *Proceedings of the National Academy of Sciences of the USA*, 89, 6818–6822.

Morris, R. G. M. (1989). Does synaptic plasticity play a role in learning and memory? In *Parallel Distributed Processing: Implications for Psychology and Neurobiology*, ed. R. G. M. Morris, pp. 248–285. Oxford: Oxford University Press.

Nottebohm, F. (1991). Reassessing the mechanisms and origins of vocal learning in birds. *Trends in Neurosciences*, 14, 206–211.

Nottebohm, F., Alvarez-Buylla, A., Cynx, J., Kirn, J. Ling, C-Y., Nottebohm, M., Suter, R. Tolles, A. & Williams, H. (1990) Song learning in birds: the relation between perception and production. *Philosophical Transactions of the Royal Society of London, Series B*, 329, 115–124.

O'Keefe, J. & Nadel, L. (1978). *The Hippocampus as a Cognitive Map*. Oxford: Oxford University Press.

Olton, D. S. (1986). Interventional approaches to memory: lesions. In *Learning and Memory: A Biological View*, ed. J. L. Martinez, Jr. & R. P. Kesner, pp. 379–397. San Diego: Academic Press.

Olton, D. S., Becker, J. T. & Handelmann, G. E. (1979). Memory, space, and the hippocampus. *Behavioral and Brain Sciences*, 2, 313–365.

Patel, S. N., Rose, S. P. R. & Stewart, M. G. (1988). Training induced dendritic spine density changes are specifically related to memory formation processes in the chick, *Gallus domesticus*. *Brain Research*, 463, 168–173.

Patel, S. N. & Stewart, M. G. (1988). Changes in the number and structure of dendritic spines, 25 h after passive avoidance training in the domestic chick, *Gallus domesticus*. *Brain Research*, 449, 34–46.

Patterson, T. A., Gilbert, D. B. & Rose, S. P. R. (1990). Pre- and post-training lesions of the intermediate medial hyperstriatum ventrale and passive avoidance learning in the chick. *Experimental Brain Research*, 80, 189–195.

Patterson, T. A. & Rose, S. P. R. (1992). Memory in the chick: Multiple cues, distinct brain locations. *Behavioral Neuroscience*, 106, 465–470.

Rogers, L. J. (1991). Development of lateralization. In *Neural and Behavioural Plasticity: The Use of the Domestic Chick as a Model*, ed. R. J. Andrew, pp. 507–535. Oxford: Oxford University Press.

Rose, S. P. R. (1991). Biochemical mechanisms involved in memory formation in the chick. In *Neural and Behavioural Plasticity: The Use of the Domestic Chick as a Model*, ed. R. J. Andrew, pp. 277–304. Oxford: Oxford University Press.

Sagar, S. M., Sharp, F. R. & Curran, T. (1988). Expression of *c-fos* protein in brain: metabolic mapping at the cellular level. *Science*, 240, 1328–1331.

Scharff, C. & Nottebohm, F. (1991). A comparative study of the behavioral deficits following lesions of various parts of the zebra finch song system: implications for vocal learning. *Journal of Neuroscience*, 11, 2896–2913.

Schoenfeld, T. A. & Hamilton, L. W. (1977). Secondary brain changes following lesions: a new paradigm for lesion experimentation. *Physiology & Behavior*, 18, 951–967.

Shallice, T. (1979). Neuropsychological research and the fractionation of memory systems. In *Perspectives on Memory Research*, ed. L.-G. Nilsson, pp. 257–277. Hillsdale, NJ: Lawrence Erlbaum.

Shallice, T. (1988). *From Neuropsychology to Mental Structure*. Cambridge: Cambridge University Press.

Sheu, F-S., McCabe, B. J., Horn, G. and Routtenberg, A. (1993) Learning selectively increases protein kinase C substrate phosphorylation in specific regions of the chick brain. *Proceedings of the National Academy of Sciences of the USA*, 90, 2705–2709.

Sokoloff, L. Reivich, M., Kennedy, C. Rosiers, M. H. D., Patlak, C. S., Pettigrew, K. D., Sakurada, O. & Shinohara, M. (1977). The ^{14}C-deoxyglucose method for the measurement of local cerebral glucose utilization: theory, procedure, and normal values in the conscious and anaesthetized albino rat. *Journal of Neurochemistry*, 28, 897–916.

Squire, L. R. (1987). *Memory and Brain*. Oxford: Oxford University Press.

Stewart, M. G. (1991). Changes in dendritic and synaptic structure in chick forebrain consequent on passive avoidance learning. In *Neural and Behavioural Plasticity: The Use of the Domestic Chick as a Model*, ed. R. J. Andrew, pp. 305–328. Oxford: Oxford University Press.

Stewart, M. G., Rose, S. P. R., King, T. S., Gabbott, P. L. A. & Bourne, R. (1984). Hemispheric asymmetry of synapses in chick medial hyperstriatum ventrale following passive avoidance learning: a stereological investigation. *Developmental Brain Research*, 12, 261–269.

Teuber, H.-L. (1955). Physiological psychology. *Annual Review of Psychology*, 6, 267–296.

Tinbergen, N. (1952). 'Derived activities': their causation, biological significance, origin and emancipation during evolution. *Quarterly Review of Biology*, 27, 1–37.

Tulving, E. (1992). Concepts of human memory. In *Memory: Organization and Locus of Change*, ed. L. R. Squire, N. M. Weinberger, G. Lynch & J. L. McGaugh, pp. 3–32. New York: Oxford University Press.

Vallortigara, G. (1992). Right hemisphere advantage for social recognition in the chick. *Neuropsychologia*, 30, 761–768.

Van Kampen, H. S. & Bolhuis, J. J. (1991). Auditory learning and filial imprinting in the chick. *Behaviour*, 117, 303–319.

Wallhäusser, E. & Scheich, H. (1987). Auditory imprinting leads to differential 2-deoxyglucose uptake and dendritic spine loss in the chick rostral forebrain. *Developmental Brain Research*, 31, 29–44.

Weiskrantz, L. (1990). Problems of learning and memory: one or multiple memory systems? *Philosophical Transactions of the Royal Society of London, Series B*, 329, 99–108.

Wisden, W., Errington, M. L., Williams, S., Dunnett, S. B., Waters, C., Hitchcock, D., Evan, G., Bliss, T. V. P. & Hunt, S. P. (1990). Differential expression of immediate early genes in the hippocampus and spinal cord. *Neuron*, 4, 603–614.

Zola-Morgan, S., Squire, L. R., Amaral, D. G. & Suzuki, W. A. (1989). Lesions of perirhinal and parahippocampal cortex that spare the amygdala and hippocampal formation produce severe memory impairment. *Journal of Neuroscience*, 9, 4355–4370.

Part two

Development of perceptual and motor mechanisms

3

The neural basis for the acquisition and production of bird song

TIMOTHY J. DEVOOGD

Avian song has received extensive attention, both as a natural behavior that is easily studied and can be related to ecology or to principles of natural selection, and as a neuroethological preparation in which it is possible to determine the neural basis for a complex motor activity in a vertebrate. This chapter is a survey of some of the central findings in both domains. It concentrates on the development of song and of brain regions that are responsible for song. It attempts to relate the behavioral and neurobiological findings in this system to more general issues of the nature of early perceptual and motor development and juvenile learning. In addition, it attempts to extend ideas on behavioral development to the neurobiological substrate for this behavior.

Overview of singing behavior

Acquisition of song – the perceptual phase

Several key findings are central to understanding avian song acquisition and performance. First, many features of song are learned. Thus, for example, many of the sounds comprising a canary's song closely resemble songs heard by the bird as a juvenile, and are distinct from sounds produced by other canaries (Marler & Waser, 1977, Waser & Marler, 1977). Zebra finches form a song by splicing elements of the songs of several individuals that they heard as juveniles, apparently favoring adults which had fed or interacted with them (Williams, 1990a). Similarly, nightingales form an elaborate song repertoire by acquiring and retaining 'packages' of sounds as a juvenile (Hultsch & Todt, 1989b), and swamp sparrows form a song by selecting from songs heard when young (Marler & Peters, 1988a). Indeed, aspects of song appear to be learned in every

49

songbird species in which this has been examined (Kroodsma, 1982). Song learning can be divided into two major subdivisions: auditory learning, in which a model is heard and retained; and motor learning, in which through practice a bird creates a song that resembles the model. In many species, these occur at distinct times and appear to follow different rules and so will be treated separately in this review.

Sensitive periods

In most songbird species studied to date, the acquisition of song from adult models occurs principally or exclusively during a restricted time period early in life. Apparently, there is a time window or sensitive period during which song learning occurs easily and other times when it is difficult or impossible. Sensitive periods have been studied by successively presenting different exemplars of song to young birds throughout development. Typically this is done using tape recordings to control for number and timing of presentations. The songs produced later by the birds are then compared to the models to which they had been exposed. Frequently, very accurate copies are produced of songs heard during a consistent, restricted interval. Thus in swamp sparrows, songs heard between the second and the fourth month after hatching are often copied in the bird's mature song, songs in the next 3 months are occasionally copied, but songs heard outside this interval are not (Marler & Peters, 1988a). In zebra finches, the sensitive period is predominantly in the second month after hatching, although some sounds heard in the third month may also be acquired (Immelmann, 1969; Price, 1979; Clayton, 1987a). Similarly, canaries readily acquire the songs to which they are exposed during months 2–4 after hatching (Waser & Marler, 1977).

Concept of template

Of perhaps equal interest in the study of song acquisition has been the discovery that there are constraints on the content of song learning as well as on its timing (reviewed by Slater, 1989; Marler, 1991). Songbirds typically do not learn to produce many of the sounds present during the sensitive period, even if these are made up of frequencies and patterns that the bird is able to hear and could produce. When presented with a choice between a song from its own species and a song from another species, a young bird typically shows a strong preference for the song of its own species. Thus, if young song and swamp sparrows are exposed to equal numbers of songs from their own and from the other species, their adult song will consist predominantly (song sparrow) or exclusively (swamp sparrow) of copies of songs from their own species (Marler & Peters, 1988b, 1989). Similarly, zebra finches exposed to the songs of zebra finches and of closely related Bengalese finches will tend to learn zebra finch song (Eales, 1987; Clayton, 1987b). Such observations have led to the concept of a template for song – an internal model of song formed in the young bird without experience of the species' song (Konishi, 1965; Marler, 1976). Young birds will learn characteristics of songs that match the template and will reject songs that do not. The richness of the template or the power of its constraints varies across species – the swamp sparrow above appears to have a more precisely defined song template than does the song sparrow. Acceptable sounds are then memorized and are used as a model upon which the bird's own song will be based. It is not clear whether this acquisition process involves altering the original internally generated template to match more precisely those sounds that are learned, or creation of additional representations for the specific acceptable sounds. These alternatives differ, for example, in whether the adult ability to identify and evaluate the songs of conspecifics is based upon the sounds learned by the bird (giving a capacity for highly precise like-different assessments) or is based on the original template (giving the ability to recognize conspecifics who learned very different songs).

Concept of irreversibility

Song learning seems quite resistant to subsequent change. For many species, auditory learning seems to end if a bird is exposed to an acceptable song model during the sensitive period. This has been shown by experiments in which a young bird is exposed to a succession of different song exemplars during development. When the bird matures, its song typically does not contain copies or variants of songs presented outside the sensitive period, even if the bird was extensively exposed to them. While there are species like canaries and red-winged blackbirds in which the content of the song may be revised from one year to the next, this need not indicate new auditory learning. It may reflect either delayed production of additional song elements learned during the sensitive period or impro-vization on sounds already being produced. However, some species such as mockingbirds or starlings are clearly capable of prolonged acquisition of new sounds. One of the intriguing unresolved questions in the neuro-ethology of avian song is whether acquisition of novel sounds in adulthood in these species is a different phenomenon from acquisition during the juvenile stage, or whether they have found a way to prolong the sensitive period for auditory learning. These alternatives are likely to be associated with very different neural mechanisms – and, indeed, may be best studied by studying the neural correlates of acquisition.

Acquisition of song – the motor phase
Behavioral stages (subsong, plastic song, full song)

The motor aspects of song acquisition have been divided into three major stages (reviewed by Marler, 1991). Subsong is first and begins early in development. It consists of soft vocalizations that lack organized structure and do not appear directed. It looks noisy in spectrograms and does not resemble songs the young bird is hearing – indeed, it does not seem to be influenced by the presence or absence of adult song. However, in spite of this amorphous quality, there are features of subsong that are consistent within a species, differ between species, and that will color the eventual song. Subsong may overlap with the sensitive period for auditory learning. It may occur in females as well as males and occurs normally in males that have been castrated (Nottebohm, 1980; Marler *et al.*, 1988). Its function is not known. Marler (1991) suggests as possibilities practicing motor sequences that will eventually become building blocks for song or matching motor signals to auditory consequences.

Subsong is followed by plastic song. Here, vocalizations are produced that are clearly derived from sounds that the young bird has heard. Execution of the sounds is initially quite variable but improves as this phase goes on. Many species-specific features occur, whether or not a bird has been exposed to songs of its species. During plastic song, birds typically practice many songs. As singing becomes more accomplished, the bird drops many of these songs and begins to settle on its eventual repertoire. It is not known what drives this selection process – possibilities will be discussed below.

Eventually plastic song gives way to crystallized or full song. For many species, this means performing a stable set of songs or song elements in a highly stereotyped way. Song is often directed, toward other males or toward females. It may be used to delineate or advertize a territory. Testosterone is required for full song to commence (Nottebohm, 1980; Heid, Güttinger & Pröve, 1985; Marler *et al.*, 1988). Indeed, it may signal the end of the period of plastic song as preventing a rise in gonadal steroids during development prevents song crystallization (Bottjer & Hewer, 1992), and raising testosterone levels prematurely results in crystallizing an abnormally small repertoire (Korsia & Bottjer, 1991). In many species, it is also required for singing to continue: decreasing testosterone either by castration or naturally as in the endocrine changes following reproduction, is associated with reduction or cessation of singing (for example, Arnold, 1975; Nottebohm *et al.*, 1987; Nowicki & Ball, 1989).

Auditory integration

Each of the motor stages of song acquisition is integrated with audition. During subsong, the young bird may learn which sounds it is making are associated with which muscular activity. It may choose to make sounds typical of its species because, in some sense, they sound right – if deafened, swamp and song sparrows no longer produce noticeably different subsongs, as do intact birds (Marler & Sherman, 1983).

Motor and auditory integration is central to plastic song. Normally, birds appear to be listening to the songs they are producing so as to match them to songs heard previously. If reared in isolation, birds also practice and improve the execution of sounds. They may copy sounds that would not normally occur in song (Price, 1979). Deafening is devastating to production of learned sounds and prevents the normal increase in stereotypy, showing the extent of the interplay between audition and execution during this phase (reviewed by Bottjer & Arnold, 1986).

After crystallization, production appears somewhat emancipated from audition. Typically, a bird's song is unaltered immediately after deafening and changes develop more or less gradually depending on the species. White-crowned sparrows appear resistant to changes in song with deafening (Konishi, 1965). In zebra finches, changes in the timing and ordering of syllables develop slowly over several months following surgery (Price, 1979; Bottjer & Arnold, 1986; K. W. Nordeen & Nordeen, 1992). While deafened canaries continue to sing all their preoperative syllables for several weeks, they rapidly begin to produce these in a hesitant, soft and variable way; their song has changed substantially by the following year (Nottebohm, Stokes & Leonard, 1976).

Sex differences and similarities in song acquisition and performance

In most songbird species, singing is done predominantly or exclusively by males. Consequently, much of the research on song and its neural basis has focused on males. In particular, there has been relatively little research on auditory learning in females, as most of the indices for measuring the timing, nature and amount of this learning depend upon eventual song production. However, attracting females is believed to be one of the major functions for song in males of many species (reviewed by Searcy & Andersson, 1986). To the extent that song is used for male to female communication, accurate perception by females is as important as accurate production by males. Several findings suggest that song processing in females is complex and interesting in its own right. Females respond to the song of their own species and not to the songs of other species (Searcy & Marler, 1981; Baker, 1983; Catchpole, Dittami & Leisler, 1984). This would suggest that, at the very least, females possess an endogenous template for song like that found in acoustically inexperienced males. Females also discriminate between conspecific songs. They appear to respond more vigorously or more quickly to an elaborate song than to a simple one (Kroodsma, 1976; Eens, Pinxten & VerHeyen, 1991). As older males often produce more elaborate songs than younger (Nottebohm, Nottebohm & Crane, 1986; Kirn *et al.*, 1989; Eens *et al.*, 1991), it has been suggested that females may evaluate the complexity of songs as an index of male age or fitness, which has led, through sexual selection, to the evolution of mechanisms that permit males to develop elaborate songs (Krebs & Kroodsma, 1980). Song ontogeny in cowbirds provides an interesting variation on the interplay between female perception and male behavior (reviewed by West & King, 1985). Cowbirds are brood parasites and so are unable to learn song from conspecific adults prior to fledging. Males later associate with female cowbirds and acquire song characteristics typical of males of the subspecies to which the female belongs. They do so by varying their early vocalizations and retaining or elaborating those which females reinforce with visual gestures (King & West, 1983a, 1983b; West & King, 1988; King & West, 1989).

Each of these aspects of female behavior could occur without learning song. Even in cowbirds, differences in female preference could derive from differences between subspecies in endogenous characteristics of the auditory template for song rather than acquired differences (although see King, West, & Eastzer, 1986, for evidence that the preferences of some females may be influenced by prior social experience). However, other data suggest that auditory learning may occur in female songbirds as well as in males. Females are able to discriminate conspecific males on the basis of song (Miller, 1979a, 1997b; Baker, Spitler-Nabors & Bradley, 1981). Female white-crowned sparrows respond more to presentations of songs from their natal dialect than from other dialects and appear to use early experience with the natal dialect to do so (Baker *et al.*, 1981; Baker, 1983; reviewed by Baker & Cunningham, 1985). When given testosterone as adults, female canaries and white-crowned sparrows begin to sing. They appear to progress rapidly through plastic song to a rendition of full song that contains male-like phonology, syntax and stereotypy (although sometimes less elaborate than in male song) (Nottebohm, 1980; Petrino-vich & Baptista, 1984, 1987). Recently, Hausberger & Black (1991) have shown that the song induced by testosterone in two female starlings overlaps almost completely with the songs of their mate. While it is not proven that songs induced in females were learned in an earlier sensitive period, it is clear that they have species-typical characteristics that would be missing in isolate song in a male. Together, such data suggest that processing of song is complex in females. It is likely to involve the learning of song characteristics early in development and learning in adulthood to distinguish individual males because of differences in song. If true, this raises issues like those being studied in males: the timing and relative extent of the early learning, and the amount of flexibility that exists in adulthood. To a large degree, these questions remain unexplored.

Interspecies variation in song learning

It is important to emphasize the extraordinary variation shown across songbird species in every aspect of song acquisition and performance. Song has been studied carefully in very few of the approximately 4200 songbird species. However, it is clear that while the ability to modify song extends across these species, the ways in which the learning works and the form of the behavior vary so much that reproductive isolation and selection based on variation in vocalization are thought to have caused much of this speciation. Thus, the acquistion pattern described above differs for each species studied in such factors as timing and duration of the sensitive period for auditory learning, as well as the extent to which boundaries for the sensitive period are rigid. Species differ in the amount of exposure needed for learning (see, for example, Hultsch & Todt, 1989a; Hultsch, 1992), and the extent to which models are learned as wholes or as collections of sound units that may be selected from or recombined (Williams & Staples, 1992). As discussed above, there are endogenous, species-specific constraints on the nature of acceptable song models; species vary as well in which characteristics are evaluated to determine suitability as a model, and in how much flexibility exists in matching these constraints. The variation between species makes it daunting to begin to study a new species; it is possible that this has limited study of the neurobiology of song to zebra finches, canaries and a handful of others. As will be seen in the discussion below, the small size, limited song repertoire and extremely rapid development of zebra finches as well as their atypical endocrinology make them a less than ideal choice for intensive study either of song learning or of song system neurobiology. Furthermore, I will suggest that interspecies variation in behavior can be a powerful tool both for understanding relations between singing behavior and neurobiology, and for inferring how these evolved.

An overview of song system neurobiology

Anatomy

In the past decade, many exciting advances have been made in understanding the neurobiology underlying song learning and production in songbirds. These began with the delineation of the neural structures that innervate the syrinx, the avian vocal organ. They have been facilitated by the discovery that song is controlled by nuclei that are clearly defined and are interconnected in a relatively simple arrangement (Figure 3.1). High vocal control center (HVC) projects to nucleus robustus archistriatalis (RA) (Nottebohm *et al.*, 1976; Nottebohm, Kelley & Paton, 1982; Gurney, 1981). Dorsal RA projects to the dorso-medial nucleus of the intercollicular complex (DM) in the midbrain, which in turn projects to the caudal portion of the hypoglossal nucleus in the brainstem (nXIIts). Ventral RA projects directly to nXIIts (Nottebohm *et al.*, 1976; Vicario, 1991). NXIIts projects to the muscles of the syrinx. Each of these structures exists bilaterally and each of these projections is exclusively unilateral (even the syrinx is essentially two structures, one at the opening of each bronchus, each of which is independently innervated). Lesions that destroy any of these nuclei or pathways bilaterally result in immediate, profound and irreversible deficits in song (Nottebohm *et al.*, 1976; Simpson & Vicario, 1990), leading to the suggestion that the motor pathway for song begins with HVC and consists of the direct projection to RA, and its projections to nXIIts.

HVC also contributes to another circuit. It projects to Area X, a large nucleus in lobus parolfactorius, which projects to DLM (medial portion of the dorsolateral thalamic nucleus), which in turn projects to LMAN (lateral portion of the magnocellular nucleus of the anterior neostriatum) and finally to RA (here termed the rostral circuit because of the placement of its two principal components). Lesions to LMAN or to Area X in mature birds that are singing result in little or no apparent deterioration in song (Nottebohm *et al.*, 1976; Bottjer, Miesner & Arnold, 1984; Sohrabji, Nordeen & Nordeen, 1990; Scharff & Nottebohm, 1991). However, lesions to either area in juvenile birds prior to song crystallization result in severe impairments in the eventual song (Bottjer *et al.*, 1984; Sohrabji *et al.*, 1990;

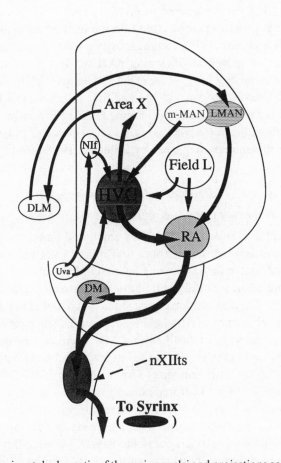

Figure 3.1 Horizontal schematic of the major nuclei and projections comprising the avian song system. All nuclei occur in both hemispheres but, with the exception of Uva → Uva, there are no known projections from any of these structures in one hemisphere to any in the other. Shading indicates relative number of cells in the nuclei that concentrate androgens. Field L is a high auditory area that projects to areas adjacent to HVC and RA. There appear to be two functionally distinct paths in the song system: caudal projections necessary to song production (HVC → RA → nXIIts), and rostral projections necessary to song acquisition (HVC → Area X → DLM → LMAN → RA). (See text for further explanation.)

Scharff & Nottebohm, 1991), suggesting that the rostral circuit may play a special role in song acquisition.

HVC and RA receive indirect projections from Field L, an area of the forebrain that contains neurons that respond to complex auditory stimuli (Kelley & Nottebohm, 1979). Neural activity can be evoked by sound in all nuclei of the song system (including nXIIts). As the latencies for this activation are shortest for HVC and RA (Williams, 1990b), this pathway from Field L probably conveys the auditory information which then is distributed throughout the song system. HVC also receives projections from several smaller nuclei whose role in song acquisition and performance is currently under study (reviewed by Williams, 1989, 1990b).

Developmental neurobiology

Formation and differentiation of song system nuclei

The telencephalon of songbirds is very small and immature at hatching. There are still large germinal zones and most of the lamina cannot be distinguished. In the days following hatching, there are high levels of cell division, migration and differentiation which are associated with rapid brain growth. These events have been studied in detail in the song system, primarily in zebra finches. At hatching, none of the telencephalic nuclei of the song system can be distinguished. Division of neurons that will constitute RA occurs primarily in the last few days before hatching and is complete by 5 days after hatching (Kirn & DeVoogd, 1989). While many HVC neurons are formed before hatching, some cell division continues to occur throughout later life (Goldman & Nottebohm, 1983; reviewed by Nottebohm et al., 1990; Nordeen & Nordeen, 1990). This late cohort includes cells projecting to RA but not to Area X (Alvarez-Buylla, Theelen & Nottebohm, 1988). In Nissl-stained tissue, RA can be distinguished at about 5 days after hatching, HVC at about 10 days and LMAN at about 2 weeks (Kirn & DeVoogd, 1989; reviewed by DeVoogd, 1991; Arnold, 1992). All three are small initially and increase rapidly in volume in the following 2–3 weeks as additional cells migrate into place and as axonal and dendritic trees are elaborated. These nuclei are large and well differentiated in zebra finches of both sexes in the fourth week after hatching. HVC and then RA become dimorphic, initially because the rate of cell death is much higher in females than in males, and later because of dendritic regression in females and continued growth of processes in males (Bottjer, Glaessner & Arnold, 1985; Konishi & Akutagawa, 1985, 1990; E. J. Nordeen, Nordeen & Arnold, 1987; Kirn & DeVoogd, 1989). It is not known how this dimorphic development is set in motion. It may follow from a pulse of estradiol in males soon after hatching that does not occur in females, in a manner similar to that found in mammals. However, while the role of a steroid pulse is theoretically plausible, direct evidence for it is ephemeral at best. It is also possible that titers of steroids are similar in the two sexes after hatching, but the hormones have divergent effects because they act on neural receptors that vary between the sexes in number or distribution, perhaps as a result of hormonal differences prior to hatching (Adkins-Regan *et al.*, 1990; personal communication). Female zebra finches given large doses of estradiol throughout the first 2–3 weeks after hatching retain large song system nuclei (Gurney, 1981; Konishi & Akutagawa, 1985; E. J. Nordeen *et al.*, 1987) and can later be induced to sing (Gurney, 1982; Pohl-Apel & Sossinka, 1984). Many of the neurons in HVC, a zone along the ventricle medial to HVC, DM and LMAN, accumulate estradiol in young zebra finches (K. W. Nordeen, Nordeen & Arnold, 1987; reviewed by DeVoogd, 1991; Arnold, 1992), thus it is possible that the steroids in the circulation affect the developmental anatomy and physiology of these regions. Whatever the means by which dimorphic development is initiated, one early consequence appears to be a sex difference in song system connectivity: axons from HVC project into RA in the fifth week after hatching in males and never do so in females (Konishi & Akutagawa, 1985). (This projection does exist in females of other species – see, for example, Alvarez-Buylla *et al.*, 1988; Volman, 1991.) Another consequence appears to be sex differences in neuro-endocrinology. The number of neurons in HVC with receptors to estradiol decreases substantially in females but not in males after 40 days of age (Gahr & Konishi, 1988). The number of cells in HVC and LMAN that accumulate androgens increases in males but not in females. Many of the neurons in HVC that concentrate steroids are projection neurons (K. W. Nordeen *et al.*, 1987). These results suggest that the song system of adult male zebra finches is able to respond directly to gonadal steroids in ways that these regions in females cannot. They suggest that the effects of steroids could easily cascade to other neurons and nuclei that do not have steroid receptors.

The development of LMAN follows a different pattern. The nucleus develops quickly in both sexes (Bottjer *et al.*, 1985). A large projection from LMAN to RA develops (Herrmann & Arnold, 1991), in which synapses are functional by 15 days after hatching, in contrast to the more slowly developing projection from HVC to RA (Mooney, 1992). While this might suggest that the rostral circuit could become functional before the direct projection from HVC to RA, the projection from DLM to LMAN does not form until 45 days after hatching (Bottjer, 1987). In the second month after hatching, the volume of LMAN then appears to regress in both sexes (Bottjer *et al.*, 1985). In both sexes, the size of neurons within LMAN and of their dendritic trees also decrease (E. J. Nordeen *et al.*, 1987; Nixdorf & DeVoogd, 1989). Up to 80% of the LMAN synapses in RA are lost during this time (Herrmann & Arnold, 1991). Aspects of the regression are dimorphic: more of the axons projecting from LMAN to RA are lost in females than in males (Nordeen *et al.*, 1992), which may contribute to the lack of definition of the nucleus in adult females. Thus in LMAN, in contrast to other major nuclei of the song system, there are major changes in morphology and connectivity that are not directly related to endocrine changes.

Little is known of the development of other connections between song system nuclei. In particular, it is not known when auditory information becomes available to each of the major nuclei in the song system.

Selectivity in cellular physiology

In mature songbirds, auditory activity can be evoked throughout the song system, including nXIIts (Katz & Gurney, 1981; McCasland & Konishi, 1981; reviewed by Margoliash, 1987; Williams, 1990b). In particular, cells can be found in HVC, Area X, DLM, LMAN and RA that respond selectively to conspecific song (Margoliash, 1983, 1986; Doupe & Konishi, 1991). Neurons in each of these areas respond better to the bird's own song than to that of another conspecific, and better to the song played forward than backward (Williams & Nottebohm, 1985; Doupe & Konishi, 1991). Intriguingly, such units are silent when the song syllable to which the unit is sensitive is produced by the bird itself (McCasland & Konishi, 1981). Apparently production of song inhibits selectivity to the song, inhibition that originates in the thalamic nucleus uva (Williams, 1989). Recently, Margoliash & Fortune (1992) have found that neural selectivity for a song in zebra finches is caused by sensitivity to the combination of two or more sound elements from the song – an exciting step toward understanding how the songbird nervous system decodes a song.

Much less is known of how the song system encodes a song for production. McCasland (1987) has demonstrated that there are neurons in HVC whose activity predicts the production of a specific syllable of zebra finch song, activity that is not evoked by hearing the syllable in a song. There are also neurons whose activity predicts all syllables that are about to be produced, and others that are excited by production of a syllable and inhibited by hearing it. Nucleus NIf is required to initiate normal singing. HVC activity associated with song production is preceded by a burst of activity in NIf, and lesioning NIf results in the loss of stereotyped phrase structure and alterations in syllable content and ordering.

To date, the development of selectivity for song has only been studied in HVC. In white-crowned sparrows singing full song or well-formed plastic song, HVC neurons that are selective to a particular song respond more strongly to ('prefer') the bird's own song to the tutor song from which it was derived, and prefer the tutored song to songs from another dialect (Margoliash, 1986; Volman, 1991). In sparrows that are 3.5–7 months old and producing subsong, neurons in HVC respond to auditory stimuli and sometimes are highly responsive to song. However, there are as many cells that prefer reversed song over normal song in these young birds as there are cells that prefer normal over reversed, and as many cells prefer the alien dialect as prefer the tutored one (Volman, 1991, unpublished data). Thus, while selectivity for complex features of auditory stimuli exists very early in HVC, the striking specificity that HVC neurons show for a bird's own song in adulthood is formed as the song is being created. Further research is needed to determine how information from afferents affects this process, and to determine how selectivity is formed in other nuclei in the song system. No research has yet systematically looked for selectivity in song system neurons of adult females; it would be very interesting to know whether similarly precise tuning exists and, if so, whether it is based on songs heard as a juvenile or songs relevant to adult life.

Plasticity in the adult song system

Most songbirds that reproduce outside the tropics restrict their breeding to a preferred season. Typically, increasing daylength induces a rise in the levels of gonadal steroids which leads to the physiological and behavioral changes associated with reproduction. For males of many species, these include territory acquisition, mate attraction – and singing. Song resumption is usually associated with a brief interval of plastic song. In many species, frequency of singing wanes late in the breeding season and may cease completely until late winter when the cycle repeats. Species vary in song organization across reproductive seasons. Some, like the white-crowned sparrow and the rufous-sided towhee, resume singing the identical song used in the prior season (Marler, 1970; Brenowitz *et al.*, 1991). Others, like canaries and red-winged blackbirds, tend to use non-reproductive times to alter or augment their repertoire (Nottebohm *et al.*,

1986; Kirn *et al.*, 1989). Females also show cyclicity in song-related behaviors, although this has been less thoroughly studied. Females become receptive to pair formation in response to song as daylength increases. Recent data suggest that song discrimination may be easier for females when days are long than when they are short (Cynx & Nottebohm, 1992).

There is a matching cyclicity in song system neurobiology (reviewed by DeVoogd, 1991). Nottebohm (1981) showed that HVC and RA appear substantially larger in the spring than in the fall in adult male canaries. Subsequent research has shown that there are also seasonal changes in these structures in several other species (Kirn *et al.*, 1989; Arai, Taniguchi & Saito, 1989; Brenowitz *et al.*, 1991). Changes occur in the volumes of Area X and nXIIts. The seasonal change in HVC is probably caused by changes in the morphology of boundary cells rather than by actual shrinkage and expansion of the nucleus, as the volume of the nucleus and the number of cells within it do not change seasonally when HVC is measured as the region that contains estrogen receptors or as the region that projects to RA or Area X (Gahr, 1990; Kirn, Alvarez-Buylla & Nottebohm, 1991). The volumes of RA and nXIIts do actually change. In RA, the volume increase in the spring is associated with longer dendrites and an increase in spine density (Hill & DeVoogd, 1991). In both RA and nXIIts, the number of transmitter vesicles per synapse increases (DeVoogd, Nixdorf & Nottebohm, 1985; Clower, Nixdorf & DeVoogd, 1989). The changes in volume are proportionately larger in females than in males (Kirn *et al.*, 1989).

It is generally believed that the rise in steroids of late winter found in cyclic songbird species leads, directly or indirectly, to the behavioral and neural changes described above. Supporting evidence has come from research on plastic changes in song system anatomy induced by ex-perimental manipulation of steroid levels. Thus, castrating male canaries results in a substantial decrease in the volumes of HVC and RA (Nottebohm, 1980). Giving testosterone to adult female canaries results in singing and in a rapid increase in the volumes of HVC, RA and nXIIts and in the size of cell bodies within these nuclei and LMAN (Nottebohm, 1980; DeVoogd *et al.*, 1985; Bottjer, Schoonmaker & Arnold, 1986; Brenowitz & Arnold, 1990; DeVoogd, Pyskaty & Nottebohm, 1991). This is associated with a variety of striking cellular changes: dendritic growth, synapto-genesis, and synaptic hypertrophy (DeVoogd & Nottebohm, 1981; DeVoogd *et al.*, 1985; Clower *et al.*, 1989). There is a massive increase in the genesis of glial and endothelial cells, the latter suggestive of increased vascularization (Goldman & Nottebohm, 1983).

These results demonstrate that the intimate relations between endocrine state and anatomy persist in the song system of adult songbirds. They raise intriguing questions about function: how do these morphological changes affect physiology and behavior? Which of the neural changes cause the changes in behavior and which are caused by them? They show that neural structure in the brain of an adult may be static but need not be fixed: even in the telencephalon, massive neural reorganization is possible in response to physiological shifts in daylength and steroid levels.

Questions and issues for further inquiry

What is the neural basis for song learning?

In spite of the extensive research described above, surprisingly little is known of the neural basis for song learning and song production. Numerous correlational studies have described the development of anatomical features in the song system and have noted which aspect of behavioral acquisition is occurring as the anatomical feature changes. This approach has been useful, but primarily in ruling out associations. For example, division of cells destined for RA is complete by 1 week after hatching (Kirn & DeVoogd, 1989). Clearly, this cell division is not a feature that is molded to encode aspects of song learning as this occurs much later. However, neurons destined for HVC (and many other parts of the forebrain) continue to be formed into adulthood (reviewed by Nottebohm *et al.*, 1990), allowing speculation that this feature could be responsible for behavioral plasticity. In this case, as in many of the developmental studies on other features of anatomy, it has proven difficult to distinguish coincidence or consequence from causation. There have been a few key findings or inferences that are likely to be productive in guiding future progress on these questions.

Special role for rostral nuclei of song system

First is the observation that Area X, LMAN and the rostral circuit of which they are part, are needed for organization of song. As described above, lesions to these rostral nuclei in juvenile males result in large impairments in the organization of the song that is subsequently formed. Lesioning LMAN results in development of a stereotyped repertoire but there are too few syllables, they are abnormally simple, and are produced in abnormally long bouts (Bottjer *et al.*, 1984). Lesioning Area X results in a song that never crystallizes but continues indefinitely to be produced in a variable, tentative way (Sohrabji *et al.*, 1990; Scharff & Nottebohm, 1991) If either is lesioned in reproductive adult males, singing continues with no apparent deficits. These observations led to hypotheses that the integrity of this rostral pathway is necessary for song learning, perhaps chiefly for the phase of auditory-motor integration (Bottjer & Arnold, 1986; Sohrabji *et al.*, 1990). Two recent observations modify and extend this perspective.

First, synapses in RA made by axons from LMAN have N-methyl-D-aspartate (NMDA) receptors (see Bolhuis, this volume) and those made by HVC axons do not (Mooney & Konishi, 1991; Kubota & Saito, 1991). In other neural systems, NMDA synapses act as comparators, transmitting a message only if they receive an action potential at a time when the postsynaptic process is already depolarized. This occurs if non-NMDA receptors, either on the same synapse or on ones nearby on the dendrite, have recently been active. In mammalian systems, NMDA transmission results in prolonged changes to the dendrite, including structural modifications and potentiated physiology. If the synapses in RA are similar, they could reinforce nearby synapses whose activity matches their own. Pushing speculation a bit further, if spontaneous activity in RA causes the unstructured sounds of subsong, selective synaptic strengthening directed by LMAN could lead to matching of auditory and motor selectivity in RA neurons.

Second, the role of LMAN does not end with initial acquisition of song. If LMAN is lesioned in adult male canaries in the fall, when they are not producing stereotyped song, their subsequent song is substantially disrupted (Suter *et al.*, 1990). As indicated above, treatment with testosterone induces singing in adult female canaries. Lesions of LMAN in such females have effects that are closely related to the timing of the lesion (Hill, Carlson & DeVoogd, 1991; unpublished data). If LMAN is lesioned before steroid treatment and the onset of singing, the eventual song has only one or two syllables that are highly stereotyped and are produced in unusually long bouts. If instead, LMAN is lesioned after testosterone treatment and the onset of stereotyped singing, no change occurs in the number of syllables and only small changes in their form. Perhaps input to RA from LMAN is needed whenever there is new motor learning. Canaries, which revise their songs from one season to the next, would then periodically reactivate the rostral circuit. Alternatively, it may be that the rostral circuit allows reference to the 'archival' version of song, which is needed after disuse for restoring fidelity to sounds already learned, providing access to sound motifs learned in plastic song but unused, as well as for the initial motor learning. This hypothesis would suggest that LMAN would continue to be needed in any species that sings phasically, whether or not it shows novel motor learning. It leaves unanswered the nature of the message being conveyed from the rostral circuit, and what role these projections play in song system function during other seasons of the year.

Correlations between anatomy and behavior

One of the guiding axioms in developmental neurobiology has been the belief that structure and function in the nervous system are associated: there are no structures without functions, and structure cannot be augmented without in some way augmenting function. Many of the findings in the song system are congruent with this point of view. For example, males that sing more elaborate songs than females have larger song system nuclei than the females. In males or females, initiation of song is associated with an increase in the volume of song system nuclei. However, two sets of observations seem paradoxical or incomplete from this perspective.

First, neurons from the song system of females have elaborate dendritic trees several weeks after hatching, and subsequently become different from those of males, primarily by regression. However, song system function in females has been studied very little, and the research that has been done has focused almost exclusively on adults. If enhanced form is associated with enhanced function, some functional attribute must be potentiated in females during this time, an attribute not yet identified or described.

Similarly, LMAN appears large and the neurons within it have elaborate dendritic trees at about 1 month after hatching. This occurs in both males and females and in both sexes it is followed by regression. In initial observations, this change was related to the apparent transient role of LMAN in song acquisition. However, this now seems unlikely. The regression happens near the beginning of the sensitive period for acquiring song characteristics – not at the end or afterward as would be predicted if temporarily enhanced anatomy is associated with this restricted period of learning. The volume of LMAN does not appear to regress in the fall in adult birds, although lesioning the nucleus at this time causes deterioration in song similar to that observed with lesions in juvenile birds. Furthermore, these initial explanations do not account for the data showing that the developmental pattern is similar in males and females, although female zebra finches never sing. How else might we explain the relations between the structure of LMAN and its function? In order for the early anatomical changes in LMAN to be related to learning, it must be a form like auditory

learning, that occurs in females as well as males. But since lesions of
LMAN can interfere with song acquisition even long after auditory
learning, it is possible that the early changes in LMAN have little to do
with learning. Alternatively, the role of LMAN may change with
development such that the song disruption caused by juvenile and adult
lesions is only superficially similar. Juvenile lesions then could disrupt
auditory learning, while later lesions might interfere with other processes
(perhaps auditory memories). Such a hypothesis would explain all results
to date except why the initial growth of LMAN occurs prior to auditory
learning. This could occur if a novel mechanism for learning is used in
LMAN.

Synaptic selection model

Learning song from an external model is similar in many ways to other
forms of early learning such as imprinting (see Bolhuis, Ten Cate, this
volume). For each, there may be a sensitive period during which acquisition
is most rapid. For each, there are constraints on the stimuli that are
acceptable for learning. Learning in both is facilitated by increasing the
richness of the stimulus as by adding visual cues to auditory (cf. Bolhuis &
Van Kampen, 1992). Furthermore, once acquired, these forms of learning
are relatively resistant to change, even though they may not be expressed
until the animal is mature. Black & Greenough (1991) have described such
forms of learning as experience-expectant, and have suggested that they
may be sufficiently distinct from other sorts of learning that they may be
associated with distinct sorts of change in the neural substrate responsible
for them. Studies of several forms of early neural plasticity suggest that
they may have in common the use of selective or subtractive neural
mechanisms, rather than adding or potentiating connections as is
commonly suggested for other forms of learning. Auditory imprinting in
chicks is a good example (reviewed by Scheich, Wallhäusser-Franke &
Braun, 1992). Prior to experience with an imprinting stimulus, there are
high numbers of spine synapses in an auditory association area. If the
chicks are deprived of experience with sound, the high spine density
(number per unit length of dendrite) persists. Experience with complex
audio-visual stimulation results in a moderate spine density (i.e. a moderate
reduction), and there is a large decrease in density after exposure to a
simple sound stimulus. The authors suggest that the neural connections
and networks responsible for this form of learning are created prior to and
independent of experience with the imprinting stimulus. When the
imprinting stimulus arrives, its characteristics are learned by preserving the
subset of synapses sufficient to encode it (but see also Bolhuis, this volume).

Preliminary data suggest that a similar process may be associated with the auditory phase of avian song learning. As noted above, complex dendritic fields and high levels of synaptogenesis occur in song system nuclei of zebra finches in the first 4 weeks after hatching, a time well before any evidence of acquisition of song features. At least in LMAN, there are data suggesting that these anatomical features are reduced in normally housed birds during the time that song features are learned (Nixdorf & DeVoogd, 1989). We have recently found that rearing male zebra finches to 55 days without exposure to song, a treatment known to prolong the sensitive period for song acquisition, results in a substantially greater synaptic density in HVC than is found in non-deprived birds (Collins, Clower & DeVoogd, 1992). Perhaps the neural representation of a sensitive period for song acquisition consists of a vast array of endogenously induced synapses potentially capable of encoding all possible songs. This hypothesis could be tested by measuring more precisely the timing of synaptogenesis and synaptic elimination in song system structures, and determining whether addition occurs before auditory learning and sub-traction occurs with the learning. Perhaps, as in chick auditory imprinting, learning consists of eliminating synapses not needed for retention of the sound features to which the young bird is exposed. This could be tested by evaluating the neural consequences of varying degrees of auditory deprivation, and relating these results to alterations in the timing or capacity for vocal learning.

Prefunctionality in the song system

Why does a young songbird readily learn the song of its own species and, typically, not learn the songs of other species even if these sounds are as common as the conspecific song? Much of the preceding discussion has considered the process of learning and its associated neurobiology. However, any consideration of the natural environment of most songbirds suggests that learning preferences or constraints on learning are critical to correct song acquisition.

Data obtained using several experimental paradigms have shown that these restrictions are powerful and extensive. They have many implications for the neurobiological substrate for song. First is the observation that if a young bird is exposed to both a conspecific and a heterospecific song, it typically learns the conspecific song. However, if presented only with the heterospecific song, a bird may acquire some or all of the alien song, showing that the selectivity is not based on perceptual or motor incapacities (Marler, 1991). Furthermore, in instances where the acquisition is partial, the 'errors' typically modify the song to be more like the species-typical song.

Second are the observations on the behavior of birds raised without exposure to any song. Such birds produce a song when they mature and the song, though very abnormal, contains sound features characteristic of the species. Thus, a male zebra finch raised by a female rather than by both sexes, ultimately forms a song in which call notes learned from the adult females are produced in a stereotyped pattern similar to that used for the syllables that make up normal zebra finch song (Price, 1979). If a young bird is raised in total isolation, it will sing a stereotyped song in which abnormal sounds are produced in a pattern like that used by individuals with normal auditory experience. Even with the extreme auditory deprivation of deafening, singing occurs and has characteristics predictable on the basis of the species' normal song.

The strong predispositions evident in song acquisition point to a high level of organization in the song system of very young birds. They suggest, for example, that in the projections between Field L, in which sound is tonotopically organized, and major nuclei of the song system, there is a neural 'filter', designed to pass on sounds satisfying the constraints shown by behavioral studies. Such a filter could be the neural instantiation of the template that is used as the reference for song production. Acceptable sounds would then preferentially activate song system nuclei. This, in turn, could strengthen connections in this network and lead to greater selectivity in the responsiveness of neurons within the song system or, equivalently on a perceptual level, would increase the preference of this song template to later sounds of self and of others that are like the learned ones. If the bird grows up without optimal stimulation, either this circuitry would be modified so as to pass more sounds, or ongoing suboptimal stimulation would have a cumulative effect such that aspects of alien sounds could be learned. Components of the song system that organize vocal output must also have substantial specificity prior to or independent of production of learned sounds. Parts of the system may form circuits prior to interaction with the auditory environment, that act as oscillators or pattern generators (see discussion by Mooney, 1992). These could encode the aspects of song that are typical of the species and that will be produced even in the absence of external song models, such as song length, number of sound units in a song, number of repetitions of a unit, or song rhythm.

These attributes in the development of avian song resemble patterns that have been called prefunctional in other systems (Hogan, 1988). Preformed or partially formed perceptual mechanisms are analogous to a template for song. The soft and non-directed vocalizations of subsong could be seen as activation of prefunctional motor units (i.e. use of a behavior prior to its adaptive context). From this framework, development can be seen as linking together of these preformed perceptual and motor units so as to enable complex behaviors in appropriate contexts. For avian song, it now seems possible to study these linkages, not merely in the growing complexity of behavioral expression, but in the literal joining of sensory and motor pathways in the brain.

Relations between structure and function in the song system

The song system is an attractive preparation in which to study the organization of a learned motor skill, in part because the circuitry which makes up the system is relatively simple. Projection patterns are clear and consistent (even between songbird species), and nuclei seldom receive projections from, or project to, more than three other loci. A major goal of research on this system has been to understand the ways in which signals are transformed within and between nuclei, leading ultimately to the muscle contractions that result in song. Not only has this not been done, but there are few plausible hypotheses on the site(s) of motor learning, or on the ways in which neural interaction at higher levels controls the muscles of the syrinx. Three experimental approaches offer opportunities for further advances in understanding these issues.

Functions of individual nuclei – physiology and lesions

First is the systematic use of conventional neuroscience tools: assessment of the electrophysiology of neurons in song system nuclei and of the behavioral and physiological consequences of lesions. We now know that subdivisions of RA project to different parts of nXIIts (Vicario, 1991), and that different zones of nXIIts project to different muscle groups in the syrinx (Vicario & Nottebohm, 1988). Further research is needed on whether there is systematic mapping of projections between Field L, HVC, and RA, and on the circuitry within any of the song system nuclei. As indicated above, many neurons in HVC become highly selective in the auditory stimuli to which they show the greatest responses, often being most responsive to the bird's own song. Understandably, most research in this area has related these responses to the songs the bird has heard and to its own emerging song. There is good evidence that prefunctional organization is important for both sensory and motor capacities of the song system. Thus it will also be interesting to determine whether neurons show a selective response to species-typical song features (perhaps like those retained by birds raised in isolation) prior to or in the absence of experience with song.

Much can also be gained from lesion studies. Such studies have shown the functional division between the rostral and caudal song system nuclei, have produced evidence for lateralization of song in canaries, and more recently, evidence that HVC may play a role in song discrimination in females (Brenowitz, 1991). Further insights on song system function could come from studies on the consequences of lesions in smaller song system nuclei such as uva or NIf, or of partial lesions of HVC or RA.

Seasonality – a tool for understanding function

A second approach to understanding song system function is through analysis of changes in song system circuitry across seasons in adult birds. We now know that large seasonal changes can occur in the morphology of neurons in HVC, RA and nXIIts without the animals necessarily losing any learned songs. This suggests that the learned song features are encoded either in the synapses that remain during the regression of neurons in these regions during fall, or in structures that are less affected by seasonal change such as LMAN. If the latter is true, neurons within HVC should lose their selectivity for the bird's own song in the fall and regain it as singing recurs in the spring. In contrast, any selectivity in the responses of neurons with LMAN should not change across seasons. Song production is organized by the caudal nuclei in the song system, and song characteristics such as stereotypy or rhythmicity are preserved or even enhanced when LMAN is damaged. Thus, it is likely that these attributes are encoded in the lasting circuitry of caudal nuclei, suggesting that any selectivity of HVC or RA neurons for these characteristics would persist across seasons. These predictions have not yet been tested.

Proportionately larger anatomical shifts occur across seasons in females. As indicated above, function of the song system in females has been minimally studied; almost nothing is known of how the function of this network would be altered by the changes in anatomy in females.

Comparative analyses of brain and behavior

Finally, comparative analyses can be used to assess structure-function relations in the song system. For example, in a group of related species, major nuclei of the song system of females are larger in the species in which females sing elaborate songs than in species in which they do not (Brenowitz & Arnold, 1986). Recently, an extensive comparative study has been completed on males from 45 songbird species (DeVoogd *et al.*, 1992). The relative volumes of HVC and Area X were compared with several sorts of song measures characterizing each species. After using statistical procedures designed to factor out effects of common phylogeny (Felsenstein, 1985), a strong association was found between relative HVC volume and the size of the song repertoire typical for individuals in the species. There were no significant correlations either between HVC volume and any other song measure, or between Area X volume and any of the song measures. Thus, variation in one behavioral attribute is associated with variation in one brain measure, suggesting that during evolution, natural selection on the behavioral attribute acted by enhancing or eroding the neural substrate most responsible for that attribute. If true, this would imply that functions of other song system structures could be determined by similar studies that examine a larger number of nuclei and a wider range of song attributes.

Developmental links between song and other neural systems

Perceiving song and producing song are closely tied together throughout development, as summarized above. There is good evidence that the union of the two behaviors is literal: single neurons in the song system are activated by song and are active in producing song. However, song is not an isolated behavior. Careful study of song shows that it is closely tied to other neural systems, both in its development and execution. There are many instances in which involvement of other perceptual systems has been demonstrated. For example, young birds are willing to learn song from a living, interactive tutor over a much longer interval than from a tape recording (Baptista & Morton, 1988). Young zebra finches, exposed to two males, choose which song to copy, apparently on the basis of how they or their mother interact with the males (Clayton, 1987b, this volume). In cowbirds, song in first year males is substantially shaped by the quality of social interactions with adult females. Initial affiliation with the females cannot be based on knowledge of cowbird song (as the babies grew up in the nests of other species), so it must be based on other auditory cues or vision. In either case, this suggests that song acquisition is tied to other perceptual systems.

It is common across many species for many songs to be learned and practiced, and for the bird to select a subset which forms the adult repertoire (Marler & Peters, 1982). It is becoming clear that this selection can be based on social factors – selecting songs that most resemble those of neighboring birds. Marler (1990) has named this form of song modification in adulthood 'action-based learning', and has pointed out that similar selective mechanisms may account for a variety of observations on variability in singing behavior. Even after the song repertoire is pared down and fixed, social factors play a large role in the execution of song. For example, the presence of a conspecific often facilitates song production (species vary in whether a male or a female is more effective). The pattern of singing may vary with the intended audience: male red-winged blackbirds organize the production of their repertoire very differently when singing to a female than when singing to a male (Searcy & Yasukawa, 1990). Also, hearing a song may induce a territorial male to respond by selecting from its repertoire and repeatedly singing the song that is most similar to that which the bird heard. Such data suggest that the song system is continually monitoring or being modulated by neural systems involved in social relations.

Singing also interacts with a variety of other motor systems and responses. At a very elementary level of description, activation of the syrinx for song is coordinated with respiration and posture. At a more complex level, song is produced as part of interactive sequences of behaviors that may be prolonged and complex and are poorly understood. Singing in mature birds often occurs in response to hearing a song. This response can become part of a cascade of agonistic behaviors, generally including approach, and potentially including direct fighting. The pattern and magnitude of response are modified by such factors as the nature of the song (neighbor or stranger) and the site of the initial song (inside or outside the territory) (Stoddard *et al.*, 1991). Such data suggest that song production is integrated with memories based on prior social interactions and with spatial knowledge about the intended receiver. It is linked to other complex behaviors: fighting with males or displaying for females, for example. Each such coordination suggests links between the song system and other neural systems. To date, little is known of projections that might be used to integrate information in these domains or to coordinate output.

Singing is also integrated with less specific neural systems. It is coordinated with many other behavioral patterns in the reproductive season of the year. It is possible that rising steroid levels activate song, female recognition, territoriality and mating behaviors in synchrony, but it also seems plausible that activation of one of these behavioral clusters can facilitate activation of others. Singing is frequently linked to increases in central 'state' factors such as arousal or motivation. Here too, behavioral data imply coordination between neural systems – but very little is known of neural projections that could bring these systems together or relay information between them.

Overview of major themes

Inevitably, a review of an active scientific area resembles an incomplete tapestry: there are loose threads that appear unconnected to the overall framework, patterns that do not seem to make sense, and uncertainty about the place at which further work will produce the most illuminating results. In spite of this, I would like to underscore some major themes from this review that are currently somewhat tangential to standard research in this area.

The first is the concept of constraints. Song acquisition is limited in time and in content. Both sorts of constraint have substantial implications for the neurobiology of the song system: much of its structure must develop prior to or independent of experience with song, and the actions of this existing structure then shape subsequent perception and learning. Experience may affect neurobiology, but within time windows and by acting on pre-existing substrates.

The second theme is related: the possibility that anatomical organization can be prefunctional, in the same sense that this concept is applied to the development of behavioral patterns. Thus, there is endogenously organized circuitry designed to perceive features that define a conspecific song, and circuitry designed to produce sound in species-typical rhythms and patterns, both of which are operational before they are integrated into a system that can produce a song. When the song system is first used in subsong, learned sound features are not linked together appropriately and the behavior is not activated by appropriate releasers. Even plastic song, though recognizably learned song, is not appropriately linked with other social behaviors. Thus, as in other systems, units of motor action have been completed and are functional before they can play their eventual role, because they have not yet been joined together.

Third are the close ties between anatomy, even at a descriptive level, and behavior. This is clearly seen across development (although interpretation of some specific anatomical shifts remains unclear). It is also seen in the seasonal changes of adulthood, in the differences between the sexes and, occasionally in differences between individuals. In each case, 'more' of a function is associated with more of a structure or its constituents. There are even powerful relations between anatomy and behavioral capacity that are evident between species, and represent accumulated shifts across evolutionary time. These suggest that essential aspects of the processing of song perception and production in song-related brain areas (and the synaptic circuitry they require) have been relatively stable for millions of generations.

Finally, there is the clear evidence that song learning is not a unitary phenomenon. Song acquisition and expression involve many forms of learning from modification of an auditory template to assembly of motor coordinations to integration of context and consequence. This behavioral evidence creates a strong presumption that changes in the song system to encode 'song learning' are also not unitary, but are likely to vary in locus and mechanism for different components of the process.

Acknowledgements

Many thanks to Joyce Reed for editorial assistance. Supported by NIH award HD21033.

References

Adkins-Regan, E., Abdelnabi, M., Mobarak, M. & Ottinger, M. A. (1990). Sex steroid levels in developing and adult male and female zebra finches (*Poephila guttata*). *General and Comparative Endocrinology*, 78, 93–109.

Alvarez-Buylla, A., Theelen, M. & Nottebohm, F. (1988). Birth of projection neurons in the higher vocal center of the canary forebrain before, during, and after song learning. *Proceedings of the National Academy of Sciences (USA)*, 85, 8722–8726.

Arai, O., Taniguchi, I. & Saito, N. (1989). Correlation between the size of song control nuclei and plumage color change in orange bishop birds. *Neuroscience Letters*, 98, 144–148.

Arnold, A. P. (1975). The effects of castration and androgen replacement on song, courtship, and aggression in zebra finches (*Poephila guttata*). *Journal of Experimental Zoology*, 191, 309–326.

Arnold, A. P. (1992). Developmental plasticity in neural circuits controlling birdsong: Sexual differentiation and the neural basis of learning. *Journal of Neurobiology*, 23, 1506–1528.

Baker, M. C. (1983). The behavioral response of female Nuttall's white-crowned sparrows to male song of natal and alien dialects. *Behavioral Ecology and Sociobiology*, 12, 309–315.

Baker, M. C. & Cunningham, M. A. (1985). The biology of birdsong dialects. *Behavioral and Brain Sciences*, 8, 85–133.

Baker, M. C., Spitler-Nabors, K. J. & Bradley, D. C. (1981). Early experience determines song dialect responsiveness of female sparrows. *Science*, 214, 819–821.

Baptista, L. F. & Morton, M. L. (1988). Song learning in montane white crowned sparrows: from whom and when. *Animal Behaviour*, 36, 1753–1764.

Black, J. E. & Greenough, W. T. (1991). Developmental approaches to the memory process. In *Learning and Memory: A Biological View*, 2nd edn, ed. J. L. Martinez, Jr. & R. P. Kesner, pp. 61–91. San Diego: Academic Press.

Bolhuis, J. J. & Van Kampen, H. S. (1992). An evaluation of auditory learning in filial imprinting. *Behaviour*, 122, 195–230.

Bottjer, S. W. (1987). Ontogenetic changes in the pattern of androgen accumulation in song-control nuclei of male zebra finches. *Journal of Neurobiology*, 18, 125–139.

Bottjer, S. W. & Arnold, A. P. (1986). The ontogeny of vocal learning in songbirds. In *Handbook of Behavioral Neurobiology*, Vol. 8, ed. E. M. Blass, pp. 129–161. New York: Plenum Press.

Bottjer, S. W., Glaessner, S. L. & Arnold, A. P. (1985). Ontogeny of brain nuclei controlling song learning and behavior in zebra finches. *Journal of Neuroscience*, 5, 1556–1562.

Bottjer, S. W. & Hewer, S. J. (1992). Castration and antisteroid treatment impair vocal learning in male zebra finches. *Journal of Neurobiology*, 23, 337–353.

Bottjer, S. W., Miesner, E. & Arnold, A. P. (1984). Forebrain lesions disrupt development but not maintenance of song in passerine birds. *Science*, 224, 901–903.

Bottjer, S. W., Schoonmaker, J. N. & Arnold, A. P. (1986). Auditory and hormonal stimulation interact to produce neural growth in adult canaries. *Journal of Neurobiology*, 17, 605–612.

Brenowitz, E. A. (1991). Altered perception of species-specific song by female birds after lesions of a forebrain nucleus. *Science*, 251, 303–305.

Brenowitz, E. A. & Arnold, A. P. (1986). Interspecific comparisons of the size of neural song control regions and song complexity in duetting bird: evolutionary implications. *Journal of Neuroscience*, 6, 2875–2879.

Brenowitz, E. A. & Arnold, A. P. (1990). The effects of systemic androgen treatment on androgen accumulation in song control regions of the adult female canary brain. *Journal of Neurobiology*, 21, 837–843.

Brenowitz, E. A., Nalls, B., Wingfield, J. C. & Kroodsma, D. E. (1991). Seasonal changes in avian song nuclei without seasonal changes in song repertoire. *Journal of Neuroscience*, 11, 1367–1374.

Catchpole, C. K. Dittami, J. & Leisler, B. (1984). Differential responses to male song in female songbirds implanted with oestradiol. *Nature*, 312, 563–564.

Clayton, N. S. (1987a). Song learning in cross-fostered zebra finches: A re-examination of the sensitive phase. *Behaviour*, 102, 67–81.

Clayton, N. S. (1987b). Song tutor choice in zebra finches. *Animal Behaviour*, 35, 714–721.

Clower, R. P., Nixdorf, B. E. & DeVoogd, T. J. (1989). Synaptic plasticity in the hypoglossal nucleus of female canaries: structural correlates of season, hemisphere and testosterone treatment. *Behavioral and Neural Biology*, 52, 63–77.

Collins, C., Clower, R. P. & DeVoogd, T. J. (1992) Development of song system nucleus HVC in juvenile male zebra finches deprived of song. Song learning and imprinting: an inquiry into mechanisms of behavioural development, Abstracts, 9. University of Groningen.

Cynx, J. & Nottebohm, F. (1992). Role of gender, season, and familiarity in discrimination of conspecific song by zebra finches (*Taeniopygia guttata*). *Proceedings of the National Academy of Sciences (USA)*, 89, 1368–1371.

DeVoogd, T. J. (1991). Endocrine modulation of the development and adult function of the avian song system. *Psychoneuroendocrinology*, 16, 41–66.

DeVoogd, T. J., Krebs, J. R., Healy, S. D. & Purvis, A. (1992). Neural correlation with song repertoire size across oscine birds. *Proceedings of the Third International Congress of Neuroethology*, 293.

DeVoogd, T. J., Nixdorf, B. & Nottebohm, F. (1985). Formation of new synapses related to acquisition of a new behavior. *Brain Research*, 329, 304–308.

DeVoogd, T. J. & Nottebohm, F. (1981). Gonadal hormones induce dendritic growth in the adult brain. *Science*, 214, 202–204.

DeVoogd, T. J., Pyskaty, D. J. & Nottebohm, F. (1991). Lateral asymmetries and testosterone-induced changes in the gross morphology of the hypoglossal nucleus in adult canaries. *Journal of Comparative Neurology*, 307, 65–76.

Doupe, A. J. & Konishi, M. (1991). Song-selective auditory circuits in the vocal control system of the zebra finch. *Proceedings of the National Academy of Sciences (USA)*, 88, 11339–11343.

Eales, L. A. (1987). Do zebra finch males that have been raised by another species still tend to select a conspecific song tutor? *Animal Behaviour*, 35, 1347–1355.

Eens, M., Pinxten, R. & VerHeyen, R. F. (1991). Male song as a cue for mate choice in the European starling. *Behaviour*, 116, 210–238.

Felsenstein, J. (1985). Phylogenies and the comparative method. *American Naturalist*, 125, 1–15.

Gahr, M. (1990). Delineation of a brain nucleus: comparisons of cytochemical, hodological, and cytoarchitectural views of the song control nucleus HVC of the adult canary. *Journal of Comparative Neurology*, 294, 30–36.

Gahr, M. & Konishi, M. (1988). Developmental changes in estrogen-sensitive neurons in the forebrain of the zebra finch. *Proceedings of the National Academy of Sciences (USA)*, 85, 7380–7383.

Goldman, S. & Nottebohm, F. (1983). Neuronal production, migration and differentiation in a vocal control nucleus of the adult female canary brain. *Proceedings of the National Academy of Sciences (USA)*, 80, 2390–2394.

Gurney, M. (1981). Hormonal control of cell form and number in the zebra finch song system. *Journal of Neuroscience*, 1, 658–673.

Gurney, M. (1982). Behavioral correlates of sexual differentiation in the zebra finch song system. *Brain Research*, 231, 153–172.

Hausberger, M. & Black, J. M. (1991). Female song in European starlings: The case of non-competitive song matching. *Ethology, Ecology, and Evolution*, 3, 337–344.

Heid, P., Güttinger, H. R. & Pröve, E. (1985) The influence of castration and testosterone replacement on the song architecture of canaries (*Serinus canaria*). *Zeitschrift für Tierpsychologie*, 69, 224–236.

Herrmann, K. & Arnold, A. P. (1991). The development of afferent projections to the robust archistriatal nucleus in male zebra finches: A quantitative electron microscopic study. *Journal of Neuroscience*, 11, 2063–2074.

Hill, K. M., Carlson, A. J. & DeVoogd, T. J. (1991). Song acquisition in adult female canaries requires n. l-MAN. *Society for Neuroscience Abstracts*, 17, 1052.

Hill, K. M. & DeVoogd, T. J. (1991). Altered daylength affects dendritic structure in a song-related brain region in red-winged blackbirds. *Behavioral and Neural Biology*, 56, 240–250.

Hogan, J. A. (1988) Cause and function in the development of behavior systems. In *Handbook of Behavioral Neurobiology*, Vol. 9, ed. E. M. Blass, pp. 63–106, New York: Plenum Press.

Hultsch, H. (1992). Time window and unit capacity: dual constraints on the acquisition of serial informtion in songbirds. *Journal of Comparative Physiology A*, 170, 275–280.

Hultsch, H. & Todt, D. (1989a). Memorization and reproduction of songs in nightingales (*Luscinia megarhynchos*): evidence for package formation. *Journal of Comparative Physiology A*, 165, 197–203.

Hultsch, H. & Todt, D. (1989b). Song acquisition and acquisition constraints in the nightingale, *Luscinia megarhynchos*. *Naturwissenschaften*, 76, 83–85.

Immelmann, K. (1969). Song development in the zebra finch and other estrildid finches. In *Bird Vocalizations*, ed. R. A. Hinde, pp. 61–74. Cambridge: Cambridge University Press.

Katz, L. C. & Gurney, M. E. (1981). Auditory responses in the zebra finch's motor system for song. *Brain Research*, 221, 192–197.

Kelley, D. B. & Nottebohm, F. (1979). Projections of a telencephalic auditory nucleus -Field L- in the canary. *Journal of Comparative Neurology*, 183, 455–470.

King, A. P. & West, M. J. (1983a). Female perception of cowbird song: A closed developmental program. *Developmental Psychobiology*, 16, 335–342.

King, A. P. & West, M. J. (1983b). Epigenesis of cowbird song: A joint endeavour of males and females. *Nature*, 305, 704–706.

King, A. P. & West, M. J. (1989). Presence of female cowbirds (*Molothrus ater ater*) affects vocal imitation and improvisation in males. *Journal of Comparative Psychology*, 103, 39–44.

King, A. P., West, M. J. & Eastzer, D. H. (1986). Female cowbird song perception: Evidence for different developmental programs within the same subspecies. *Ethology*, 72, 89–98.

Kirn, J. R., Alvarez-Buylla, A. & Nottebohm, F. (1991). Production and survival of projection neurons in a forebrain vocal center of adult male canaries. *Journal of Neuroscience*, 11, 1756–1762.

Kirn, J. R., Clower, R. P., Kroodsma, D. E. & DeVoogd, T. J. (1989). Song-related brain regions in the red-winged blackbird are affected by sex and season but not repertoire size. *Journal of Neurobiology*, 20, 139–163.

Kirn, J. R. & DeVoogd, T. J. (1989). The genesis and death of vocal control neurons during sexual differentiation in the zebra finch. *Journal of Neuroscience*, 9, 3176–3187.

Konishi, M. (1965). The role of auditory feedback in the control of vocalization in the white-crowned sparrow. *Zeitschrift für Tierpsychologie*, 22, 770–783.

Konishi, M. & Akutagawa, E. (1985). Neuronal growth, atrophy and death in a sexually dimorphic song nucleus in the zebra finch brain. *Nature*, 315, 145–147.

Konishi, M. & Akutagawa, E. (1990). Growth and atrophy of neurons labeled at their birth in a song nucleus of the zebra finch. *Proceedings of the National Academy of Sciences (USA)*, 85, 3538–3541.

Korsia, S. & Bottjer, S. W. (1991). Chronic testosterone treatment impairs vocal learning in male zebra finches during a restricted period of development. *Journal of Neuroscience*, 11, 2362–2371.

Krebs, J. R. & Kroodsma, D. E. (1980). Repertoires and geographical variation in bird song. *Advances in the Study of Behavior*, 11, 143–177.

Kroodsma, D. E. (1976). Reproductive development in a female songbird: Differential stimulation by quality of male song. *Science*, 192, 574–575.

Kroodsma, D. E. (1982). Learning and the ontogeny of sound signals in birds. In *Acoustic Communication in Birds*, Vol. 2, ed. D. E. Kroodsma & E. H. Miller, pp. 5–23. New York: Academic Press.

Kubota, M. & Saito, N. (1991). NMDA receptors participate differently in two synaptic inputs in neurons of the zebra finch robust nucleus of the archistriatum in vitro. *Neuroscience Letters*, 125, 107–109.

Margoliash, D. (1983). Acoustic parameters underlying the responses of song-specific neurons in the white-crowned sparrow. *Journal of Neuroscience*, 3, 1039–1057.

Margoliash, D. (1986). Preference for autogenous song by auditory neurons in a song system nucleus of the white-crowned sparrow. *Journal of Neuroscience*, 6, 1643–1661.

Margoliash, D. (1987). Neural plasticity in birdsong learning. In *Imprinting and Cortical Plasticity*, ed. J. P. Rauschecker & P. Marler, pp. 289–309. New York: Springer-Verlag.

Margoliash, D. & Fortune, E. S. (1992). Temporal and harmonic combination-sensitive neurons in the zebra finch's HVc. *Journal of Neuroscience*, 12, 4309–4326.

Marler, P. (1970). A comparative approach to vocal learning: Song development in white-crowned sparrows. *Journal of Comparative and Physiological Psychology Monographs*, 71, 1–25.

Marler, P. (1976). Sensory templates in species-specific behavior. In *Simpler Networks and Behavior*, ed. J. Fentress, pp. 314–329. Sunderland, Mass.: Sinauer.

Marler, P. (1990). Song learning: The interface between behaviour and neuroethology. *Philosophical Transactions of the Royal Society of London, Series B*, 329, 109–114.

Marler, P. (1991). Differences in behavioural development in closely related species: birdsong. In *The Development and Integration of Behaviour*, ed. P. Bateson, pp. 41–70. Cambridge: Cambridge University Press.

Marler, P. & Peters, S. S. (1982). Developmental overproduction and selective attrition: new processes in the epigenesis of birdsong. *Developmental Psychobiology*, 15, 369–378.

Marler, P. & Peters, S. S. (1988a). Sensitive periods for song acquisition from tape recordings and live tutors in the swamp sparrow, *Melospiza georgiana*. *Developmental Psychobiology*, 15, 369–378.

Marler, P. & Peters, S. S. (1988b). The role of song phonology and syntax in vocal learning preferences in the song sparrow, *Melospiza melodia*. *Ethology*, 77, 125–149.

Marler, P. & Peters, S. S. (1989). Species differences in auditory responsiveness in early vocal learning. In *The Comparative Psychology of Audition: Perceiving Complex Sounds*, ed. S. Hulse & R. Dooling, pp. 243–273. Hillsdale, NJ: Lawrence Erlbaum.

Marler, P., Peters, S., Ball, G. F., Dufty, A. M. & Wingfield, J. C. (1988). The role of sex steroids in the acquisition and production of birdsong. *Nature*, 336, 770–772.

Marler, P. & Sherman, V. (1983). Song structure without auditory feedback: Emendations of the auditory template hypothesis. *Journal of Neuroscience*, 3, 517–531.

Marler, P. & Waser, M. S. (1977). Role of auditory feedback in canary song development. *Journal of Comparative and Physiological Psychology*, 91, 8–16.

McCasland, J. S. (1987). Neuronal control of bird song production. *Journal of Neuroscience*, 7, 23–39.

McCasland, J. S. & Konishi, M. (1981). Interaction between auditory and motor activities in an avian song control nucleus. *Proceedings of the National Academy of Sciences (USA)*, 78, 7815–7819.

Miller, D. B. (1979a). The acoustic basis of mate recognition by female zebra finches (*Taeniopygia guttata*). *Animal Behaviour*, 27, 376–380.

Miller, D. B. (1979b). Long-term recognition of father's song by female zebra finches. *Nature*, 280, 389–391.

Mooney, R. (1992). Synaptic basis for developmental plasticity in a birdsong nucleus. *Journal of Neuroscience*, 12, 2464–2477.

Mooney, R. & Konishi, M. (1991). Two distinct inputs to an avian song nucleus activate different glutamate receptor subtypes on individual neurons. *Proceedings of the National Academy of Sciences (USA)*, 88, 4075–4079.

Nixdorf, B. E. & DeVoogd, T. J. (1989). Developmental changes in nucleus magnocellularis of the anterior neostriatum (MAN) in zebra finches before and during song acquisition. In *Dynamics and Plasticity in Neuronal Systems*, Proceedings of the 17th Göttingen Neurobiology Conference, ed. N. Elsner & W. Singer, p. 221. Stuttgart: Georg Thieme Verlag.

Nordeen, E. J., Grace, A., Burek, M. J. & Nordeen, K. W. (1992). Sex-dependent loss of projection neurons involved in avian song learning. *Journal of Neurobiology*, 23, 671–679.

Nordeen, E. J. & Nordeen, K. W. (1990). Neurogenesis and sensitive periods in avian song learning. *Trends in Neuroscience*, 13, 31–36.

Nordeen, E. J., Nordeen, K. W. & Arnold, A. P. (1987). Sexual differentiation of androgen accumulation within the zebra finch brain through selective cell loss and addition. *Journal of Comparative Neurology*, 259, 393–399.

Nordeen, K. W. & Nordeen, E. J. (1992). Auditory feedback is necessary for the maintenance of stereotyped song in adult zebra finches. *Behavioral and Neural Biology*, 57, 58–66.

Nordeen, K. W., Nordeen, E. J. & Arnold, A. P. (1987). Estrogen accumulation in zebra finch song control nuclei: implications for sexual differentiation and adult activation of behavior. *Journal of Neurobiology*, 18, 569–582.

Nottebohm, F. (1980). Testosterone triggers growth of brain vocal control nuclei in adult female canaries. *Brain Research*, 189, 429–437.

Nottebohm, F. (1981). A brain for all seasons: Cyclical anatomical changes in song-control nuclei of the canary brain. *Science*, 214, 1368–1370.

Nottebohm, F., Alvarez-Buylla, A., Cynx, J., Kirn, J., Ling, C. Y. & Nottebohm, M. (1990). Song learning in birds: The relation between perception and production. *Philosophical Transaction of the Royal Society of London, Series B*, 329, 115–124.

Nottebohm, F., Kelley, D. B. & Paton, J. A. (1982). Connections of vocal control nuclei in the canary telencephalon. *Journal of Comparative Neurology*, 207, 344–357.

Nottebohm, F., Nottebohm, M. E. & Crane, L.A . (1986). Developmental and seasonal changes in canary song and their relation to changes in the anatomy of song-control nuclei. *Behavioral and Neural Biology*, 46, 445–471.

Nottebohm, F., Nottebohm, M. E., Crane, L. A. & Wingfield, J. C. (1987).

Seasonal changes in gonadal hormone levels of adult male canaries and their relation to song. *Behavioral and Neural Biology*, 47, 197–211.

Nottebohm, F., Stokes, T. M. & Leonard, C. M. (1976). Central control of song in the canary. *Journal of Comparative Neurology*, 165, 457–468.

Nowicki, S. & Ball, G. F. (1989). Testosterone induction of song in photosensitive and photorefractory male sparrows. *Hormones & Behavior*, 23, 514–525.

Petrinovich, L. & Baptista, L. F. (1984). Song dialects, mate selection and breeding success in white-crowned sparrows. *Animal Behaviour*, 32, 1078–1088.

Petrinovich, L. & Baptista, L. F. (1987). Song development in the white crowned sparrow: modification of learned song. *Animal Behaviour*, 35, 961–974.

Pohl-Apel, G. & Sossinka, R. (1984). Hormonal determination of song capacity in females of the zebra finch: critical phase of treatment. *Zeitschrift für Tierpsychologie*, 64, 330–336.

Price, P. (1979). Deveolpmental determinants of structure in zebra finch song. *Journal of Comparative and Physiological Psychology*, 93, 260–277.

Scharff, C. & Nottebohm, F. (1991). A comparative study of the behavioral deficits following lesions of various parts of the zebra finch song system: Implications for vocal learning. *Journal of Neuroscience*, 11, 2896–2913.

Scheich, H., Wallhausser-Franke, E. & Braun, K. (1992). Does synaptic selection explain auditory imprinting? In *Memory: Organization and Locus of Change*, ed. L. R. Squire, N. M. Weinberger & J. L. McGaugh, pp. 114–159. Oxford: Oxford University Press.

Searcy, W. A. & Andersson, M. (1986). Sexual selection and the evolution of song. *Annual Review of Ecology and Systematics*, 17, 507–533.

Searcy, W. A. & Marler, P. (1981). A test for responsiveness to song structure and programming in female sparrows. *Science*, 213, 926–928.

Searcy, W. A. & Yasukawa, K. (1990). Use of the song repertoire in intersexual and intrasexual contexts by male red-winged blackbirds. *Behavioral Ecology and Sociobiology*, 27, 123–128.

Simpson, H. B. & Vicario, D. S. (1990). Brain pathways for learned and unlearned vocalizations differ in zebra finches. *Journal of Neuroscience*, 10, 1541–1556.

Slater, P. J. B. (1989). Bird song learning: causes and consequences. *Ethology, Ecology, & Evolution*, 1, 19–46.

Sohrabji, F., Nordeen, E. J. & Nordeen, K. W. (1990). Selective impairment of song learning following lesions of a song control nucleus in juvenile zebra finches. *Neural and Behavioral Biology*, 53, 51–63.

Stoddard, P. K., Beecher, M. D., Horning, C. L. & Campbell, S. E. (1991). Recognition of individual neighbors by song in the song sparrow, a species with song repetoires. *Behavioral Ecology and Sociobiology*, 29, 211–215.

Suter, R., Tolles, A., Nottebohm, M. & Nottebohm, F. (1990). Bilateral lMAN lesions in adult male canaries affect song in different ways with different latencies. *Society for Neuroscience Abstracts*, 16, 1249.

Vicario, D. S. (1991). Organization of the zebra finch song control system: II. Functional organization of outputs from nucleus robustus archistriatalis. *Journal of Comparative Neurology*, 309, 486–494.

Vicario, D. S. & Nottebohm, F. (1988). Organization of the zebra finch song control system: I. Representation of syringeal muscles in the hypoglossal nucleus. *Journal of Comparative Neurology*, 271, 346–354.

Volman, S. F. (1991). Development of auditory responses in nucleus HVC with reduced auditory feedback during song acquisition. *Society for Neuroscience Abstracts*, 17, 1050.

Waser, M. S. & Marler, P. (1977). Song learning in canaries. *Journal of Comparative and Physiological Psychology*, 91, 1–7.

West, M. J. & King, A. P. (1985). Learning by performing: An ecological theme for the study of vocal learning. In *Issues in the Ecological Analysis of Learning*, ed. T. D. Johnston & A. Pietrewicz, pp. 245–272. Hillsdale, NJ: Lawrence Erlbaum.

West, M. J. & King, A. P. (1988). Female visual displays affect the development of male song in the cowbird. *Nature*, 334, 244–246.

Williams, H. (1989). Multiple representations and auditory–motor interactions in the avian song system. *Annals of the New York Academy of Sciences*, 563, 148–164.

Williams, H. (1990a). Models for song learning in the zebra finch: fathers or others? *Animal Behaviour*, 39, 745–757.

Williams, H. (1990b). Bird song. In *Neurobiology of Comparative Cognition*, ed. R. P. Kesner & D. S. Olton, pp. 77–126. Hillsdale, NJ: Lawrence Erlbaum.

Williams, H. & Nottebohm, F. (1985). Auditory responses in avian vocal motor neurons: A motor theory for song perception in birds. *Science*, 228, 279–282.

Williams, H. & Staples, K. (1992). Syllable chunking in zebra finch (*Taeniopygia guttata*) song. *Journal of Comparative Psychology*, 106, 272–286.

4

Sexual imprinting as a two-stage process
HANS-JOACHIM BISCHOF

There is now a considerable literature concerning the phenomenon known as sexual imprinting and the mechanisms underlying it (for reviews see Bateson, 1966; Immelmann and Suomi, 1981; Kruijt, 1985), However, recent findings by Immelmann, Lassek, Pröve & Bischof (1991) and by Kruijt & Meeuwissen (1991) suggest that earlier concepts of imprinting-like learning have to be revised. In this chapter I will analyse this new evidence and discuss its implications for some of the presumed characteristics of imprinting, such as the existence of a sensitive period and the stability of preferences. Further, I will consider some important questions such as stimulus selection and the reasons for stability of preferences. Many of the ideas I will present here are speculative and have little experimental backing. However, they may help us discard some of the old ideas concerning imprinting and so allow for the generation of new ones.

I will start with a brief description of the findings which prompted this chapter. Then I will propose an interpretation of these findings in terms of a two-stage process. The period in early development where information about the appearance of the parents is stored is called 'acquisition phase' here. Subsequently, there is a 'consolidation process' which takes place when the animal becomes sexually mature. In the final section of this chapter, I summarize the main features of the two-stage process and try to evaluate how the ideas presented here can be generalized to other learning paradigms.

Sexual imprinting in zebra finches

Immelmann (1969) was the first to describe sexual imprinting in zebra finches. In his experiments, young zebra finches were cross-fostered by Bengalese finches, birds which readily adopt and rear young of other species. The song of cross-fostered adult zebra finch males resembled that of the Bengalese finch (Immelmann, 1968). Further, the males showed a strong preference for the foster species if tested in a simultaneous choice test, in which they could choose between a female of their own and of the foster species (Immelmann, 1969, 1972; Clayton, 1987a). Immelmann also presented evidence suggesting that sexual imprinting was possible only during a sensitive phase, which started at about 10 days. Imprintability was high until about 20 days and then decreased asymptotically to zero (Immelmann, 1972; Immelmann & Suomi, 1981).

Once a significant sexual preference was established, it was thought to be stable (Lorenz, 1935). Immelmann (1979; Immelmann &. Suomi, 1981) tested this hypothesis by exposing adult zebra finch males, which were raised by Bengalese finches and showed a sexual preference for the foster species, to a zebra finch female for 7 months or longer. The males courted these females irrespective of their previously shown preference for Bengalese finch females, and eventually reared clutches of young. However, when the birds were separated from the females, after some time they again preferred females of the species of their foster parents. Immelmann (1979) concluded from these results that imprinting on the foster parents was stable, and that the preference for the conspecific female was superimposed on the original preference, but did not erase it. However, there were some indications in his experiments that the pretest, which served to measure the initial preference of the bird, influenced the final results: if Immelmann omitted this test, some of the males developed a new and stable preference for zebra finch females. On the basis of these findings, I postulated a 'second sensitive period', in the course of which the final preference of the bird was established (Bischof, 1979).

Although at present I would not defend all of the features of the model that I presented in 1979, subsequent research by Immelmann *et al.* (1991) and Kruijt & Meeuwissen (1991; cf. Kruijt, 1991) confirmed the notion of a two-stage process. These authors reared male zebra finch young with Bengalese finch foster parents for 40 days and subsequently isolated them until day 100. Half of the males were then given a preference test and showed a strong preference for females of the foster species. All the birds then received breeding experience with a female of their own species for 7 months (Immelmann *et al.*, 1991) or 3 months (Kruijt & Meeuwissen, 1991). Thereafter, the birds were tested in two series of preference tests, one immediately after the end of the breeding experience, the other some months later (at different times in both studies). Although there were slight differences in design between the two experiments, the results were remarkably similar. Whereas in the group that received a pretest, most of the birds retained their original preference for the foster species, the majority of the males of the group without pretest preferred females of their own species. The second test, up to 1 year later, showed that this preference remained stable. This similarity of results in two independent studies shows that the effect of the pretests is very reliable.

The authors of both studies concluded that the initial preference which the bird acquires during the first 40 days of its life is not stable, but has to be validated by later experience. Immelmann *et al.* (1991), following Bischof (1979), described the effect as 'consolidation', presuming that under natural conditions the information which was acquired during the early sensitive period, and that which is stabilized or consolidated by the first sexual experience with a female of the parent species will be the same. Kruijt & Meeuwissen (1991), more exactly, point out that in the experiments described above, the effects of the original experience are modified if the first and the second experience differ. Thus, they claim that, depending on the similarity or dissimilarity of the two subsequent experiences, the process involved is described as 'consolidation' or 'modification', respectively.

In addition, both studies showed that the acquired preference remained stable for 7 months or more than a year, respectively. Besides these general findings, they reveal some details of the two-stage process. Immelmann et al. showed that the timing of the experiment is important for the final result: more of the animals exposed to the foster parents for 35 instead of 40 days showed a preference for their own species, irrespective of previous choice tests. Surprisingly, longer experience (50 days) with foster parents also diminished the consolidation effect. This may be due to the presence of conspecific siblings, which has been shown by Kruijt, ten Cate & Meeuwissen (1983) to affect later sexual preferences significantly. Kruijt & Meeuwissen (1991) showed, in addition, that the effect of the breeding experience on preferences is not due to physical interactions between the male and the female, nor is it dependent on breeding itself. The effect of keeping the experimental bird in visual contact with a female, but separated by wire, for 3 months was not different from that of keeping the male in direct contact with the female. The same was true for the preference tests: the stimulus females in both studies were separated by wire from the experimental male, showing that direct physical contact was not necessary for the stabilizing effect of the preference test.

Bischof & Clayton (1991), in an experiment similar to those described above, showed that at least in an experimental situation where the birds were isolated between day 40 and day 100, it was the first exposure to a female or 'the first courtship' which established the preference. They used a more balanced design where the birds were raised by Bengalese finch parents and isolated between day 40 and day 100, until they were exposed for 1 week to a Bengalese female, then for an equal time to a zebra finch female or vice versa. Whereas all birds which were first exposed to the Bengalese finch showed a 100% preference for females of this species, the birds first confronted with zebra finch females showed preferences either for the one or the other species. Raising the birds with their own parents, however, resulted in all cases in a preference for zebra finch females, irrespective of the sequence of exposure to females as adults. Bischof & Clayton interpreted this result as suggesting an 'own species bias' in the formation of sexual preferences. However, the alternative explanation that differential social interactions with the parents (ten Cate, 1982) or the influence of siblings (Kruijt *et al.*, 1983) may contribute to the difference between Bengalese and zebra finch-reared animals cannot be excluded. In a recent study, Kruijt & Meeuwissen (1993), using a design similar to Bischof & Clayton (1991), compared the preferences of zebra finch males raised by conspecific parents with those of cross-fostered zebra finch males in their earlier study (Kruijt & Meeuwissen, 1991). The comparison did not reveal an asymmetry in the preferences of conspecific-reared and cross-fostered males. Because the main difference between the two experiments was the number of siblings with which the experimental males were raised (0–1 in the Kruijt and Meeuwissen studies, and 2–4 in the Bischof and Clayton experiment), this may be the crucial factor for the development of such asymmetries (see Kruijt & Meeuwissen, 1993, for further discussion).

Bischof and Clayton (1991) also showed that the outcome of the experiments depends on the interactions between the young and their parents, and the male and the female in the 'first courtship' situation, respectively. Comparing brothers within clutches, the one that begged and was fed more by its foster parents developed a stronger preference for Bengalese finch females. The more song phrases a male directed to the zebra finch female during the first exposure period after isolation, the stronger was the sexual preference for zebra finch females in the choice tests.

Although the experiments described above support the concept of sexual imprinting as a two-stage process, they provide relatively little information concerning mechanisms underlying it. The evidence suggests that interactions between the experimental males and their parents and also their mate, respectively, might be important. In the next section, I shall discuss some of the features of the phase during which the birds learn about the environment. I call this period, which is identical with the classical 'sensitive phase' of sexual imprinting, the 'acquisition phase'.

The acquisition phase

It is logical and generally accepted that sensitive phases for acquisition of external information cannot start before the sensory system has developed to a stage that sensory input can reach the central nervous system. The question is then which sensory systems are involved and at which time they become functional (Gottlieb, 1971; see also Balsam & Silver, this volume). Because in sexual imprinting the characteristics of the object for sexual behaviour are learned (Lorenz, 1935), it is useful to ask which sensory information is mainly used in courtship behaviour. For sexual behaviour in zebra finches, visual information seems to be the most important by far. Male zebra finches court stuffed dummies of zebra finch females, but do not respond with sexual behaviour to female calls, even if there is a chance for the male and female to communicate acoustically. However, the reaction to the visual image of a female is enhanced by acoustic cues. Most probably, acoustical features of the female arouse the male to a certain degree, but do not alone release courtship behaviour (Bischof, 1985a).

In the course of development, young zebra finches react with gaping responses to acoustic and mechanosensory cues from the time of hatching (Bischof & Lassek, 1985). Reactions to stuffed dummies are observed not earlier than day 10, although the eyes of the birds open at about 6 days. Thus, for the first few days the gaping reaction is elicited only by acoustical and mechanosensory cues, and these stimuli may, after the visual system starts to develop, help to guide the animal's attention towards the visual stimuli. The appearance of the first reaction to visual stimuli could be interpreted as showing that the sensitive phase for the acquisition of visual cues in zebra finches starts at about 10 days from birth. However, at 15 days of age, the young birds react to stuffed dummies with fear instead of gaping reactions. The birds obviously cannot differentiate between dummies and the parents (which had free access to the young except during experiments) earlier than this day. Thus, it is possible that the visual system needs another 5 days to allow the perception of complex stimuli, and one might therefore set the starting point for the sensitive phase at 15 days of age. The data from Immelmann's consecutive rearing experiments (see above) show that the truth lies somewhere in between. By 13 days of age, 10% of the birds have already acquired enough information about the parents to show a stable preference for them. However, if the parents are replaced by foster parents at 19 days of age, about 85 % of the birds retain the initial preference. Therefore it appears as if, under normal conditions, most of the information is stored between 13 and 19 days of age (Immelmann & Suomi, 1981). However, the acquisition phase may be extended if, as mentioned above, the birds are not reared by their own parents. That raises the question of how the storage of information is accomplished, what factors are responsible for the acquisition of information, and what are the factors that end the acquisition period.

The first attempt to explain sensitive periods was a model in which the sensitivity to the environment was switched on and off, by an internal, perhaps genetically determined factor (e.g. Scott, 1962) At present, most theoretical considerations favour the view that a store of limited capacity is filled with information, starting at the time when the sensory system becomes functional. Information storage is fast at the beginning (but perhaps hindered initially by an incomplete development of the sensory system), and slows down dependent on the filling of the store (Bateson, 1981, 1987; Bischof 1983, 1985b; Boakes & Panter, 1985; Bolhuis & Bateson, 1990).

To prevent misinterpretations, it should be mentioned here that the notion of 'limited store' could be taken literally. That is, there may really be some limited space within the brain which is capable of storing the information which comes in during development. It is more probable, however, that the limitations are set by the information itself: if the bird has learned the main features of, say, its parents, there is little opportunity for acquiring new information. The storage process concerning the parents may come to an end because new features appear only very rarely. If at this time the foster parents are replaced by conspecific parents (or *vice versa*) the representation of the characteristics of the parent is already complete, but details of the 'new' parent image may be added. Ten Cate (1986a, 1986b) suggested that such consecutive rearing (as well as simultaneous rearing with two species) results in a mixed representation of both kinds of parent features. Our results (Bischof & Lassek, 1985) on the development of the gaping reaction suggest a third mechanism for limiting the acquisition of information. They show that acquisition of the parent features may simultaneously result in a development of fear towards other, new objects that prevents the birds, at least partly, from learning features other than those with which they are familiar already. This has often been mentioned as a factor limiting learning in filial imprinting (Sluckin & Salzen, 1961; Salzen, 1962; Hoffman & Ratner 1973; Bateson 1981).

However, if the fear reaction is overcome, acquisition of new information may be possible.

So far, nothing has been said about what kind of information is stored during the acquisition period. When sexual imprinting was supposed to be a one-stage process, it was thought that the bird acquired information about its later sexual partner. The problem with this idea was that the system selecting the input for the store had to 'know' at the time of input by what features such information is characterized. Because the bird at that time has no sexual experience, this could only be some sort of prefunctional knowledge about what might be relevant features for finding a sexual partner later in life.

The two-stage model offers another interpretation. It is conceivable that at the time of the aquisition phase the young bird does not know anything about sexual partners and sexual behaviour. However, there are other things which have to be learned, for example 'who is feeding me' or 'who is competing with me for food'. These are examples of acquisition processes that include interactions between the individual that learns and the object of learning. For filial imprinting there is some evidence that associative learning may be the basis for this kind of acquisition (for reviews see Hoffman & Ratner, 1973; Bolhuis, de Vos & Kruijt, 1990). On the other hand, zebra finches may learn about their inanimate environment, for example, without conventional reinforcement, as has been shown for filial imprinting (reviewed in Bolhuis, 1991). However, this issue will not be discussed here since other chapters in this volume (Clayton, ten Cate; see also Clayton, 1987b) deal with the factors influencing information storage in imprinting and song learning.

Whatever the mechanisms of acquisition may be, one can infer that some sort of internal representation – a 'neuronal model', as Salzen (1962) put it, or a 'template', as it is usually called in the song learning literature (Konishi, 1965) – is formed in the brain of the bird. This suggestion implies that the part of the brain or store which acquires information about the environment has to possess some sort of prestructured knowledge. Information about the inanimate environment, such as the walls of the nest box, is put in a different store from, for instance, information about the parents. On the other hand, if birds are reared consecutively by two species, a mixed representation of the features of both species is formed (ten Cate, 1986a, 1986b; see also Clayton, 1988). This means that the system has to 'know' which incoming information has to be stored where. As mentioned above, this could be accomplished by associative processes; for example, categories could be built like 'these are the subjects which are always feeding me' or 'these are subjects which are competing with me for food'. However, there have to be other ways of categorization; it is not easy to understand, for example, how information about the colour of the walls is put into an appropriate store.

It must be emphasized here that it is not only visual information that is stored during the acquisition period. The song learning literature shows that acoustic information is stored as well (e.g. Böhner, 1990). This storage may start even earlier than the storage of visual information. As described above, zebra finches react with gaping to acoustic stimuli from the first day of their life (Bischof & Lassek, 1985). It cannot be excluded that they can perceive acoustic information even in the egg, as has been shown for ducklings by Gottlieb and colleagues (Gottlieb, 1971).

Thus, although acoustic features are not the most important for the elicitation of sexual behaviour (Bischof, 1985a), there is no reason to believe that these features are not learned during development. It has been shown that features of the calls of zebra finches are learned, although the basic structure is inherited (Zann, 1985). It may then be possible that in cross-fostering experiments the birds learned acoustic features from their natural parents before the eggs or the very young hatchlings were transferred to their foster parents. This may also contribute to the apparent 'own species bias' which is observed in many experiments of sexual imprinting, and may also have implications for the consolidation process.

The consolidation process

Neural representations

At the end of the acquisition period, the bird has a set of neural representations of individuals which are important during this period of its life. It may already distinguish between father, mother, and siblings. The 'siblings' category is probably updated several times because the siblings grow older and change their appearance (Bateson, 1981). The 'sibling' category may later also be divided into several subdivisions, allowing the bird to identify each individual brother or sister.

The idea I wish to present here is that the information acquired in the course of the acquisition period is used also in situations other than those which led to its storage. In the experiments by Kruijt & Meeuwissen (1991) and Immelmann *et al.* (1991), the birds were, after reaching independence, isolated for about 60 days. If they are exposed to a female after this isolation period this is a new situation in two respects. First, the male zebra finch sees another bird after a long time of isolation, which may lead to high levels of arousal (Bischof & Herrmann, 1986), and second, the male has become sexually mature in the meantime.

It is therefore conceivable that the bird's attention towards the new stimulus animal(s) is very high. If there are two females from different species, as is the case in the choice test, the male most probably attends to the one similar to the birds to which it has been exposed previously. If there is only one bird, and this is not of the species with which he has been reared, he may nonetheless try to court it, probably because the strange bird shares some features, like having wings or the general shape of a bird, with his parents or siblings. On the other hand, he may court the strange female because his courtship motivation is so strong that he courts almost anything, or he may not court at all.

The two-stage hypothesis predicts what happens if the female that the young male sees first after isolation is of the species he was reared with. There is a stored representation which resembles closely the image of the female to which the male is exposed: that of the mother. Thus the male is familiar with this sort of stimulus and tries to direct the new behaviour which he has developed towards it. As a consequence of courtship behaviour directed towards the female and some contingent behaviour of hers, the young bird learns that the image in the store is good not only for getting food, but also for courtship behaviour. The new association of a stored representation and a new behaviour system (sex) is then stably installed (see also Hogan, 1988).

How can a fostered bird change its preference towards zebra finch females during the consolidation period? In this case no zebra finch representation would be stored, and this should result in a lack of interest in this species, as it is not known. There are different ways to explain this problem, and at present there is no possibility to decide which is the right one. The simplest solution would be to state that the acquisition phase has not yet ended. Immelmann & Suomi (1981) showed that, under certain conditions, the preference of a male zebra finch can be changed very late, an extreme case being reversibility after 70 days. This means that new features may be added to the recognition system even in late stages of development.

An alternative explanation for the shift towards a zebra finch preference is that there is already some representation in the recognition system. This may be some sort of predetermined structure, as I preferred to think (Bischof, 1979; Bischof & Clayton, 1991). However, the objections against such a view, formulated frequently by Kruijt and his colleagues, have somewhat changed my mind. It may well be that, for example, the presence of zebra finch siblings in the nest has affected the formation of a representation of zebra finch features (see Kruijt & Meeuwissen, 1993, for further discussion), or that the young birds, before being transferred to the foster parents, had acquired some acoustic features of their natural parents (see above). As in many other cases, both factors may be involved. Gottlieb (1971, 1981), for example, demonstrated that young ducklings need some sort of acoustic stimulation before hatching to develop a preference for the mother's call, but this does not have to be the call itself. Likewise, Horn and colleagues (e.g. Horn, 1985, 1990; Johnson & Bolhuis, 1991; cf. Bolhuis, this volume) showed that there is some sort of predetermined preference (a 'predisposition') for more natural objects such as a stuffed junglefowl in filial imprinting in chicks. Thus, a prefunctionally determined structure of the recognition system cannot be excluded.

Some of the arguments presented above can probably be tested. If there is no representation in the recognition system at the time where consolidation occurs and a new one can be added at this time, the zebra finch males should also be able to court a third species, for example silverbills (*Euodice cantans*), another finch species, and probably develop a stable preference for this species. If the representation stems from influence of siblings, the cross-fostered male should, given a choice between young and adult females, probably prefer the young one, and if the representation involves acoustic cues from the zebra finch parents, mute zebra finch females should not be courted.

Linking the appropriate behaviour

In any case, in the course of the consolidation process, one (or more) stored representations will influence sexual behaviour. Independently of my first account (Bischof, 1979) of the results discussed here (Immelmann *et al.*, 1991; Kruijt & Meeuwissen, 1991), Bateson (1981, 1987) developed a model of imprinting based on results of Immelmann (1969) and Cherfas & Scott (1981). As mentioned above, Immelmann found that male zebra finches that were reared early in life with Bengalese finches, and after a choice test were kept for 7 months with conspecifics, bred successfully with female zebra finches. However, in a subsequent choice test, the males courted Bengalese finches in preference to zebra finches. Thus, for the time of breeding with conspecific females the preference for Bengalese finch females was masked by a new one, but resurfaced when a choice was possible. Bateson stated that this could be explained by proposing two separate systems, a recognition system and an executive system (for example the one for sexual behaviour).

The recognition system, according to Bateson, is organized as presented above in this paper. Bateson suggested that the recognition system gains access to the executive system in the course of the sensitive phase. If more than one image (for example a zebra finch and a Bengalese finch image) 'captures' a behavioural subsystem of the executive system (for example sexual behaviour), the one which has stronger access controls the behaviour as long as the stimulus is available. However, if the bird does not have a choice, the other image can also control behaviour.

So far, this is very similar to the view presented here. In contrast to Bateson's proposals, however, it is assumed here that the acquisition process and the consolidation process are guided by different and independent mechanisms. Whereas, as outlined above, the recognition system is shaped during the 'classical' sensitive phase and does not have anything to do with sexual imprinting directly, the consolidation process takes place during the second stage, when connections are made with the executive system (Kruijt & Meeuwissen, 1991). The most important feature of this consolidation process is that it occurs only if sexual behaviour is performed and the validity of the learned image for sexual behaviour can be tested; thus, it depends on a certain sexual maturity of the bird.

First courtship attempts in young zebra finch males can be observed between 30 and 35 days of age (Kalberlah, 1980). At this time, there is a peak in plasma testosterone levels (Pröve, 1983). When this peak ends, the plasma concentration of 17β-oestradiol, which is the metabolite acting in the brain (e.g. Gurney & Konishi, 1980; Harding, 1983), is raised substantially (Pröve, 1983). If courtship is a necessary prerequisite for the consolidation process, it cannot occur earlier than 30 to 35 days. As the results of Kruijt & Meeuwissen (1991) and Immelmann *et al.* (1991) show, it can be delayed even until the bird is 100 days of age. However, it is conceivable that under natural conditions consolidation occurs earlier and that it may not be restricted to a single courtship bout.

Direct evidence that raised levels of testosterone are a prerequisite for consolidation has been provided by Pröve (1990). He reared zebra finch males with Bengalese finch parents until they were 35 days old. During this time the birds received silastic implants containing testosterone. From day 35 to 95 the birds were transferred to a zebra finch female. Half of these birds received silastic implants containing Cyproteronacetate (CyA), an antiandrogen which has been shown to suppress reversibly sexual behaviour (Pröve & Immelmann, 1982). Subsequent choice tests showed that the birds which had received CyA implants retained the preference towards the foster species, whereas the non-treated animals preferred their own species. Unfortunately, no observations of the behaviour of the birds were performed in these experiments. They suggest that stabilization occurs only if testosterone is available. It is conceivable, however, that because of the lack of testosterone, the birds did not court their cagemates between 35 and 95 days (cf. Bolhuis, 1991). In that case, the experiments would support the suggestion by Immelmann *et al.* (1991) that sexual behaviour is necessary to stabilize or modify the preference.

In the experiments performed by Immelmann *et al.* (1991) and Kruijt & Meeuwissen (1991), the birds appeared highly aroused. Observations by Bischof & Herrmann (1986), using ^{14}C-2-deoxyglucose (see also Bolhuis, this volume) may be relevant to this issue. These authors found that four areas of the forebrain of zebra finch males were activated strongly if the bird was aroused by chasing it round the cage, or under the same conditions that Immelmann *et al.* (1991) used in their experiments. It may well be that this high arousal level is a precondition for consolidation. This suggestion receives some support from anecdotal evidence presented by Bateson (1983).

We cannot define exactly what factors may contribute to the level of arousal. Obviously, the male is aroused by the appearance of the female. Probably, the behaviour of the female towards the male is an important factor. However, Kruijt & Meeuwissen (1991) have shown that a female separated from the test male by wire has the same effect as a female that can be accessed directly by the male. Bischof & Clayton (1991) demonstrated that the preference for the test female is greater when the male's courting activity is high. Whether this, in turn, depends on the behaviour of the female, cannot yet be decided. Both findings can also be interpreted as showing that the male's behaviour itself may be sufficient to stabilize his preferences. This idea has been put forward in filial imprinting, e.g. in the form of a 'law of effort' (Hess, 1973). Kruijt & Meeuwissen (1991) argued that the high level of arousal that the birds show in the given experimental situation may be dependent on the long period of isolation preceding exposure to a female. This is supported by findings of Bischof & Herrmann (1988), who showed that the activation of the four brain areas mentioned above, which was used as an indicator for arousal, was dependent upon the time of isolation preceding the experiment.

However, it is conceivable that under normal conditions the arousal level of a male is enhanced if a female appears. The enhancement may be small, but sufficient for consolidation to occur. According to this view, the experiments by Kruijt & Meeuwissen (1991) and Immelmann *et al.* (1991) did not measure an artifact, but drove one of the factors involved (namely the level of arousal) to its limits. To be really sure about the role of arousal, however, it is necessary to find ways to test this hypothesis more directly.

Conclusions

From the experiments described in this chapter, sexual imprinting can now be characterized as occurring in two stages: First, an 'acquisition phase' during which the bird learns about its environment, for example about its nest, its siblings, and its parents. When the bird is able to perform sexual behaviour, this previously acquired representation is used when choosing a sexual partner. If the representation matches the courtship partner, it is used for sexual behaviour subsequently. If it does not fit, the existing representation is altered or a new representation influencing sexual behaviour may be formed. In either case, the representation remains stable after the consolidation process.

It may be too early to generalize these ideas to other imprinting-like paradigms. However, as has been mentioned above (see also Clayton, ten Cate, this volume), song learning shares many features with sexual imprinting. It is generally accepted that song learning in birds is a two-stage process. With slight modifications, it is assumed by all authors working in this field that birds learn their song during early development in the course of a sensitive phase. The songs which are acquired during this sensitive phase are stored as a 'template' which can later be recalled. In the so-called 'sensorimotor phase', this template is used as a guide for the young bird to develop its own song by matching the own song output with the neuronal model stored in the template. As yet, it is not easy to decide whether the similarities end here and whether dissimilarities show up with closer examination. Further research may clarify this issue.

Comparisons with other imprinting paradigms are also difficult at this stage. Filial imprinting, for example, obviously does not have a time lag between acquisition and stabilization, although formally the two stages can be separated (Bischof, 1979). An idea that might be interesting to investigate is that each behaviour which is triggered by a releaser, the characteristics of which are learnt, is coupled to this releaser by a consolidation process. However, at present one can only speculate that the two-stage process that has been found in sexual imprinting may be a common feature of all imprinting-like processes.

Acknowledgements

I thank Johan Bolhuis and Jerry Hogan for their critical comments on earlier versions of the manuscript. I am grateful to Jaap Kruijt for his continuing interest in our work and his constructive critique over the years. The many discussions with him and the members of his group have been very stimulating.

References

Bateson, P. P. G. (1966). The characteristics and context of imprinting. *Biological Reviews*, 41, 177–220.

Bateson, P. P. G. (1981). Control of sensitivity to the environment during development. In *Behavioral Development*, ed. K. Immelmann, G. W. Barlow, L. Petrinovich, M. Main, pp. 432–453, Cambridge: Cambridge University Press.

Bateson, P. P. G. (1983). The interpretation of sensitive periods. In *The Behavior of Human Infants*, ed. A. Oliverio & M. Zapella, pp 57–70. New York: Plenum Press.

Bateson, P. (1987). Imprinting as a process of competitive exclusion. In *Imprinting and Cortical Plasticity*, ed. J. P. Rauschecker, P. Marler, pp. 151–168. New York: John Wiley & Sons.

Bischof, H.-J. (1979). A model of imprinting evolved from neurophysiological concepts. *Zeitschrift für Tierpsychologie*, 51, 126–139.

Bischof, H.-J. (1983). Imprinting and cortical plasticity: a comparative review. *Neuroscience and Biobehavioral Reviews*, 7, 213–225.

Bischof, H.-J. (1985a). Der Anteil akustischer Komponenten an der Auslösung der Balz männlicher Zebrafinken (*Taeniopygia guttata castanotis*). *Journal für Ornithologie*, 126, 273–279.

Bischof, H.-J. (1985b). Environmental influences on early development: a comparison of imprinting and cortical plasticity. In *Perspectives in Ethology*, Vol. 6: Mechanisms, ed. P. Bateson, P. Klopfer, pp. 169–217. New York: Plenum Press.

Bischof, H.-J. & Clayton, N. (1991). Stabilization of sexual preferences by sexual experience in male zebra finches *Taeniopygia guttata castanotis*. *Behaviour*, 118, 144–155.

Bischof, H.-J. & Herrmann, K. (1986). Arousal enhances 14C-2-Deoxyglucose uptake in four forebrain areas of the zebra finch. *Behavioural Brain Research*, 21, 215–221.

Bischof, H.-J. & Herrmann, K. (1988). Isolation-dependent enhancement of 2–14C-deoxyglucose uptake in the forebrain of zebra finch males. *Behavioral and Neural Biology*, 49, 386–397.

Bischof, H.-J. & Lassek, R. (1985). The gaping reaction and the development of fear in young zebra finches (*Taeniopygia guttata castanotis*). *Zeitschrift für Tierpsychologie*, 69, 55–65.

Boakes, R. & Panter, D. (1985). Secondary imprinting in the domestic chick blocked by previous exposure to a live hen. *Animal Behaviour*, 33, 353–365.

Böhner, J. (1990). Early acquisition of song in the zebra finch, *Taeniopygia guttata*. *Animal Behaviour*, 39, 369–374.

Bolhuis, J. J. (1991). Mechanisms of avian imprinting: a review. *Biological Reviews*, 66, 303–345.

Bolhuis, J. J. & Bateson, P. (1990). The importance of being first: A primacy effect in filial imprinting. *Animal Behaviour*, 40, 472–483.

Bolhuis, J. J., De Vos, G. J. & Kruijt, J. P. (1990). Filial imprinting and associative learning. *Quarterly Journal of Experimental Psychology*, 42B, 313–329.

Cherfas, J. J. & Scott, A. M. (1981). Impermanent reversal of filial imprinting. *Animal Behaviour*, 29, 301.

Clayton, N. S. (1987a). Mate choice in male zebra finches: some effects of cross-fostering. *Animal Behaviour*, 35, 596–622.

Clayton, N. S. (1987b). Song tutor choice in zebra finches. *Animal Behaviour*, 35, 714–722.

Clayton, N. S. (1988). Song learning and mate choice in estrildid finches raised by two species. *Animal Behaviour*, 36, 1589–1600.

Gottlieb, G. (1971). *Development of Species Identification in Birds*. Chicago: University of Chicago Press.

Gottlieb, G. (1981). Roles of early experience in species-specific perceptual development. In *Development of Perception*, Vol. I, ed. R. Aslin, J. R. Alberts & M. R. Petersen, pp. 5–44. New York: Academic Press.

Gurney, M. & Konishi, M. (1980). Hormone induced sexual differentiation of brain and behavior in zebra finches. *Science*, 208, 1380–1383.

Harding, C. F. (1983). Hormonal specificity and activation of social behaviour in the male zebra finch. In *Hormones and Behaviour in Higher Vertebrates*, ed. J. Balthazart, E. Pröve & R. Gilles, pp. 275–289. Berlin: Springer Verlag.

Hess, E. H. (1973). *Imprinting: Early Experience and the Developmental Psychobiology of Attachment*. New York: Van Nostrand Reinhold.

Hoffman, H. S. & Ratner, A. M. (1973). A reinforcement model of imprinting. Implications for socialisation in monkeys and men. *Psychological Review*, 80, 527–544.

Hogan, J. A. (1988). Cause and function in the development of behavior systems. In *Handbook of Behavioral Neurobiology*, Vol. 9, ed. E. M. Blass, pp. 63–106. New York: Plenum Press.

Horn, G. (1985). *Memory, Imprinting, and the Brain*. Oxford: Clarendon Press.

Horn, G. (1990). Neural bases of recognition memory investigated through an analysis of imprinting. *Philosophical Transactions of the Royal Society of London*, Series B, 329, 133–142.

Immelmann, K. (1968). Zur biologischen Bedeutung des Estrildidengesangs. *Journal für Ornithologie*, 109, 284–299.

Immelmann, K. (1969). Über den Einfluß frühkindlicher Erfahrungen auf die geschlechtliche Objektfixierung bei Estrildiden. *Zeitschrift für Tierpsychologie*, 26, 677–691.

Immelmann, K. (1972). The influence of early experience upon the development of social behaviour in estrildine finches. *Proceedings of the 15th International Ornithological Congress.*, The Hague 1970, pp 316–338. Leiden: E. J. Brill.

Immelmann, K. (1979). Genetical constraints on early learning: a perspective from sexual imprinting in birds and other species. In *Theoretical Advances in Behavior Genetics*, ed. J. R. Royce & P. Mos, pp. 121–136. Alphen aan de Rijn: Sijthoff & Noordhoff.

Immelmann, K., Lassek, R., Pröve, R. & Bischof, H.-J. (1991). Influence of adult courtship experience on the development of sexual preferences in zebra finch males. *Animal Behaviour*, 42, 83–89.

Immelmann, K. & Suomi, S. J. (1981). Sensitive phases in development. In *Behavioral Development*, ed. K. Immelmann, G. W. Barl, L. Petrinovich & M. Main, pp. 395–431. Cambridge: Cambridge University Press.

Johnson, M. H. & Bolhuis, J. J. (1991). Imprinting, predispositions and filial preferences in chicks. In *Neural and Behavioural Plasticity*, ed. R. J. Andrew, pp. 133–156. Oxford: Oxford University Press.

Kalberlah, H. H. (1980). Quantitative Untersuchungen zur Ontogenese des Sexualverhaltens beim Zebrafinken (*Taeniopygia guttata castanotis*). Dissertation, Universität Bielefeld.

Konishi, M. (1965). The role of auditory feedback in the control of vocalization in the white-crowned sparrow. *Zeitschrift für Tierpsychologie*, 22, 770–783.

Kruijt, J. P. (1985). On the development of social attachment in birds. *Netherlands Journal of Zoology*, 35, 45–62.

Kruijt, J. P. (1991). The possible role of courtship experience in the consolidation of sexual preferences in zebra finch males. *Acta XX Congressus Internationalis Ornithologici*, 1068–1073.

Kruijt, J. P. & Meeuwissen, G. B. (1991). Sexual preferences of male zebra finches: effects of early and adult experience. *Animal Behaviour*, 42, 91–102.

Kruijt, J. P. & Meeuwissen, G. B. (1993). Consolidation and modification of sexual preferences in adult male zebra finches. *Netherlands Journal of Zoology*, 43, 68–79.

Kruijt, J. P., Ten Cate, C. J. & Meeuwissen, G. B. (1983). The influence of siblings on the development of sexual preferences of male zebra finches. *Developmental Psychobiology*, 16, 233–239.

Lorenz, K. (1935). Der Kumpan in der Umwelt des Vogels. *Journal für Ornithologie*, 83, 137–213, 289–413.

Pröve, E. (1983). Hormonal correlates of behavioural development in male zebra finches. In *Hormones and Behaviour in Higher Vertebrates*, ed. J. Balthazart, E. Pröve & R. Gilles, pp. 368–374. Berlin: Springer Verlag.

Pröve, E. (1990). Haben Steroidhormone einen Einfluß auf die sexuelle Prägung? Untersuchungen zu physiologischen Korrelaten eines frühontogenetischen Lernvorgangs bei männlichen Zebrafinken (*Taeniopygia guttata*). *Die Vogelwarte*, 35, 329–340.

Pröve, E. & Immelmann, K. (1982). Behavioral and hormonal responses of male zebra finches to antiandrogens. *Hormones & Behavior*, 16, 121–131.

Salzen, E. A. (1962). Imprinting and fear. *Symposia of the Zoological Society of London*, 8, 199–217.

Scott, J. P. (1962). Critical periods in behavioral development. *Science*, 138, 949–958.

Sluckin, W. & Salzen, E. A. (1961). Imprinting and perceptual learning. *Quarterly Journal of Experimental Psychology*, 8, 65–77.

ten Cate, C. (1982). Behavioural differences between zebra finch and bengalese finch (foster) parents raising zebra finch offspring. *Behaviour*, 81, 52–172.

ten Cate, C. (1986a). Sexual preferences in zebra finch males raised by two species: I. A case of double imprinting. *Journal of Comparative Psychology*, 100, 248–252.

ten Cate, C. (1986b). Sexual preferences in zebra finch males raised by two species: II. The internal representation resulting from double imprinting. *Animal Behaviour*, 35, 321–330.

Zann, R. (1985). Ontogeny of the zebra finch distance call: I. Effects of cross-fostering to bengalese finches. *Zeitschrift für Tierpsychologie*, 68, 1–23.

5

The influence of social interactions on the development of song and sexual preferences in birds

NICKY S. CLAYTON

Bird song and sexual imprinting in birds are ideal systems for studying behavioural development, as we have seen already (DeVoogd, Bischof, this volume). They are also ideal systems to compare because of the many parallels between them (see also ten Cate, this volume). In this chapter I compare aspects of song learning and sexual imprinting in birds to show how the relationship between them contributes to our general understanding of how behavioural processes interact during development. I shall focus on the importance of social interactions for both song learning and sexual imprinting, discussing the circumstances under which social interactions can override two key features of imprinting-like processes, namely sensitive phases and stability, and describing some experiments that demonstrate which features of social interaction seem to be important (see ten Cate, this volume; Bolhuis, 1991, for the role of social interaction in filial imprinting). Finally, I shall adopt an interdisciplinary approach by linking the behaviour with its underlying neural substrate. An understanding of the role social interaction plays in learning and memory is important for two reasons: it contributes towards our general understanding of the complex process of behavioural development, and may help to elucidate the fascinating problem of how memory is stored and processed in the brain.

Sexual imprinting

Although the extraordinary phenomenon of newly hatched precocial birds following humans and subsequently developing sexual preferences for them had been described long before, the term imprinting, or 'Prägung', was popularised by Konrad Lorenz (1935), who drew attention to its biological significance. Lorenz believed imprinting to be a unique

98

learning process because its two most striking features are, firstly, that the learning is restricted to a brief period early in life, the so-called 'critical' or 'sensitive phase' (see Hinde, 1970; Immelmann & Suomi, 1981), and secondly, that once learnt, the information is stable, not to be forgotten and replaced by information which might be acquired before or after that time.

However, recent studies demonstrate that these two hallmarks of imprinting are not as cut and dried as had been thought. Firstly, imprinting does not occur during a sharply defined time window of fixed duration, with a step-like onset and termination. Instead, the readiness to learn begins, and ends, more gradually and varies both in time and duration, as a result of the individual's own experiences. Secondly, the information acquired during the sensitive phase is not always permanent and it is possible, under certain circumstances, to learn at other times (see Bolhuis, 1991; Bischof, this volume).

Evidence obtained with zebra finch males that have been raised by another species, such as the Bengalese finch (*Lonchura striata*), illustrate these points. If zebra finch males are raised by Bengalese finch foster-parents, they develop a sexual preference for Bengalese finch females. However, sexual imprinting takes longer in these cross-fostered zebra finches than in zebra finches raised by conspecific parents. Whereas the normally raised birds will only court zebra finch females, those raised by Bengalese finch foster-parents may court both species of females, and the preference for Bengalese and zebra finch females will depend on the extent of the interaction with other zebra finch conspecifics and with their Bengalese finch foster-parents (e.g. Immelmann, 1972; ten Cate, Los & Schilperoord, 1984; Clayton, 1987a; Immelmann *et al.*, 1991; Kruijt & Meeuwissen, 1991; Bischof & Clayton, 1991). Despite this flexibility in what is learnt and when, the degree of plasticity does decline over the course of development, and the fact remains that there are temporal constraints which can impose strong limitations on the readiness to learn.

Song learning

Of the 8500 living species of birds, almost half are classified in the songbird suborder Oscines. As their name suggests, songbirds produce long, complex vocalizations – songs – in addition to short, simple calls. In most of these species, singing is more or less exclusively a male activity, although both males and females are able to learn to discriminate between the songs of different individuals (McGregor & Avery, 1986; Miller, 1979; Clayton,

1988a). One common feature of all the songbirds studied to date is that they need to hear other birds singing in order to develop a song typical of their species. This means that learning must play a key role in song development.

Most species of songbird, such as the white-crowned sparrow, *Zono-trichia leucophrys*, do not change their songs seasonally: they learn their songs early in development and these remain remarkably stable throughout the individual's life. A few species, like the canary, *Serinus canaria*, learn new songs each year. Marler (1987) coined the term 'age-limited learners' to refer to species such as the white-crowned sparrow that do not change their song repertoires in later life.

The pattern of song learning varies greatly between species, but for both 'age-limited' and 'non-age-limited' learners, song learning can be viewed as consisting of two parts: a memorization phase, during which the songs that are heard are stored; and a motor phase, when the young bird develops its own song by perfecting its motor output and matching this to sounds that it has heard previously. These two phases can occur at different times. For example, under natural conditions, young male white-crowned sparrows spend the first few months of life in an environment in which the adult males are singing their individually distinct, territorial song. However, a young male does not usually begin to sing until he is several months old. The first attempts consist of a twittering subsong that bears but a vague similarity to an adult male's fully crystallized song. Within a period of about 2 months, as the bird practises, this develops into plastic song and this in turn becomes more and more structured until it closely resembles a typical adult white-crowned sparrow song (see Thorpe, 1961; Marler, 1976; DeVoogd, this volume).

Song learning as an imprinting-like process

There are a number of similarities between song learning and sexual imprinting. Firstly, learning is important for the development of both song and sexual preferences in all the true songbirds that have been studied in any detail. Secondly, the information is usually learnt during a sensitive phase early in the bird's life, and the songs and sexual preferences established during this time are remarkably stable from year to year. A third similarity between song learning and sexual imprinting is that of 'own-species bias'. In both the birdsong and sexual imprinting literature, there have been several studies suggesting that a preference for conspecifics or for their song may be established more easily than one for another species (e.g. Slater, Eales & Clayton, 1988; Marler, 1991a, 1991b, for bird song; Immelmann & Suomi, 1981, for sexual imprinting).

Moreover, as with sexual imprinting, there is a flexibility in the timing and duration of the sensitive phase for song learning which is influenced by social experience as well as age. In several species, the young males will not copy their songs from those heard on a tape-recorder, and even in those that will do so, they are more likely to learn from a live song tutor. For example, Baptista & Petrinovich (1984) found that social interactions with an adult male can have a dramatic effect on what a young bird learns and when, even if that adult is of the wrong species. Baptista & Petrinovich found that if a fledgling white-crowned sparrow is placed in a cage next to a strawberry finch, *Amandava amandava*, in an aviary where it can hear members of its own species but can only interact with the strawberry finch, it will copy the song of its social tutor, the strawberry finch. This is the case even if the experiment does not begin until after day 50, i.e. after the supposed sensitive phase for song learning. This is but one example of the importance of social interaction in governing what a young bird learns and when.

Given these parallels, it should be enlightening to compare the development of sexual preferences and song in the same species. Unfortunately, studies of sexual imprinting have been confined to only a few species, including only one songbird, the zebra finch. Nonetheless, the zebra finch has proved to be an ideal subject for studying both these aspects of development. In the rest of this chapter, I shall explore the analogies between sexual imprinting and birdsong in the zebra finch.

Song learning in zebra finches

Much of the recent work on song learning in zebra finches has been carried out by Slater and his colleagues at the University of St Andrews (reviewed in Slater *et al.*, 1988). To examine whether zebra finches might learn their fathers' songs or those of other males, the young were separated from their parents at 35 days, the age of independence, at 50 days, in about the middle of the juvenile period, and at about 65 days, when the birds are reaching sexual maturity. The birds were divided into two groups and, on removal from the parents, they were housed with their clutch mates either in visual and acoustic isolation from adults or with a different adult pair so that they had the opportunity to learn song from another male zebra finch.

The results of both groups suggest that young zebra finches learn their song during the period from 35 to 65 days. The percentage of song elements that the young males shared with their fathers increased with the time they spent together, from 0 % if the males were removed from the father at 35 days to 100 % if they remained with him until day 65. There was no significant difference in the amount of the father's song that was learnt between males that were isolated and those that were housed with a new tutor at 35, 50 or 65 days of age. This suggests that elements learnt from the new tutor are added to those memorized from the father rather than overriding them (Slater *et al.*, 1988).

In contrast, Böhner (1990) found that young males that were housed with their father until about 35 days post-hatch developed songs as similar to their father's song as birds that were not separated from him until day 100. Slater & Mann (1990) suggested one explanation for the discrepancy between the two results. In Böhner's study, the young males were placed in visual and partial auditory isolation after they had been separated from their father. Those birds that had been isolated at day 35 lacked stimulation from other birds after this time, and this may explain why they recalled and sung the songs that they had heard from the father, perhaps rather earlier in life than those they would usually copy.

However, an even more recent study by Slater & Richards (1990) suggests that re-nesting may be an important factor in guiding whether a young male learns before or after 35 days post-hatch, the time at which he becomes nutritionally independent from his parents. Slater & Richards found that, in captivity, young zebra finches are more likely to learn song from their father if the nest box has been removed from the cage once the nestlings have fledged. Slater & Richards suggest that the reason for this is that the young birds have closer contact with their father if he does not re-nest – males that re-nest probably spend more time building a new nest, courting the female and helping to incubate the eggs than those that do not re-nest – and that this would explain why their songs are less likely to be copied. An alternative explanation is that the young birds are forced to mature more quickly if their father re-nests and therefore would reach independence, begin courting and develop their songs earlier. Presumably this could be tested by measuring the development of secondary sexual characteristics and testosterone levels in the two groups.

Is there any field evidence to support any of these claims? In the wild, zebra finches become independent from their parents at about 35 days of age and form small flocks within the colony where they are likely to hear songs from a variety of different adult males. Furthermore, if conditions remain favourable for breeding, at least some of the parents will already have a fresh clutch in a different nest by this time and so these fathers may not be available for the young birds to learn from after day 35 (Immelmann, 1962). One problem with working with zebra finches in the wild is that they usually disperse after independence and are very difficult to relocate. This means that it is rather difficult to study song development in zebra finches under natural conditions. Nonetheless, Zann (1990) overcame this difficulty by catching young zebra finches from day 36 onwards and housing them not with their father but with other adult zebra finch song tutors. Zann kept the birds in aviaries in the middle of his wild colony until he was able to record their songs, as well as those of their fathers and the song tutors with whom they were housed during this period. He found that most of the young birds did copy their fathers' songs. However, the tendency to do so was less in those that had been confined early, i.e. the ones that were deprived of any social interaction with their fathers during the time immediately after independence. Zann's results also demonstrate that when adult, the young zebra finches were most likely to sing those songs that they had heard from day 36. In other words, Zann's results suggest that wild zebra finches also tend to base their songs largely on those heard in the period immediately after day 35, although some males may base their songs on those heard before this time.

To return to the laboratory studies, subsequent work has shown that zebra finches do actually memorize the songs that they hear before day 35, even if they do not sing those songs (Clayton, 1988a; Slater *et al.*, 1988). Furthermore, if males are exposed after this time to a tutor whose song is in some way inadequate for learning, but sufficient to stimulate song development, they may sing a song which was heard before day 35. There appear to be three situations in which this occurs: firstly, when the young males can hear the song tutor but cannot see him because they are separated by an opaque screen; secondly, when young males are housed at independence with a tutor which belongs to a different species from their father or foster-father; and thirdly, when young males are raised by mixed species parents, i.e. with a zebra finch male and a Bengalese finch female or *vice versa* (Slater *et al.*, 1988). In all three cases, the auditory and/or visual stimulation after day 35 did not match in quality that received earlier. Such a mismatch seems to cause the bird to recall and reproduce song learned before 35 days and also to block further modification later. For young zebra finches, which have been observed in mixed-species flocks (Immelmann, 1962), this mechanism might act as a safeguard to ensure that the young birds only sing songs typical of their own species.

The experiments described in the previous paragraph demonstrate that social interaction is important in determining both the duration and the nature of song learning. However, the first situation in which the young males can hear but not see the song tutor is particularly striking: the finding that some song learning may occur if the young zebra finch can interact vocally, but not visually, with the song tutor suggests that sight is not crucial for song learning in zebra finches. On the basis of two further findings by Slater *et al.* (1988), namely that young zebra finches that have been exposed only to a tape-recorded song, and those that have been denied the opportunity to interact visually and vocally with the song tutor, fail to sing that song tutor's song, it has been concluded that song learning is stimulated by the presence of a living song tutor.

Additional evidence that exposure to the visual components of the song tutor and/or his display are not essential for song learning in young zebra finch males comes from the results of a study by Patrice Adret (personal communication). During the sensitive phase for song learning (35–65 days post-hatch period), young males were blindfolded by glueing a small, plastic cone-shaped eyecap to the feathers around each eye, and housed in a cage with adult male song tutors with whom they could physically interact. These young blindfolded males produced accurate copies of the songs of tutors sharing their cage, suggesting that visual contact between the tutor and young male is not necessary for song copying to occur. In contrast, control siblings that did not wear eye caps and could see normally, but were separated from the song tutor by an opaque screen, showed very poor, if any, imitation of the song tutor's song. One effect of the visual deprivation in the blindfolded group may have been to increase the auditory sensitivity of these birds. It cannot be concluded, therefore, that visual stimuli do not play any role in song learning in young males with normal vision. Nonetheless, the results do suggest that sight is not crucial for song learning to occur in zebra finches. These results also suggest that, in addition to the auditory stimulus of the song tutor, physical interactions with the tutor are important, at least in the blindfolded group. This raises the question of whether social interactions are also important in zebra finches that have not been subject to visual deprivation.

In order to address this question, and to look in more detail at what features of social interaction are important, I observed the behavioural interactions between young zebra finches and their song tutors in captivity and correlated these with the young males' subsequent choice of song tutors.

How do zebra finches choose their song tutors?

Two experiments were performed (Clayton, 1987b). In one experiment, young zebra finches were raised only by their mother, and at independence were transferred to a cage with two live male tutors. Social interactions between the young and the tutors were observed and the songs of both the young birds and their tutors were recorded after 4 months. Each of the 11 birds in this experiment was found to learn song from the tutor that was most aggressive towards him.

In a second experiment, I analysed the role of the father's song in the young male's choice of which tutor to learn from. Chicks were raised by both parents until independence and then placed in a cage with two tutors. The young birds were separated from both tutors by wire mesh so that physical attacks, shown to be important in the first experiment, could not influence which tutor the young bird chose. The tutors were selected so that one had a song similar to the young bird's father and the other did not. Nine of the ten birds copied the song of the tutor whose song was similar to that of their father, and the tenth bird learnt from both tutors. This suggests that the early experience of hearing the father's song influences the young bird's subsequent choice of song tutor.

One criticism of these experiments is that the birds were housed in cages and therefore were more or less forced to interact with the song tutors. Caution must be applied when interpreting these results and attempting to extrapolate them to the situation found in the wild. For example, it is unclear whether aggression *per se* is the crucial factor guiding a male's choice of song tutor or whether any form of social interaction would suffice, aggression merely being the most prevalent form of interaction in captive birds housed in cages.

A recent laboratory experiment by Williams (1990) is more realistic in the sense that the birds were housed in aviaries rather than in cages. She placed 12 male and female zebra finches in an aviary along with 12 nest boxes. The birds were left to form pairs and 10 of these bred, producing 35 young. She recorded and analysed the songs of all the males in the aviary, both young and adult, and observed the social interactions between them. Unfortunately the observations were made only until the fledglings reached day 40, i.e. at a time when learning would not yet have been completed (e.g. Immelmann, 1969; Slater *et* al., 1988). In spite of this, Williams' observations did allow her to see how the songs that these young zebra finch males developed related to the interactions that they had had with particular adult males. All the birds were left in the aviary until the young were sexually mature and had fully developed songs which could be recorded.

Williams found that all the adult males were copied by some young birds, and most of them used song elements from more than one of these song tutors. Interestingly, the two males whose songs were copied the most were the two that interacted most with the young. These males were not more aggressive but they did feed the young more frequently than the other adult males did.

Unfortunately, these results cannot be interpreted as easily as might appear at first (Slater & Mann, 1990). One problem is that most of the young formed a single 'crèche' in one corner of the aviary and therefore all the fledglings had a rather similar, and rather limited, experience with the song tutors. An additional problem is that, even for a social species, the aviary was tiny (20 m) given the number of birds involved. Perhaps this is why the birds formed a 'crèche'. The closer proximity, and more intense exposure, to the song tutors may also explain why these birds learnt from more than one adult whereas those in the previous studies usually based their song on that of only one male song tutor. Thirdly, because the observations were made only until day 40, it is difficult to compare the two studies and at present, it is unclear as to which of the features of living tutors are important stimuli for song learning. Perhaps feeding is the crucial social interaction prior to independence whereas aggressive interactions assume importance after this time. Alternatively, the story may not be so clear cut. Perhaps all types of social interaction are important, in different ways and at different times, depending on the social situation, and it is social interactions *per se* that guide song learning.

One common feature of these conditions may be the presence of a contingency between the young zebra finch's behaviour and the behaviour of the song tutor, at least in terms of his singing rate. For example, the way in which a song tutor responds to a young male may make that young male more attentive to the song, or in some way heighten his arousal, so that his song learning is enhanced. It has been suggested that behaviour-contingent exposure to song may act as a reinforcer to young zebra finches because there have been indications that exposure to song of poor quality and/or quantity might increase a young male's motivation to expose himself to song (ten Cate, 1986).

Two experiments have been designed to test this hypothesis (ten Cate, 1991; Adret, 1993). In both experiments, young zebra finch males were raised normally by both parents and then separated from all other birds by housing them in single cages in separate rooms during the sensitive phase

(35–65 days post-hatch). Each cage had a loud-speaker in front of it, connected to a cassette recorder loaded with an endless tape containing zebra finch song. The birds were divided into two groups. In the experimental group the birds were given control over tape-recorded song, in the case of ten Cate's experiment by hopping on one of the perches, and in Adret's study by pecking a key by one of the perches. In both cases, the perch or key giving exposure to song was alternated every couple of days to randomize any position bias. The other males served as yoked controls: each was housed in an adjacent room to an experimental male and was exposed to song when the experimental one was.

Interestingly, the results of the two studies differ. Ten Cate (1991) reports that, although some males visited a perch more often when it provided exposure to song, and that these males tended to have a greater preference for the reinforced perch than their yoked controls, only two birds in the experimental group appeared to have learnt the contingency. ten Cate found copying from the tape to be poor in both groups and there was no significant difference between experimental and yoked control males. On the basis of this, he wrote 'the conclusion that behaviour-contingent exposure to song does not improve song learning can be no more than a tentative one. However, what can be concluded is that the mechanism giving rise to the reinforcing value of song is not inextricably linked to the one controlling access to the memory for songs to be produced later in life'.

In contrast, Adret (1993) did find evidence that behaviour-contingent exposure to song improves song learning. The experimental birds imitated the tape-recorded songs to a much greater extent than the males in the yoked control group, and not only copied more of the elements from the tape-recordings, but also copied the syntax of the songs more precisely. He found that this operant conditioning procedure with song as a reward strongly influenced song learning during the sensitive phase. However, it did not affect song preferences in adulthood: in choice tests using the same technique, both groups of males showed strong preferences for the training song over a novel song. Clearly, both groups had learnt to recognize the training song but only the experimental birds tended to copy that song.

It is not clear why the two studies yielded such different results. Adret (1993) listed a number of differences between the two studies that might be responsible for the difference in results. Further experiments are required to elucidate why the birds in Adret's experiment showed behaviour-contingent song learning while those in ten Cate's study failed to do so. One possibility is that the birds in ten Cate's study did not actually learn the contingency, in contrast to the males in Adret's study. It may be that the ability to learn the behaviour-contingent response depends upon the operant used. That is, hopping on a perch may be more difficult for the bird to associate with hearing song than the positive key-pecking response.

What is evident from these studies is that social interaction plays an important role in the acquisition of song in zebra finches. The studies by Baptista & Petrinovich (1984) on white-crowned sparrows which were described earlier in the chapter lend further support to the hypothesis that song learning may be stimulated by using a live song tutor with whom the young birds can interact. These examples show how specific kinds of social experience can guide both the timing and the nature of song learning. Given the similarities between song learning and sexual imprinting, this raises the question of how important social factors are in guiding the development of sexual preferences.

How do social interactions guide sexual imprinting in zebra finches?

One of the most important techniques for studying the extent to which early social experience influences the development of sexual preferences is that of cross-fostering, in which the young are removed from their natural parents and raised by a pair of foster-parents either of the same or of a different species. Much of the work has focused on the development of sexual preferences in young zebra finch males that have been raised either by zebra finches or by Bengalese finches. When adult, the males are given a mate-choice test, usually with two females, one of each species. By observing the amount of time and sexual displays towards the two females, the sexual preference of the young male can be determined.

Such cross-fostering experiments demonstrate that the social conditions that a bird experiences early in life play a key role in guiding both timing and the nature of sexual imprinting. Young zebra finch males that have been raised by zebra finch parents or foster-parents develop a stable preference for zebra finch females when adult, whereas those that have been raised by Bengalese finch foster-parents, although later strongly preferring Bengalese finch females, will sometimes also court zebra finch females. Young males that were raised by a mixed-species pair, i.e. one zebra finch and one Bengalese finch parent, strongly preferred zebra finch females when adult, in spite of the fact that they had been reared by both species (Immelmann, 1972).

In the following chapter, ten Cate discusses how, by examining the behavioural interactions between the adults and their young, he and his colleagues (ten Cate, 1982; ten Cate, 1984; ten Cate *et al.*, 1984) were able to assess how social factors can influence sexual imprinting. Ten Cate argues that behavioural interactions with the parents and with other birds may explain why young males reared by a mixed-species pair develop a preference for zebra finch females rather than for Bengalese finch females. The study by ten Cate *et al.* (1984) throws further light on the role that social experience plays in sexual imprinting. Young male zebra finches were exposed to zebra finches and Bengalese finches successively, rather than simultaneously. The zebra finches were raised by zebra finch parents for 1 month. Each young male was then put in with a separate group of Bengalese finches for another month and their behaviour observed. After a period of subsequent isolation, all the birds were tested to see whether they preferred Bengalese finch or zebra finch females.

There was a considerable amount of variation between young males: about two-thirds of the birds preferred zebra finch females, whereas the other third preferred Bengalese finch females. In order to attempt to account for this individual variation in sexual preferences, ten Cate *et al.* (1984) compared the social interaction between the young zebra finches and the adult Bengalese finches with their sexual preferences. They found that the social behaviour of the young zebra finch male to Bengalese finches was not correlated with the sexual preference. However, a higher number of behavioural initiatives of Bengalese finches, both aggressive and non-aggressive ones, directed towards the young male was related to a stronger sexual preference for Bengalese finch females. Other young males which had spent their second month of life with a group of stuffed Bengalese finches developed much stronger preferences for the zebra finch female.

These results suggest that exposure to the visual appearance of Bengalese finches has little effect on the development of sexual preferences. What seems to be important are the social interactions between the Bengalese finches and the young males. Thus, in parallel with the results from the song-learning studies, the findings of ten Cate and his colleagues suggest that a live stimulus strongly influences the sexual development of the young birds and that it is the social interactions, rather than simply the visual stimulus, that are crucial. However, these results leave open questions about the precise nature of the social experience and the mechanism by which the sexual preferences develop.

In the previous chapter, Bischof discussed recent studies by Immelmann *et al.* (1991) and Kruijt & Meeuwissen (1991) that shed light on these matters. The results of both studies are remarkably similar, and demonstrate that young zebra finch males that had been reared by Bengalese finches until about the time of independence, isolated until adulthood and then housed with a zebra finch female, may switch their preference from a Bengalese finch female to a zebra finch one. However, this switch in preference can be prevented by two brief preference tests with a Bengalese finch and a zebra finch female, immediately after the males have been isolated. The effect is so striking that even after subsequent exposure to the zebra finch female for several months, more than half the males preserved their initial preference for Bengalese finch females. These results cast doubt upon the original ideas about sensitive phases and stability of preferences formed during imprinting because both groups of researchers show that the sexual preferences can be altered even in adult birds.

Immelmann *et al.* (1991) and Bischof (this volume) raise the point that the view of sexual imprinting being a two-stage process is rather similar to that developed in the song-learning literature. As outlined briefly in the section on song learning, song acquisition is thought to consist of two steps, a memorization and a motor phase (cf. DeVoogd, this volume). In the memorization phase of song learning, the young male develops a 'template' based on the songs that he hears. Step two consists of a so-called motor phase in which the young male matches his own song with the template that he has developed during the first phase (see Konishi & Nottebohm, 1969). Thus, the memorization phase of song learning may be comparable to the sensitive phase for imprinting, as originally defined, and the motor phase could be compared with the act of consolidation.

Following the studies of Immelmann *et al.* (1991) and Kruijt & Meeuwissen (1991), Bischof & Clayton (1991) carried out their experiment in which they found that the order in which the young males were exposed to the two females has a strong influence on the final outcome of the development of their sexual preferences: it is the first exposure to a female of the same species as the one that raised them which consolidates the information, rather than any exposure to a female of that species (see Bischof, this volume). In interpreting the results of Bischof & Clayton (1991), it should be noted that the birds were isolated from all other birds between day 40 and day 100, a scenario that would be extremely unlikely to occur under natural conditions. This period of isolation may have had some important effects on the results and, in particular, the impact of the first courtship experience may have been enhanced by this preceding isolation. In the wild, consolidation of the previously acquired sexual preference may result from a series of sexual encounters, rather than just the first one. However, although the experimental design used by Bischof

& Clayton (1991) probably enhances the effect of the first sexual encounter, it seems unlikely that the mechanism for the consolidation of the previously acquired sexual preference would be different.

What all these results demonstrate is that social interactions with a live bird are extremely important in guiding the development of mate preference, and that some types of interaction seem to be more important than others. The study by Bischof & Clayton (1991) suggests, at least for zebra finches, that parental feeding prior to nutritional independence, and initial courtship experience with females are two crucial factors governing the timing and nature of the sexual preferences that are established during the process of sexual imprinting.

How are song learning and sexual imprinting related in zebra finches?

One of the points emerging from the studies of sexual imprinting is that it appears to occur earlier than song learning, at least for zebra finches that have been raised by conspecific parents (see also Immelmann, 1972; Slater *et al.*, 1988). This difference in timing might have an important consequence for song tutor choice: a young male might learn song from the tutor of the same species as he has sexually imprinted upon.

Is there any evidence to support this hypothesis? In a study by Clayton, (1988b), zebra finches were exposed to both zebra finches and Bengalese finches during their development. The young male zebra finches were raised with their siblings in a large cage together with either a zebra finch and a Bengalese finch pair of foster-parents, or a pair of foster-parents of their own species, or a pair of Bengalese finch foster-parents. At 35 days after hatching they were transferred to another cage with their brothers and given two song tutors, a zebra finch and a Bengalese finch male. From 70 to 120 days post-hatch, the males in each brood were housed together but in visual isolation from all other birds.

Once the birds reached sexual maturity (day 120) their songs were recorded and analysed to see whether they sang zebra finch, Bengalese finch or a mixture of both types of song elements. To measure their sexual preference for both species of females, each male received three, 20-minute simultaneous choice tests with three different pairs of zebra finch and Bengalese finch females in order to calculate the mean number of song phrases that each male directed towards zebra finch and Bengalese finch females. For each test, each male was placed individually in the central compartment of a test cage and separated by wire mesh from the two side compartments, one of which contained a zebra finch female and the other a Bengalese finch female.

Some males that had been raised by both species sang mixed species song and some courted females of both species. At first sight, this would seem to support the hypothesis that young males might be learning their songs from the species that they have imprinted upon. However, comparing the responses of individual males showed that males do not necessarily choose song tutors who are of the same species as those they have sexually imprinted upon. These results suggest that in spite of the striking similarities between the two systems, the development of choice of song tutor and sexual preference are not directly related, and that sexual imprinting and song learning in zebra finches are two separate systems. One challenge for future work in this field will be to determine in more precise detail the way in which the two systems are linked and the extent to which these results for zebra finches can be generalized to other species.

One way in which these questions can be tackled is to investigate the underlying neural mechanisms of the two systems and then to attempt to make links between the brain and behavioural findings. What is the relationship between behavioural changes and changes in the brain? For example, do changes in the neuroanatomical structures of the brain precede or accompany the behavioural development? In order to answer questions of this nature, an understanding of the neuroanatomical structures is required.

Despite the abundant ethological literature on sexual imprinting and song learning, and the ease with which their behavioural attributes can be studied in the laboratory, the neurobiological analysis is considerably less straightforward because of the complexity of the behaviour. To date, there have been no reports on the neural basis of sexual imprinting (but see Bolhuis *et al.*, 1989), and although the neural basis of song learning is fairly well established (e.g. reviews by Nottebohm, 1989; Nottebohm *et al.*, 1990; Clayton & Bischof, 1990; Nordeen & Nordeen, 1990), virtually nothing is known about the neural mechanisms of how social interactions can affect sensitive phases and stability of song learning and sexual preferences. There do appear to be certain forebrain nuclei which are activated in arousing situations (Bischof & Herrmann, 1986). Although merely speculative, these nuclei may play a role during development in determining which stimuli are likely to be memorized. Determining how motivation and arousal influence song learning and sexual imprinting at the neural level is likely to be a difficult task (Clayton & Bischof, 1990). Nevertheless, the song control system of birds is a promising starting point for such research and may point the way for future investigations into the neural basis of sexual imprinting.

Acknowledgements

Many of the ideas for this chapter came from numerous discussions with Peter Slater and Hans-Joachim Bischof. I am extremely grateful to both of them for their continued support and for their critical comments on the manuscript. Thanks also to Patrice Adret for allowing me to cite his unpublished work, to Richard Zann for comments and to Johan Bolhuis and Jerry Hogan for editing the chapter. Financial support for the research came from an Imperial Chemical Industries Educational Trust, an Alexander von Humboldt Stiftung Fellowship and from a Postdoctoral Fellowship funded by the Science and Engineering Research Council; support during the writing of this manuscript came from the Agricultural and Food Research Council and Linacre College, Oxford.

References

Adret, P. (1993). Operant conditioning, song learning and imprinting to taped song in the zebra finch. *Animal Behaviour*, 46, 149–159.

Baptista, L. F. & Petrinovich, L. (1984). Social interaction, sensitive phases and the song template hypothesis in the white-crowned sparrow. *Animal Behaviour*, 32, 172–181.

Bischof, H.-J. & Clayton, N. S. (1991). Stabilization of sexual preferences by sexual experience in male zebra finches, *Taeniopygia guttata castanotis*. *Behaviour*, 118, 144–155.

Bischof, H.-J. & Herrmann, K. (1986). Arousal enhances 14-C-2 deoxyglucose uptake in four forebrain areas of the zebra finch. *Behavioural Brain Research*, 21, 215–221.

Böhner, J. (1990). Early acquisition of song in the zebra finch, *Taeniopygia guttata*. *Animal Behaviour*, 39, 369–374.

Bolhuis, J. J. (1991). Mechanisms of avian imprinting: a review. *Biological Reviews*, 66, 303–345.

Bolhuis, J. J., Johnson, M. H., Horn, G. & Bateson, P. (1989). Long-lasting effects of IMHV lesions on social preferences in domestic fowl. *Behavioral Neuroscience*, 103, 438–441.

Clayton, N. S. (1987a). Mate choice in male zebra finches: some effects of cross-fostering. *Animal Behaviour*, 35, 596–622.

Clayton, N. S. (1987b). Song tutor choice in zebra finches. *Animal Behaviour*, 35, 714–722.

Clayton, N. S. (1988a). Song discrimination learning in zebra finches. *Animal Behaviour*, 36, 1016–1024.

Clayton, N. S. (1988b). Song learning and mate choice in estrildid finches raised by two species. *Animal Behaviour*, 36, 1589–1600.

Clayton, N. S. (1989). Song, sex and sensitive phases in the behavioural development of birds. *Trends in Ecology and Evolution*, 4, 82–84.

Clayton, N. S. & Bischof, H.-J. (1990). *Neurophysiological and behavioural development in birds: song learning as a model system. Naturwissenschaften,* 77, 123–127.

Hinde, R. A. (1970). *Animal Behaviour: A Synthesis of Ethology and Comparative Psychology.* New York: McGraw-Hill.

Horn, G. (1985). *Memory, Imprinting, and the Brain.* Oxford: Clarendon Press.

Immelmann, K. (1962). Beiträge zu einer vergleichenden Biologie australischen Prachtfinken (*Spermestidae*). *Zoologisches Jahrbuch Systematische Bedeutung,* 90, 1–196.

Immelmann, K. (1969). Song development in the zebra finch and other Estrildine finches. In *Bird Vocalizations,* ed. R. A. Hinde, pp. 61–74. Cambridge: Cambridge University Press.

Immelmann, K. (1972). The influence of early experience on the development of social behaviour in Estrildine finches. *Proceedings of the XI International Ornithological Congress, Den Haag,* 1970, 316–338.

Immelmann, K., Pröve, R., Lassek, R. & Bischof, H.-J. (1991). Influence of adult courtship experience on the development of sexual preferences in zebra finch males. *Animal Behaviour,* 42, 83–90.

Immelmann, K. & Suomi, S. J. (1981). Sensitive phases in development. In *Behavioural Development,* ed. K. Immelmann, G. W. Barlow, L. Petrinovich & M. Main, pp. 395–431. Cambridge: Cambridge University Press.

Konishi, M. & Nottebohm, F. (1969). Experimental studies in the ontogeny of avian vocalizations. In *Bird Vocalizations,* ed. R. A. Hinde, pp. 29–48. Cambridge: Cambridge University Press.

Kruijt, J. P. (1985). On the development of social attachments in birds. *Netherlands Journal of Zoology,* 35, 45–62.

Kruijt, J. P. & Meeuwissen, G. B. (1991). Sexual preferences of male zebra finches: Effects of early adult experience. *Animal Behaviour,* 42, 91–102.

Lorenz, K. (1935). Der Kumpan in der Umwelt des Vogels. *Journal für Ornithologie,* 83, 137–213, 289–413.

Marler, P. (1976). Sensory templates in species-specific behavior. In *Simpler Networks and Behavior,* ed. J. C. Fentress, pp. 314–329. Sunderland, Massachusetts: Sinauer.

Marler, P. (1987). Sensitive periods and the roles of specific and general sensory stimulation in birdsong learning. In *Imprinting and Cortical Plasticity,* ed. J. P. Rauschecker & P. Marler, pp. 99–136. New York: John Wiley & Sons.

Marler, P. (1991a). Song-learning behavior: the interface with neuroethology. *Trends in Neurosciences,* 14, 199–206.

Marler, P. (1991b). The instinct for vocal learning: songbirds. In *Plasticity of Development,* ed. S. E. Brauth, W. S. Hall & R. J. Dooling, pp. 107–125. Cambridge, Mass.: MIT Press.

McGregor, P. K. & Avery, M. I. (1986). The unsung songs of great tits (*Parus major*): Learning neighbours' songs for discrimination. *Behavioural Ecology and Sociobiology,* 18, 311–316.

Miller, D. B. (1979). Long-term recognition of father's song by female zebra finches. *Nature,* 280, 389–391.

Nottebohm, F. (1989). From bird song to neurogenesis. *Scientific American,* 260, 74–79.

Nottebohm, F. , Alvarez-Buylla, A., Cynx, J., Kirn, J., Ling, C-Y., Nottebohm, M, Suter, R., Tolles, A. & Williams, H. (1990). Song learning in birds: the relation between perception and production. *Philosophical Transactions of the Royal Society of London, Series B,* 329, 115–124.

Nordeen, E. J. & Nordeen, K. W. (1990). Neurogenesis and sensitive periods in avian song learning. *Trends in Neurosciences*, 13, 31–36.

Schutz, F. (1965). Sexuelle Prägung bei Anatiden. *Zeitschrift für Tierpsychologie*, 22, 50–103.

Slater, P. J. B., Eales, L. A. & Clayton, N. S. (1988). Song learning in zebra finches (*Taeniopygia guttata*): progress and prospects. *Advances in the Study of Behavior*, 18, 1–34.

Slater, P. J. B. & Mann, N. I. (1990). Do male zebra finches learn their fathers' songs? *Trends in Ecology and Evolution*, 12, 415–417.

Slater, P. J. B. & Richards, C. (1990). Renesting and song learning in the zebra finch, *Taeniopygia guttata*. *Animal Behaviour*, 40, 1191–1192.

ten Cate, C. (1982). Behavioural differences between zebra finch and Bengalese finch (foster) parents raising zebra finch offspring. *Behaviour*, 81, 152–172.

ten Cate, C. (1984). The influence of social relations on the development of species recognition in zebra finch males. *Behaviour*, 91, 263–285.

ten Cate, C. (1986). Listening behaviour and song learning in zebra finches. *Animal Behaviour*, 34, 1267–1269.

ten Cate, C. (1991). Behaviour-contingent exposure to taped song and zebra finch song learning. *Animal Behaviour*, 42, 857–859.

ten Cate, C., Los, L. & Schilperoord, L. (1984). The influence of differences in social experience on the development of species recognition in zebra finch males. *Animal Behaviour*, 32, 852–860.

Thorpe, W. H. (1961). *Bird-Song*. Cambridge: Cambridge University Press.

Williams, H. (1990). Song learning in the zebra finch: do fathers supply the models for their sons' songs? *Animal Behaviour*, 39, 745–757.

Zann, R. A. (1990). Song and call learning in wild zebra finches in south-east Australia. *Animal Behaviour*, 40, 811–828.

6

Perceptual mechanisms in imprinting and song learning

CAREL TEN CATE

At birth, young animals are confronted with a variety of stimuli in various modalities. Individuals of precocial species, in particular, have to respond adequately within hours in order to survive. Altricial individuals may have more time available to explore the world around them, but eventually they face the same problem. Of course, the change in stimulation from before until after birth is not one from zero input to a bewildering complexity. Several senses can be stimulated before birth, and learning about features of the external world does occur before birth in both mammals and birds. Also, not all senses may be functioning at full strength at birth, making the increase in incoming stimulation less abrupt. Nevertheless, the task of making sense of the world is not an easy one. One may argue that parents may provide some guidance, but then this requires that the animal recognizes its parent, reducing, but not eliminating the problem. At the same time, ducklings, only a few hours after hatching, and fowl chicks a bit later will follow their mother as though they have known her for a long time; adult zebra finches will court conspecifics as though no other species has ever been around, and many songbirds will sing their father's song as though they had never been exposed to any other sound. However, in all these cases there is abundant evidence that interfering with normal rearing conditions of the birds may severely disrupt the natural outcome of the processes. So, although we are dealing with flexible learning processes, the natural variation in outcome seems limited. Such developmental processes might be called 'canalized' or 'buffered' and one of the aims of ontogenetic research is to examine how this is achieved. A concept which is frequently used in the context of perceptual development is that of 'templates' or '(pre)dispositions'. These terms refer to the fact that the young bird does not seem to face the world unequipped, but shows, instead, some differentiation in the reaction towards different stimuli. It is as though it possesses some 'prestructured' or 'prefunctional' (Hogan, 1988) knowledge of the world, which guides it towards responding to, or learning about, the right stimuli. More recently, it has become clear that another factor may have at least as powerful an influence on the developmental process, namely the context in which the stimuli to learn about are presented. In particular, social interactions with the stimuli (as exemplified by song learning from, and imprinting with, live tutors) strongly affect the learning process (see also Clayton, this volume). This factor also contributes strongly to the canalization of development, as, at an early age, social interactions are most likely to occur with parents or siblings. Predispositions and social interactions can be seen as two different dimensions and be analyzed separately for their contribution to development (e.g. ten Cate, 1989a). However, this does not imply that the ways in which they obtain their effects are based on very different mechanisms, although this has been suggested (e.g. Pepperberg, 1991). My aim in this chapter is to analyze both the nature of predispositions and social interactions and how they contribute to development. I will concentrate on the processes of imprinting and song learning in birds. Not only do they provide some of the best examined instances of canalized development, but they also provide the classic examples of the way in which predispositions and social interactions may affect development. In doing so, I will distinguish between the question of the stimulus input appealing to the predisposition or necessary to give rise to the effect of social interactions and the question of how the relevant experience affects further development. I hope to demonstrate that the way in which the perceptual sensitivities underlying predispositions and those underlying social interactions both obtain their effects, as well as the ways in which they affect the learning mechanisms, may not be radically different. Functionally they might be complementary in their effects, and their interaction may make the young bird highly sensitive to parental stimulation and may lead to strongly canalized, but flexible, development.

I will first address the traditional view of imprinting and song learning as examples of learning guided by predispositions. This brings me to a discussion on the nature of 'predispositions'. Next I will concentrate on the effects of social interactions on imprinting and song learning, which will be illustrated with a few case studies. Finally, I return to the theme of canalization and I will discuss how the active interchange between the developing individual and its environment affects the developmental process.

Imprinting and song acquisition in birds

The phenomena of filial and sexual imprinting were discussed by Bischof and Clayton in Chapters 4 and 5 of the present volume (see also Bolhuis, 1991, for an extensive review). Lorenz (1935) suggested that the young bird's early experience may also have consequences for later sexual preferences. As later research revealed, filial and sexual imprinting are not just two consequences of a single process, but they differ in various respects and are quite separate processes (Bateson, 1979; Kruijt, 1985).

Many songbirds acquire the knowledge about the song to be sung as a result of exposure to this song at an early age (see also Bischof, Clayton, this volume). So, both song learning and sexual imprinting are characterized by a powerful effect of early experience on later behaviour. Both processes also share an age-related sensitivity to this external stimulation, and the acquired preference or song is relatively resistant to change at a later age. In addition, for sexual imprinting and, in a species such as the chaffinch for song learning, the effects of early exposure need not become overt until long after the end of the exposure period. There are, of course, limits to the similarity between song development and imprinting, as song development does not only involve the formation of some sort of internal representation of the song to be sung, but also involves the development of the motor output to produce that song later in life. I will not deal with this latter issue, but instead concentrate on the processes involved in the acquisition of the internal representation of the song to be sung. I will first analyze the role of 'predispositions' in canalizing development, starting with the traditional view of the mechanisms thought to underlie imprinting and song learning.

Programmed learning

The apparent absence of clearly identifiable reinforcers for both imprinting and song learning has led to categorizing them as a special form of learning: 'template learning' or 'programmed learning'. This terminology refers not only to the seemingly programmed nature of the time course of the learning processes, but also to another similarity between the two; the constraints there seem to be with respect to what is acceptable as a model about which to learn. Some songs are more attractive for copying than others and some models seem more likely to become the object for social or sexual attachment than others. A clear example illustrating this phenomenon is song acquisition in the white-crowned sparrow (Marler, 1970). White-crowned sparrows will copy songs from a tape if exposed to these songs between about 10 and 50 days. However, Marler's experiments suggested that they will only do so when exposed to a dialect of their own species and not when exposed to song of another species. This finding led Gould (1982) to conclude about white-crowned sparrows that 'evolution has provided them with filters – IRMs – by which they distinguish their own song from those of other species.... When they hear the appropriate cues, they turn on their mental tape recorders'. These 'appropriate cues' might be a characteristic such as song elements of a specific nature, a particular arrangement of elements, tonal qualities, etc. which may serve as markers or 'releasing stimuli' for its acceptance as a suitable song to learn about. This idea has been corroborated by later experiments demonstrating that, in this case, song sparrows will learn swamp sparrow elements if these are presented in the syntax specific for conspecific songs (Marler, 1987). The syntax may thus operate as the cue for the 'mental tape recorder' to switch on. Staddon (1983), arguing along the same lines as Gould, uses the metaphor of a 'slot' or 'template', waiting to be filled with the appropriate matching input. Although both the concept of 'template' (Johnston, 1988) and the evidence for selective learning (Petrinovich, 1988) have been criticized, the observation that white-crowned sparrows exposed to taped song will learn conspecific but not allospecific songs still suggests that, at least under certain conditions, the songs of conspecifics are processed differently from those of other species.

It has been suggested that certain stimulus features enhance filial imprinting in a similar way. In spite of the fact that the objects for filial attachment can be so varied, including other bird species, humans, flickering or rotating lights, conspicuously coloured artificial objects, etc. (see Sluckin, 1972), not everything will do, but some objects are preferred to others. An example of this, to which I shall return later on, is provided by a number of studies by Horn and co-workers (see Horn, 1985). Their experiments suggest that domestic chicks have a disposition to respond to certain features that resemble those of a (stuffed) hen (Horn, 1985).

From metaphors to mechanisms

In spite of the compelling terminology of 'templates' or empty 'maternal slots', these are metaphors which do not necessarily provide understanding of the underlying mechanisms (see also Johnston, 1988; Bolhuis & Johnson, 1991). Yet, it is understanding of these mechanisms which we want to obtain. The effects observed both in the white-crowned sparrow example and the chick example give rise to a number of questions. First one might ask for the precise nature of the underlying learning or processing mechanism. For instance, does imprinting involve a specialized learning mechanism (Lorenz, 1935), some form of associative learning (Hoffman & Ratner, 1973), or is it exposure learning (Sluckin, 1972) or 'released image' recognition learning (Suboski, 1989)? This issue has been addressed recently (Bolhuis, de Vos & Kruijt, 1990; Hollis, ten Cate & Bateson, 1991; Bolhuis, 1991) and I will not go into it in much detail in this chapter. For the moment, I take as a common denominator of the various views that both song learning and imprinting seem to proceed without obvious external rewards (food, warmth), but that aspects of a visual or acoustic stimulus may be inherently attractive and that, by learning, other aspects of the stimulus object or sound may become associated with these inherently attractive aspects. Taking this general model for the learning process, at least two other questions can be posed. First, we can ask what the young bird is sensitive to, i.e. analyze the crucial features of the stimulus input required for a song to be acceptable or a hen to be attractive. The second question concerns the influence that these features exert on the learning process. In this case we take it for granted that there is some aspect of the stimulus responsible for the effect, but ask how this experience affects the processing of stimuli.

The nature of perceptual sensitivity

As mentioned above, for a song to be learned by a song sparrow, syllables have to be presented in the 'right' syntax. 'Right' means the syntax used by the avarage song sparrow. Unknown at present is what characterizes the right syntax; is it, for instance, the fact that songs consist of different parts, or does the time structure require specific features? Further experimentation on this issue and on the question of whether species differ in their sensitivity, as seems to be the case with the song sparrow and the swamp sparrow (Marler, 1987, 1990), is required. This may also provide some insight into whether perceptual sensitivity is tuned to the characteristics of the average conspecific song, or to those features which make conspecific songs different from those of other species, and hence into the function of the sensitivity with respect to maintaining species-specificity of songs.

The nature of the sensitivity of the chick to respond to a hen has been addressed by Johnson & Horn (1988). By cutting up a stuffed hen and by scrambling and re-combining the pieces in a number of ways, it was found that a scrambled model remained as attractive as a normal one for as long as it contained a head and neck. This finding was extended in experiments indicating that a head alone was at least as likely to be responded to as a whole hen (Johnson & Horn, 1988). Although this suggests that a chick may have some sensitivity for the head features of a hen, it need not be so specific, as a gadwall duck and even a stuffed pole cat appeared to be as attractive as a hen model (Johnson & Horn, 1988). This finding suggests that what the chicks are sensitive to might be a more general property, for instance a vertical symmetric arrangement of a few features such as circular dots ('eyes'), but this issue has not yet been investigated.

Apart from this disposition for what most likely is a specific configuration of stimulus elements, earlier research on filial imprinting had already revealed sensitivities for certain types of much simpler, artificial, models over others. When, for instance, individual chicks are exposed to two identical objects, one red, the other yellow, most chicks will, after some time, have become attached to the red object (e.g. Bateson & Jaeckel, 1976; de Vos & Bolhuis, 1990). Similarly, certain shapes, sizes or patterns are preferred over others (see Sluckin, 1972). The relation of these types of sensitivities to those making a hen attractive is, as yet, not quite clear (cf. Johnson & Bolhuis, 1991).

The effects of perceptual stimulation

What these examples of predispositions have in common is that emphasis is placed on some existing (pre-functional) knowledge as the prime factor for the observed canalization during the acquisition process. However, the similarity is based primarily upon the observation of a similar outcome: birds are more likely to end up with a particular song or to show an attachment to a particular object rather than another. The caveat here is that a similar outcome can be achieved in various ways. For instance, the song learning example given above suggests that songs lacking specific acoustic features are rejected as a model, i.e. the features act upon some sort of 'switch', allowing specific songs to enter memory. The features of a song thus affect the processing directly. For imprinting, on the other hand, the present evidence suggests that biases in later preferences may arise in various ways and, as I will argue below, to lump them all under the heading of 'predispositions' may hamper rather than increase our insight in the mechanisms which may underlie the expression of a particular preference.

In filial imprinting, most authors refer to predispositions when dealing with preferences for certain features shown by a chick which has not been exposed to these features before. This means that we are dealing with a predisposition to *respond* to some features rather than others. It does not tell us anything about whether such a disposition affects further processing of these features or others associated with it. This is most clearly illustrated by some experiments by Kovach (1983a, 1983b, 1983c). In separate experiments he showed that naive quail chicks prefer blue lights over red ones; lights with vertical lines over lights with dots (Kovach, 1983a, 1983b) and flickering lights over stationary ones (Kovach, 1983c). In all these cases the preference for the first over the second stimulus is about 60 % versus 40 %. If chicks are exposed to the various stimuli, a remarkable phenomenon appears. As expected, chicks exposed to red or blue lights increase their preference for the colour to which they have been exposed (Kovach, 1983a, 1983b, 1983c). However, chicks exposed to vertical lines or dots do not, or marginally, alter the strength of their preference as a result of exposure (Kovach, 1983b). The same applies to exposure to ficker or non-flicker (Kovach, 1983c). So, although for colour, pattern and flicker a similar initial preference is present for one type of stimulus over the other, additional learning about the feature seems to occur for colour only. Thus, one might conclude that the predisposition to respond to pattern or flicker only affects the further development of a preference in that it might help to guide the young bird towards objects having these characteristics. Next, the young bird may learn to recognize these objects by learning other features of them. Nevertheless, the attraction towards objects having vertical lines or showing flicker remains intact in chicks having learned about other features (Kovach, 1983b, 1983c), so it will still contribute to the later preference if the chick has developed a preference towards objects possessing the preferred traits (see also Klopfer, 1967). A comparable mechanism may underlie the finding in sexual imprinting that birds of the rearing type were preferred more strongly if they possessed certain features not present, or less obvious, in the parents (ten Cate & Bateson, 1989).

Is there an interaction between the disposition to respond to certain colours and learning about colours? Several experiments indicate that chicks need not learn more about the preferred than about the non-preferred colour. Bateson & Jaeckel (1976) showed that naive chicks preferred a red light to a yellow one. Other chicks were exposed to either a red or a yellow light. In turned out that red-exposed chicks showed a stronger preference for red than the yellow-exposed chicks showed for yellow. Yet, both groups differed from the naive controls to the same extent, i.e. yellow-exposed chicks had shifted their preference as much as red-exposed ones when compared with the initial preference, indicating a similar amount of learning. The experiments by Kovach allow us to go one step further. Kovach (1979) selected for colour preferences in naive chicks. After ten generations the selected lines were clearly different from each other as well as from the control line for their naive preference. Similar to Bateson & Jaeckel's experiment, chicks that were selected for a preference for red learnt as much about blue as unselected chicks, while blue-line chicks learnt as much about red as unselected chicks. Thus, strength of the naive preference did not seem to affect learning about the non-preferred feature. A possible interpretation of these findings is that yellow and red, or blue and red, are equally 'learnable', but that the final outcome is a combination of the naive preference and the learnt one. This interpretation is, in fact, the same as the one above, with the difference that in this case, in contrast to the preference for lines over dots or flicker over non-flicker, the disposition to respond is for a feature about which learning may also take place.

A final example along the same line is provided by the work of Horn and co-workers. The preference for hen-like objects emerges in naive chicks, provided they are handled at some time between about 20 and 40 hours after hatching (Johnson, Davies & Horn, 1989). If the preferences of chicks at 24 hours after hatching are compared to those at 48 hours, it turns out that, irrespective of whether chicks are exposed to (and have developed a preference for) a hen, a red rotating box, or no stimulus object at 24 hours, the preferences at 48 hours is always shifted to the hen. More importantly, the shift is always by the same magnitude. This indicates, again, that the emerging disposition is added to the acquired preference. However, the phenomenon allows us to address the question whether chicks in which the preference has emerged show signs of being hampered in learning about the red box, compared to chicks in which the preference has not emerged as a result of not handling these chicks. So far, experiments (Bolhuis, Johnson & Horn, 1989) have not revealed any evidence that an emerged preference for hen-like objects hampers learning about an alternative object. Thus, as for the examples on colour learning discussed above, a disposition to respond to hen- like objects seems not to affect the learning itself, but adds to the preference in trained birds.

All these examples point to a similar mechanism: birds may have naive preferences which mean that they may direct themselves more, and/or attend more, to certain objects and hence learn about these objects in favour of others. At the same time, the feature for which a preference is present does not have to be the same as the one about which learning can take place, nor does the preference imply that more learning takes place about a preferred rather than a non-preferred trait. Such dispositions to respond to certain features seem to remain present (and to affect the preference) after training with an object, whether or not it possesses these features. Whether the conclusions can be read to suggest that the degree of learning about a preferred trait is *never* different from that about a non-preferred trait remains to be seen. That there may still be something special about preferred features compared to others is indicated by a recent experiment by van Kampen & de Vos (1991). They demonstrated that a chick exposed to a red object learned more about *another* feature of the object, in this case its shape, than chicks exposed to a (less preferred) yellow object. This example indicates that interactions may occur between a disposition to respond to, or to learn about, one feature and learning about another. However, the better learning about the red object may have been induced by the red-exposed chicks spending more time close to the object and hence receiving more exposure to the shape of the object, rather than by an effect on the processing mechanism.

A final point concerns the emergence of dispositions. The finding that very young birds may have clear biases in their preference has too often been interpreted to imply that such dispositions were not affected by earlier experience. However, dispositions have a developmental history which might have been affected by external stimulation just as much as learned preferences. For instance, the preference for 'hen-like objects' emerges only after some degree of visual or tactile stimulation, while prior exposure to a complex pattern may speed up its development (Bolhuis, Johnson & Horn, 1985). In addition, the non-specific stimulation needs to occur within a certain time after hatching in order for the disposition to emerge (Johnson *et al.*, 1989). With respect to the disposition to respond to certain patterns, Klopfer & Hailman (1964; Klopfer, 1967) demonstrated that both chicks and ducklings which had been exposed to either a plain white stimulus object or a strikingly patterned one during training preferred the striking model later on, whereas untrained individuals did not discriminate. So, in this case the emergence of a disposition for patterned objects was dependent upon having received some sort of visual stimulation earlier on. The importance of prior visual experience for the development of a preference for patterned over unpatterned stimuli was also demonstrated by Berryman, Fullerton & Sluckin (1971), while Bateson & Wainwright (1972) demonstrated that such prior experience might affect subsequent learning. On the other hand, exposure to achromatic light prevented the emergence of colour preferences in naive quail chicks (Kovach, 1971). So, underlying a perceptual preference for certain patterns over others, or for 'hen-like' features over others at one stage in development, might be a sensitivity to some other type of stimulation at an earlier age. In addition, stimulation in one modality may affect the emergence of sensitivity for another. This issue has been examined extensively by Lickliter *et al.* (Lickliter & Virkar, 1989; Lickliter, 1990; Lickliter & Stoumbos, 1991; Lickliter & Hellewell, 1992). They demonstrated that experimentally induced visual stimulation before hatching interfered with the later responsiveness of bobwhite quail (*Colinus virginianus*) hatchlings to the species-specific maternal call, while enhanced prenatal auditory experience facilitated species-specific visual responsiveness.

Programmed learning revisited

The conclusion at the end of this section on 'programmed learning' might be that young birds do have various perceptual sensitivities, differing with age or preceding stimulation, which can affect further processing. This effect may range from altering sensitivities to learning about specific features. The final preference which a chick might show for, say, a stuffed hen to which it has been exposed for some time, has thus been brought about by it receiving certain types of stimulation before hatching, certain types of stimulation after hatching and some exposure to the features of the hen. The final preference is based on one for traits for which the chick needed no exposure to the object to respond to it, but for which it may have needed some other type of stimulation earlier on; on one for traits for which it also needed no exposure to prefer the trait, but for which exposure has reinforced the preference; on one for features to which it would not develop a preference had it not been exposed to them; and on one for features which it may have learned about because the learning was stimulated as a result of the presence of other features in the same object. This complexity will not exactly cover what most people have in mind when they mention filial imprinting as an example of 'template' or 'programmed' learning, nor is it covered by metaphors of 'slots' waiting to be filled. In contrast, the apparent programmed nature of the developmental process originates from an intricate interplay between organism and environment, and the outcome is dependent on environmental stimulation as much as on changing properties of the organism itself. In addition, this dynamic nature of the developmental process makes it doubtful whether we should make such a strong distinction between developing dispositions and learning in their contribution to perceptual development (see also Hogan, 1988). However, I hope to have made clear that the program underlying 'programmed learning' should itself be the subject for developmental study if we want to understand predictability and canalization of development.

I started this chapter by drawing parallels between imprinting and song learning (see also Bischof, Clayton, this volume). Returning to this issue for a moment, the emergent complexity out of the studies on filial imprinting does raise a number of questions with respect to the way in which we might look at song development. For instance, little effort has so far been put into analyzing the nature of the sensitivity underlying the preference for certain songs over others and how this affects the learning process. It has been recognized that analyzing song development in terms of 'templates' interacting with 'learning' is an approach which may have limitations (Johnston, 1988; West, King & Duff, 1990). Nevertheless, the possibility that certain types of stimulation may give rise to perceptual sensitivities of a seemingly different nature seems a neglected area of research, but one which may be useful in interpreting the results obtained by rearing birds in acoustic and visual isolation. Could it, for instance, be that the lack of proper development which some species show when tape-tutored may also be caused by lack of some other type of stimulation which the bird needed at an earlier stage to induce the sensitivity to respond to the song?

Social context

Above, I concentrated on the factors underlying the role of so-called 'predispositions' in perceptual development. In this section will follow a similar approach to gain further understanding of the factors underlying the role of social interactions.

To illustrate the powerful effect of social interactions on perceptual development, I will return briefly to the white-crowned sparrow. In 1981, Baptista & Morton published a paper with a sonagram of a white-crowned sparrow singing an allospecific song differing in many features from white-crowned sparrow song. The song was a near-perfect copy of the song of a strawberry finch (*Amandava amandava*). The classic model of song learning outlined above was obviously contradicted; this song somehow got stored, in spite of its dissimilarity to white-crowned sparrow song. The crucial factor which Baptista & Morton (1981) suggested to be responsible for this finding was the way in which the bird was exposed to the song. It was housed with the strawberry finch in the same aviary for some time and it might be that this exposure to a live tutor had a powerful influence on the processing mechanism. A series of systematic studies followed up this anecdotal observation (e.g. Baptista & Petrinovich, 1986; Petrinovich & Baptista, 1987). They not only confirmed that exposure to allospecific tutors in a social context could induce the copying of a variety of other species' songs, but also indicated that the effect could be obtained outside the period which had been identified previously, by tape-tutoring experiments, as the sensitive phase. So, we have to conclude that the way in which the song to be learned is presented to a young white-crowned sparrow may have at least such a powerful effect on whether it is accepted as a model as do the acoustic qualities of this song. The superiority of live tutoring has been demonstrated for other species as well, and for a species like the zebra finch even conspecific song will usually not be copied from a tape, whereas not only conspecific, but also allospecific song will be copied from a live tutor (Immelmann, 1969a; Clayton, 1989).

For imprinting, an intriguing demonstration of the effects of the social context is provided by the work of Gottlieb and co-workers (Lickliter & Gottlieb, 1985, 1986, 1988; Johnston & Gottlieb, 1985). They were interested in the effects of broodmates on filial imprinting in ducklings. Intuitively one would think that isolating the subject together with the imprinting stimulus provides the best rearing condition if one wants to imprint a duckling. Isolated ducklings do, indeed, imprint with artificial objects such as a red light. When trained with a stuffed mallard, individually raised ducklings preferred the mallard over a pintail. However, when presented with a choice between a stuffed mallard and a stuffed redhead female, they failed to discriminate, suggesting that the latter discrimination is more difficult. The surprising finding emerged when ducklings were kept in small groups from training to testing. Under these conditions one would expect interference of imprinting on the stuffed model, as age mates might act as competing imprinting objects. Such interference does indeed occur when groups of ducklings, rather than individuals, are trained with a stuffed mallard (Lickliter & Gottlieb, 1986). However, ducklings exposed to age mates *after* exposure to the stuffed mallard not only discriminated a mallard female from a pintail, but also a mallard from a redhead (Lickliter & Gottlieb, 1988). So, somehow social experience affected the processing of the visual information acquired earlier on.

A final example concerns the role of social experience in sexual imprinting. It serves to illustrate that social factors can provide an alternative explanation to that of a 'template' when canalized development has been observed. Zebra finches raised by the related Bengalese finch show, when adult, a preference for partners of the foster species over their own species (Immelmann, 1969b; Kruijt, ten Cate & Meeuwissen, 1983; Sonnemann & Sjölander, 1977; ten Cate & Mug, 1984; see also Bischof, this volume). At the same time, Immelmann and others (Immelmann 1972a, 1972b; Sonnemann & Sjölander, 1977; Bischof, 1979; ten Cate & Mug, 1984) noticed that mate preferences tended to be biased towards the own species. That is to say, the preference for the foster species in cross-fostered zebra finches did not become as strong as the preference for conspecifics in zebra finches that were reared normally. This was partly due to the presence of conspecific siblings, which push the sexual preference shown later in the direction of the own species (Kruijt *et al.*, 1983). However, when zebra finch males are raised by a mixed pair (one zebra finch and one Bengalese finch parent), males later on prefer to mate with conspecifics over Bengalese finches (Immelmann, 1972a; ten Cate, 1982), a finding which holds if sibling effects are minimized or when Bengalese finch siblings are used (ten Cate, 1984). Following the results of experiments on the development of filial attachments, some interpreted these findings as evidence for the presence of a predisposition to respond to conspecific features. However, observations of mixed pairs showed that zebra finch parents did most of the feeding and were also more aggressive towards the young birds, a difference which was present irrespective of which species was male or female in such a pair (ten Cate, 1982). Thus, young zebra finches were exposed to more behavioural interactions with their own species than with the foster species. This opened the possibility that the bias towards the own species was due to social factors rather than to some pre-programmed preference for features of the own species. This was tested in a series of experiments in which the possibility for the zebra finch parent to interact with the young birds was reduced systematically, while maintaining visual exposure to the zebra finch (ten Cate, 1984, 1989a). One of these manipulations involved an exchange of the zebra finch parent for another zebra finch, which was not rearing offspring. As a result the Bengalese finch parent now did most of the feeding. Other behavioural interactions between zebra finch parents and young birds were not affected by this particular manipulation (ten Cate, 1984). The outcome of this and other manipulations was that the preference of males became more Bengalese finch directed with increasing interactions with the parent of this species. So, the behaviour of the stimulus birds was related to the strength of the later preference, supporting the hypothesis that the own species bias present in males raised by mixed pairs may have been due to social factors, making a predisposition for conspecific features less likely.

'What?' and 'how?' for the social context

The above example and those given before illustrate the powerful effect of social experience on perceptual development. As imprinting and song learning will normally occur when a bird is raised by conspecific parents and in the presence of siblings, such social experience may thus contribute considerably to the canalized development observed in the natural situation. The next step is to understand how this may have been brought about, just as for 'programmed learning' in the previous section. For social interactions too, we have the question of what the bird is sensitive for, i.e. the stimulus input necessary for producing the effect and, secondly, the question of the impact which the exposure may have on the learning mechanism.

Social experience: what is important?

Before examining the sensitivity to aspects of live stimuli in some detail, we have to consider the possibility that the causes for improved learning could be trivial. For instance, in the zebra finch example discussed above, the live stimulus bird moves around. Therefore it will sometimes be closer to the young bird than at other times. It might be that zebra finch parents in non-manipulated mixed pairs spent more time close to the young birds than Bengalese finch parents and that manipulations of the behaviour of the finch parent led to it spending more time at some distance. As a consequence, the visual input from this zebra finch will be decreased and, if amount of exposure determines the amount of learning, this decrease may have led to a corresponding decrease in learning about the stimulus. However, although such a cause cannot be excluded, it is not the full explanation. For instance, one of the manipulations involved exposure to a stuffed zebra finch instead of a living one. The stuffed bird was visible from inside the nest box and placed on a perch in a conspicuous position in the cage. Young zebra finches were observed to spend considerable time sitting next to this stuffed bird and also showed other behaviour (including precocial sexual behaviour) to it. The visual input received from this stuffed bird is therefore unlikely to have been less than that of a living zebra finch in other experimental conditions. Nevertheless, hardly a trace of an effect of exposure to the stuffed bird could be found (ten Cate, 1984; similar results were obtained by ten Cate, Los & Schilperoord, 1984). These experiments suggest that mere quantity of visual input is unlikely to be responsible for how much is learned about a stimulus object. Rather, the effect may have been obtained because young are sensitive to parents showing particular behaviour patterns, or to parental responses to their own behaviour. Some further insight into these latter factors has been obtained in various studies on filial imprinting and I will now concentrate on these.

For filial imprinting, living stimulus birds can be superior stimuli (Beaver, Shrout & Hess, 1976; Boakes & Panter, 1985; but cf. Reese *et al.*, 1972). Further, several experiments have demonstrated that moving artificial stimuli are more effective than non-moving ones (e.g. Salzen & Sluckin, 1959; Moltz, Rosenblum & Stettner, 1960; Salzen, 1969). It suggests that one possible factor making living stimuli attractive is that they move. But is this the only factor? This question led to an experiment in which a direct comparison was made between the effects of exposure to a moving stuffed bird and to a living one (ten Cate, 1989b). During exposure to the stimulus, various parameters of a developing preference, such as a chick's orientation or its tendency to approach the stimulus, indicated that a moving stuffed hen was more attractive than a non-moving one, as was to be expected. In addition, however, the living hen was significantly more attractive than the moving stuffed one. Also, removal of the stimulus object led to the strongest response in chicks exposed to the living hen, suggesting that the exposure had resulted in a greater attachment to the living hen. What could have been responsible for the effect? As live hens were spending considerable amounts of time sitting quietly, it is unlikely that their attractiveness originated from a higher frequency of movement *per se*. So, a first conclusion must be that movement as such was not the only factor making the living hen attractive. This was supported by a further analysis in which it was examined whether, among chicks exposed to the living hen, the strength of the response to removal of the hen was correlated with variation in the hen's behaviour. This analysis confirmed that movement of the hen was likely to have been of some importance, as correlation coefficients between virtually all measured activities of the hen and the chicks' attachment to it were positive. Nevertheless, the effects were not significant, apart from two remarkable exceptions: 'pecking' by the hen (directed at food particles) and 'chick-directed behaviour' (directed pecks, threat postures and binocular focusing) showed highly significant correlations with the distress calls caused by removal of the hen. It is of interest and probably not without meaning that, of all behaviour shown by the hen, these are clearly the ones which are also of the greatest functional significance to the chick: pecking of the hen may indicate the presence of food; chick-directed behaviour may require a response. An effect of chick-directed behaviour is also implied in some other studies (Gray, 1964; Shapiro & Agnew, 1975). The latter authors exposed ducklings to living females of various species and mentioned that stimulus females: 'did emit some threat displays toward the ducklings and occasionally they would peck at the subjects. Interestingly enough, a subject could be pecked at severely but it would not change its preference. In fact, it seemed that the subjects' preference was enhanced'. In the quail study, pecking and chick-directed behaviour formed only a small proportion of what the hen was doing. Therefore, a second, tentative conclusion may be that chicks are more sensitive to particular kinds of movements or postures than they are to others; i.e. the *quality* of the movement may be important.

The above finding brings us one step further, but leads to the question of the nature of this sensitivity to particular behaviour patterns. One possiblity is that chicks are sensitive to some specific posture which is adopted when a hen performs the pecking behaviour, like a bill pointing down. Another possiblity is that the pattern of movement as such was affecting the chick. This question has not yet been addressed for imprinting, but it has been in respect to the development of food recognition in chicks. Turner (1964) and, later on, Suboski (Bartashunas & Suboski, 1984; Suboski, 1989) demonstrated that a pecking hen stimulates chicks to peck at the same food particles as the hen does. In further experiments on the stimulus elements responsible for this effect they used a two-dimensional cardboard model of a hen. When the model was in a stationary position, either upright or with the bill pointing down, this did not give rise to a substantial response from the chicks. When, however, the model was moved to mimic pecking, alternating between the two positions, the chicks responded strongly, indicating that it was the movement rather than a specific posture that was essential for arousing the responsiveness of the chicks. With respect to imprinting, these data suggest that there too the dynamics of the movement, rather than adopting some posture, may have been the factor responsible for the formation of a filial attachment. If so, it would also imply that two different learning processes (food recognition and imprinting) can be stimulated by the same behavioural act. In terms of stimulus input, the sensitivity underlying the response could be a combination of one for an edge (provided by the bill) and one for a downward movement (pecking). In the same way, the strong effect of chick-directed behaviour may not so much be caused by a sensitivity for certain specific postures, but by a combination of sensitivity for some crude shape linked to one for stimuli moving closer. The latter sensitivity is indicated by an experiment by Schulman, Hale & Graves (1970), who showed that chicks are more attracted to approaching than to retreating objects. The attraction by different types of movement has also been examined by Smith & Hoyes (1961) and Gossop (1974), but it is not clear whether this affected the formation of an attachment to the moving objects. Sensitivities for combinations of shapes and movement might also underlie the effects obtained in the above-mentioned experiments on sexual imprinting in the zebra finch. These suggested that a whole variety of behaviour patterns might affect the imprinting process, ranging from feeding to aggression. Rather then attributing their effects to subtle or higher order effects of each pattern separately, it could be that they all originate from one or two basic propensities, making a bird respond to behaviour showing a certain directedness or involving some type of approaching movement. Such an explanation fits in with the finding that siblings or allospecific foster parents, which differ in details of their behaviour from conspecific parents, may contribute to the formation of preferences.

The perceptual sensitivity to a combination of shape and movement may be relatively simple, comparable to the mechanism which enables a toad to discriminate 'prey' from 'predators' on the basis of perceptual sensitivities for different combinations of shapes and movement (Ewert & Traud, 1979; Ewert, 1980). It suggests that underlying the sensitivity to live tutors are some dispositions, similar in nature to those discussed in the previous section, to respond to, or to learn about, objects moving in certain ways. If conspecific siblings or parents show these patterns more frequently, this would induce a preference for the own species. However, although a sensitivity to certain types of movement might have been a factor in creating the powerful effect of a living model, is this all there is to say about the factors making a living stimulus bird attractive?

In applying the reductionist, 'top-down', approach followed above, we have lost sight of several features of our living stimulus bird. Apart from adopting certain postures or moving in a specific way, the behaviour of a stimulus bird might bear a relationship to that of the young individual, for instance by responding to the young bird's behaviour. Also, behaviour of the stimulus may show some (ir)regularity over time; i.e. sometimes the behaviour at one moment will predict the behaviour to follow, but at other times it may change quite unexpectedly. Would such aspects of behaviour also affect the formation of filial attachments? These aspects might be examined with living stimuli, for instance, by using one-way vision screens to prevent a hen from responding to a chick. Such an approach has been applied by Friedman (1977), to analyze which aspects of male ring dove behaviour stimulate ovulation in females, but to my knowledge such experiments have not been done in the context of imprinting. However, we can also start at the other end, using a 'bottom-up' approach; i.e. we can add aspects of interactions or predictability to a moving stimulus and assess whether these enhance imprinting, and a number of studies followed this approach.

In the past, several authors (e.g. Campbell & Pickleman, 1961; Hoffman & Kozma, 1967) have demonstrated that an imprinted chick or duckling will perform an operant to get access to the stimulus to which it is imprinted. Bateson & Reese (1969) showed that chicks need not have been imprinted with an attractive object in order to work for exposure to it. The behaviour will also be shown by naive birds. At the same time these birds will develop an attachment to the stimulus. However, the experiment by Bateson & Reese (1969) did not indicate whether chicks for which the exposure to the stimulus was contingent upon their behaviour showed a stronger, or earlier attachment to it than controls. This was examined in two subsequent studies. The first one (ten Cate, 1986a) used calling of Japanese quail chicks as the operant to make a stuffed hen move. These calls will, under natural conditions encourage hens to give attention to their chicks. For each experimental chick there was a yoked control that was exposed to the same stimulus, but had no control over it. A preference test indicated that the experimental chicks had formed the strongest attachment to the stimulus. However, the effect was weak and only significant in comparison with another control group, exposed to a non-moving stimulus bird. Also, it could not be demonstrated that the experimental chicks differed from the controls in the level of performing the operant, making it unclear whether the effect was caused by perceived contingency or because, for example, the movement was more likely to coincide with periods of activity in the experimental than in the control chick. A similar experiment was carried out by Bolhuis & Johnson (1988). Their experimental design was different in that they used pedal pressing as operant and, also, their imprinting stimulus was visible only when the (domestic) chick performed the operant. The outcome was clear cut: experimental and control chicks did not differ in filial preference. This outcome could be taken to imply that there is no 'contingency effect' on imprinting, as the first experiment was not very convincing anyway. However, differences in procedure might have an impact on whether effects do or do not appear, and the hypothesis that the appearance of a contingency effect is dependent on factors like the operant used or the way of presenting the stimuli cannot be discarded yet.

Although the experiment by Bolhuis & Johnson (1988) obtained no evidence for a contingency effect, it did reveal another factor. Initially, they used two control groups, one containing yoked control chicks, the other containing chicks which had been exposed to a stimulus object for fixed durations alternated with fixed periods of non-movement. Surprisingly, the second control group developed a much weaker preference for the stimulus than the first one. A further experiment demonstrated that chicks exposed to the stimulus on a variable interval/variable duration schedule developed a stronger preference for the stimulus than chicks exposed to the stimulus on a fixed duration/fixed interval schedule, even when number of exposures and total time of exposure was the same for both groups. This finding suggests an effect of predictability of stimulus exposure on the imprinting process. It is not unlikely that the stimulation provided by a living stimulus bird might be more of a variable type than following a fixed pattern, and hence this type of sensitivity will be another factor contributing to enhancement of filial imprinting with living stimulus objects. The above experiment concentrated on predictability with respect to timing and duration of exposure. So far, it has not been examined whether predictability with respect to *type* of movement has a similar effect, i.e. whether it makes a difference when the movement is always the same or when different types of movement are performed in randomized order.

In terms of underlying mechanisms, the effects of predictability or contingency cannot be based on a purely perceptual sensitivity, but we need to infer some involvement of memory as the effect of exposure at a given moment depends on the bird's history of exposure up to that moment.

Is there any evidence to suggest that the types of sensitivities mentioned above may also play a role in song learning? To my knowledge, no extensive studies have been carried out on, for instance, degree of copying in relation to manipulations of the number or type of interactions between young and the tutor(s), although there is some correlational evidence that interactions might affect tutor choice in the zebra finch (Clayton, 1987, this volume; Williams, 1990). Nevertheless, the finding that several species will copy songs from a live tutor of a different species, or copy vocalizations heard from, or in the presence of human, interacting, caretakers (Todt, Hultsch & Heike, 1979; West, Stroud & King, 1983) might imply that the stimulation needed to produce the effect is not limited to the occurrence of some precise species-specific display, but might be of a more general nature. This could imply the presence of similar sensitivities for combinations of shape and movement as seem present for imprinting. It seems worthwhile to examine whether a taped song, played when a visual stimulus approaches a young bird or mimics some behaviour pattern of a parent, increases the likelihood of the song being copied.

The experiments on filial imprinting raised the question of whether part of the stimulation making live song tutors effective is also due to their interactive or (un)predictable nature. This question was recently examined in zebra finches for which hearing a taped song was made contingent upon the young bird performing either perch hopping (ten Cate, 1991) or key pecking (Adret, 1993; see also Clayton, this volume). In line with some earlier work on chaffinches (Stevenson-Hinde, 1973), both studies revealed that zebra finches increased the performance of the operant to get exposure to the song; i.e. hearing song may act as a reinforcer in young males, in a similar way as exposure to a visual imprinting object is reinforcing for a duckling or a chick. The young males in my own study showed no indication of copying songs from the tape (ten Cate, 1991). However, the experimental birds in the study of Adret (1993) generally copied the songs, whereas his yoked controls did not, or did so much less. Adret's findings make it clear that zebra finches which have some control over stimulus presentation may copy songs from tape, where mere exposure to taped songs fails. The reasons for the differences between the two studies might be several, ranging from the operant chosen to the precise details of the exposure schedule (see Adret, 1993) and a further study of these differences may reveal more about what is essential for learning to occur. Nevertheless, it has become clear that adding some aspect of 'interaction' to a tape may bring copying at the same level as that present with live tutors. It suggests, tentatively, that one factor involved in the effectiveness of live tutors may be a responsiveness to behaviour of juveniles. With respect to the other factor which seemed of importance in imprinting – unpredictability – no experiments on this are available. However, Petrinovich (1988) has suggested that one reason why white-crowned sparrows learn more from live tutors than from taped ones is that the stimulation provided by a live tutor is much more variable with respect to direction, timing, structure and volume of the sound. This might prevent habituation to the stimulus. Such habituation was demonstrated to be present in the responses which free-living white-crowned sparrows showed in play-back experiments of taped songs (e.g. Petrinovich & Patterson, 1982). In summary, it seems likely that the effectiveness of live song tutors might originate from similar propensities to respond to rather simple patterns of stimulation to those involved in imprinting.

So far, my way of arguing may be taken to imply that the sensitivity to 'social interaction' is nothing special, but rests on simple dispositions to respond to some types of stimulation which happen to be present frequently between interacting individuals. Whether or not this is the case remains to be seen. My prime aim has been to make clear that we should not overlook simpler explanations where more complex ones might be more appealing. How far this will get us is a matter for empirical research, but it may be that effects now attributed to some higher order cognitive capacities in fact have a simpler basis. For instance, Pepperberg, following up earlier studies by Todt (1975) and others, has been able to teach an African Grey parrot, 'Alex', an amazing vocabulary, which it uses in an appropriate context. Pepperberg considers the fact that she has been training Alex by a social modelling technique, involving interaction between two trainers, as a key factor for her success (Pepperberg, 1985, 1991), and has suggested that social modelling might also be involved in producing the effects of live tutoring in a species like the white-crowned sparrow (Pepperberg, 1985). The technique assumes that the trainees, in this case the birds, are capable of higher order mental processes such as 'referentiality' and that the effect of live tutoring is due to observing vocalizations being used in their appropriate context. However, her technique contains a number of procedures which might optimize the occurrence of learning by stimulating the perceptual sensitivities discussed above. For instance, training involves long-lasting and intensive interaction with the parrot at close distance, with responses to vocalizations or behaviour it performs. This should provide, at least, a basis for learning the vocalizations given by tutors at the same time.

Social experience and learning

Earlier I emphasized that the sensitivities for stimulation underlying the effects of social interactions might not be that different from those involved in the programmed learning under the influence of dispositions. If so, can the effect of social interaction or of stimulus movement on the processing also be based on similar mechanisms? To some extent, it might. It has been suggested that stimulus movement is a crucial feature for successful imprinting (e.g. Hoffman & Ratner, 1973). However, it has become clear that non-moving objects may provide enough stimulation for imprinting (e.g. Bateson, 1964; Salzen, 1969; Eiserer, 1980; de Vos & Bolhuis, 1990). One effect of movement seems to be that it helps to focus the attention to an object by increasing its conspicuousness (Eiserer, 1980). If this conspicuousness is increased by other means, for instance by putting a spotlight on a static object in a darker environment, learning may also be increased (Eiserer, 1980). Thus, the way in which social interactions obtain their effect may be quite similar to the way in which the colour of an object increases learning about other features of the object, or, to phrase it more generally, to the way in which predispositions might obtain their effect. Movement may not only stimulate the young bird to look at certain objects and thereby increase its visual input, but it also seems to affect the internal processing of the object providing the stimulation. For both social interactions and stimulation provided by certain conspicuous static patterns, this can be achieved by affecting the 'selective attention' which is a crucial feature of associative learning processes (Dickinson, 1980). The effects of contingency and predictability, found both for imprinting and song learning, may also originate from effects on attentional mechanisms.

The data from the studies by Gottlieb and co-workers, discussed above, suggest that the effects of social stimulation may also operate at an even higher level of processing, in which later experience retroactively interferes with the processing of earlier acquired information. The latter phenomenon seems one of the few indications that predispositions and social interactions may differ in their effects on information processing, as there is no clear indication that a similar phenomenon occurs upon exposure to attractive static stimuli.

Presence of movement may thus add to the interest which naive young birds show for the stimulus object and to the processing of information about the object. For filial imprinting, this notion has been around for some time, albeit in a cruder form than expressed here (cf. Matthews & Hemmings, 1963). As we have seen for predispositions to respond to colours or patterns, the preference for a particular feature may or may not lead to learning about the feature. Whether exposure to movement results in learning about this movement or whether it adds to the preference without learning about it is not yet clear. Several experiments indicated that exposure to moving stimulus objects during a test increases the response to them, but no experiment seems to have addressed the question of whether exposure to one type of movement made the bird more responsive to that type of movement rather than to another later on. Such an experiment would be critical to demonstrate whether movement is a part of what the bird learned about the stimulus (cf. Green, 1982). What has been demonstrated is that training with a moving object may lead to a preference for that object even when it is not moving during the test (Klopfer, 1965; Hoffman & Segal, 1983). So, the presence of movement which stimulated interest in the object is not essential for showing attachment during a test later on, although it may add to the responsiveness towards the object. In this case the effect of movement is comparable to that of pattern or flicker obtained in the studies by Kovach (1983b, 1983c) discussed above.

The conclusion at the end of this section on which aspects of the social context affect the developmental processes and how they do so, might be that the effect of the social context seems mediated by stimulation of certain perceptual biases ('predispositions') which may then affect further processing of various aspects of the object providing the stimulation. At present we know very little about how the sensitivity to certain patterns of movement develops or whether it changes as development proceeds. However, given the similarities with the way in which predispositions affect development, it seems likely that sensitivities to various types of movements may also be affected by preceding stimulation and alter as development proceeds. Under natural cicumstances, social interactions will usually be most frequent or intense with parents and/or siblings. Also, these will be the most likely ones to show the behaviour patterns for which particular sensitivities seem to exist. As a result, learning about their features will be promoted over those of others. In this way the interactions play their own role in producing canalization of the developmental process.

Development as a dynamic process

As I hope to have demonstrated, the development of filial and sexual preferences and songs may superficially look like a process of 'programmed learning' stimulated by a 'social context'; in reality, the situation is much more dynamic. Programmed learning depends heavily on external stimulation to induce, maintain or alter specific perceptual sensitivities, while social interactions obtain their impact to a large extent by acting upon similar sensitivities and affecting the learning mechanism. Although the appearance (or vocal characteristics) of an object (or song tutor) and its behaviour may each be sufficient to guide the developing preference, when operating together the effect may be even more powerful. For instance, the attraction to pecking movements observed in quail and fowl chicks may help to draw attention to the head and neck region of the hen, for which a perceptual sensitivity seems present (Horn, 1985; Johnson & Horn, 1988; Johnson & Bolhuis, 1991).

During filial imprinting, chicks do not just learn visual characteristics of the stimulus, but may also show 'auditory imprinting' (see Bolhuis & van Kampen, 1992, for a recent review). Live stimulus birds may vocalize in addition to showing behaviour. A broody hen utters special vocalizations (e.g. Collias & Joos, 1953; Kent, 1987) to attract the chick and, in line with this, various studies demonstrated that young chicks are more attracted to a visual object providing acoustic stimulation than to a silent one (e.g. Gottlieb & Klopfer, 1962; Smith & Bird, 1963; Storey & Shapiro, 1979). Recently, van Kampen & Bolhuis (1993) demonstrated that adding a vocalization to an imprinting object led to the formation of a stronger attachment to the object. Interestingly, the process works both ways, as chicks exposed to the combined stimulus also developed a stronger preference for the acoustic stimulus than chicks which were trained on this stimulus without a visual one (van Kampen & Bolhuis, 1991). Thus, combined stimulation via different sensory modalities may result in enhanced processing of stimulus features, which may be caused by increased arousal or by an attentional effect (Bolhuis & van Kampen, 1992). Again, living stimulus objects are more likely to show such a combination, and thus the sensitivity to combined stimulation is another factor likely to canalize the formation of the attachment towards the right object. All the separate propensities discussed so far, when combined, are likely to give rise to the powerful effect which living stimuli may have on filial imprinting. So, rather than being based on some unique or higher order stimulation, the effect might originate from a joint stimulation of various not very complicated perceptual properties, which might primarily affect the attentive mechanism involved in information processing. This notion is attractive as it enables us to reduce the effects of various treatments to some common denominators. It does, however, not imply that 'everything is of equal importance at all times'. At some stage in development, one type of stimulation may have a larger impact than another, and the dominant modality at one time may be different from that at another. Also, changes in sensitivities might be contingent upon earlier stimulation in the same or a different modality. Given this sort of complexity, it is no wonder that different experimental paradigms may produce different outcomes: artificial stimulation may capture one or more aspects of the natural, living stimulus object, but it will be hard to accommodate for the multitude of aspects of the natural objects which seem to be involved in promoting the establishment of social bonds with, or the learning of songs from, these objects.

Discussing the propensities of the young birds may have given the impression of a passive role for the youngsters; they may sit and wait for their parent to do the right thing at the right time. This view would be wrong. For both imprinting and song learning, there is evidence to suggest that the young may play a very effective role to obtain the stimulation needed at a certain moment (Bateson, 1971; ten Cate, 1986b, 1989a). In this way they may compensate for variation in parental behaviour. At the same time their interactions with the parents may help to modify their parents' behaviour in such a way that this behaviour becomes more suited to their needs. Flexibility in behaviour of young and parents will in this way produce predictability and stability in outcome of the developmental process, and so produce the canalization which seems so characteristic for the development of many aspects of the behaviour of an individual.

Acknowledgements

I dedicate this chapter to Jaap Kruijt on the occasion of his 65th birthday. He put me on to the subject of behavioural development, provided me with the tools to get some hold on the issue, and supported my work in many ways. His broad interest, combined with a critical attitude, definitely had an impact on my own scientific development. It was a great pleasure to have been one of the members of his group for a long time. Many of the ideas in this paper are influenced by discussions with Jaap and other members of that group: Gerrit de Vos, Ton Groothuis, Hendrik van Kampen and Dave Vos; with the editors of this book: Johan Bolhuis and Jerry Hogan; and with Pat Bateson. I thank them all for their contribution. Johan, Jerry and Hendrik provided helpful criticisms on earlier versions of this chapter.

References

Adret, P. (1993). Operant conditioning, song learning and imprinting to taped song in the zebra finch. *Animal Behaviour*, 46, 149–159.

Baptista, L. F. & Morton, M. L. (1981). Interspecific song acquisition by a white crowned sparrow. *The Auk*, 98, 383–385.

Baptista, L. F. & Petrinovich, L. (1986). Song development in the white-crowned sparrow: Social factors and sex differences. *Animal Behaviour*, 32, 172–181.

Bartashunas, C. & Suboski, M. D. (1984). Effects of age of chick on social transmission of pecking preferences from hen to chicks. *Developmental Psychobiology*, 17, 121–127.

Bateson, P. P. G. (1964). Relation between conspicuousness of stimuli and their effectiveness in the imprinting situation. *Journal of Comparative and Physiological Psychology*, 58, 407–411.

Bateson, P. P. G. (1971). Imprinting. In *Ontogeny of Vertebrate Behavior*, ed. H. Moltz, pp. 369–378. New York: Academic Press.

Bateson, P. (1979). How do sensitive periods arise and what are they for? *Animal Behaviour*, 27, 470–486.

Bateson, P. P. G. & Jaeckel, J. B. (1976). Chicks' preference for familiar and novel conspicuous objects after different periods of exposure. *Animal Behaviour*, 24, 386–390.

Bateson, P. P. G. & Reese, E. P. (1969). The reinforcing properties of conspicuous stimuli in the imprinting situation. *Animal Behaviour*, 17, 692–699.

Bateson, P. P. G. & Wainwright, A. A. P. (1972). The effects of prior exposure to light on the imprinting process in domestic chicks. *Behaviour*, 42, 279–290.

Beaver, P. W., Shrout, P. E. & Hess, E. H. (1976). The relative effectiveness of an inanimate stimulus and a live surrogate during imprinting in Japanese quail, *Coturnix coturnix japonica*. *Animal Learning & Behavior*, 4, 193–196.

Berryman, J., Fullerton, C. & Sluckin, W. (1971). Complexity and colour preferences of chicks of various ages. *Quarterly Journal of Experimental Psychology*, 23, 255–260.

Bischof, H.-J. (1979). A model of imprinting evolved from neurophysiological concepts. *Zeitschrift für Tierpsychologie*, 51, 126–137.

Boakes, R. & Panter, D. (1985). Secondary imprinting in the domestic chick blocked by previous exposure to a live hen. *Animal Behaviour*, 33, 353–365.

Bolhuis, J. J. (1991). Mechanisms of avian imprinting: a review. *Biological Reviews*, 66, 303–345.

Bolhuis, J. J. & Johnson, M. H. (1988). Effects of response-contingency and stimulus presentation schedule on imprinting in the chick (*Gallus gallus domesticus*). *Journal of Comparative Psychology*, 102, 61–65.

Bolhuis, J. J. & Johnson, M. H. (1991). Sensory templates: mechanism or metaphor? *Behavioral and Brain Sciences*, 14, 349–350.

Bolhuis, J. J., Johnson, M. H. & Horn, G. (1985). Effects of early experience on the development of filial preferences in the domestic chick. *Developmental Psychobiology*, 18, 299–308.

Bolhuis, J. J., Johnson, M. H. & Horn, G. (1989). Interacting mechanisms during the formation of filial preferences: the development of a predisposition does not prevent learning. *Journal of Experimental Psychology: Animal Behavior Processes*, 15, 376–383.

Bolhuis, J. J., De Vos, G. J. & Kruijt, J. P. (1990). Filial imprinting and associative learning. *Quarterly Journal of Experimental Psychology*, 42 B, 313–329.

Bolhuis, J. J. & van Kampen, H. S. (1992). An evaluation of auditory learning in filial imprinting. *Behaviour*, 122, 195–230.

Campbell, B. A. & Pickleman, J. R. (1961). The imprinting object as a reinforcing stimulus. *Journal of Comparative and Physiological Psychology*, 54, 592–596.

Clayton, N. S. (1987). Song tutor choice in zebra finches. *Animal Behaviour*, 35, 714–722.

Clayton, N. S. (1989). The effects of cross-fostering on selective song learning in estrildid finches. *Behaviour*, 109, 163–175.

Collias, N. E. & Joos, M. (1953). The spectographic analysis of sound signals in the domestic fowl. *Behaviour*, 12, 175–187.

De Vos, G. J. & Bolhuis, J. J. (1990). An investigation into blocking of filial imprinting in the chick during exposure to a compound stimulus. *Quarterly Journal of Experimental Psychology*, 42B, 289–312.

Dickinson, A. (1980). *Contemporary Animal Learning Theory*. Cambridge: Cambridge University Press.

Eiserer, L. A. (1980). Development of filial attachment to static visual features of an imprinting object. *Animal Learning and Behavior*, 8, 159–166.

Ewert J. P. (1980). *Neuroethology – An Introduction to the Neurophysiological Fundamentals of Behavior*. Berlin: Springer Verlag.

Ewert, J. P. & Traud, R. (1979). Releasing stimuli for anti-predator behaviour in the common toad *Bufo bufo L. Behaviour*, 68, 170–180.

Friedman, M. B. (1977). Interactions between visual and vocal courtship stimuli in the neuroendocrine response of female ring doves. *Journal of Comparative and Physiological Psychology*, 91, 1408–1416.

Gossop, M. R. (1974). Movement variables and the subsequent following response of the domestic chick. *Animal Behaviour*, 4, 982–986.

Gottlieb, G. & Klopfer, P. H. (1962). The relation of developmental age to auditory and visual imprinting. *Journal of Comparative and Physiological Psychology*, 55, 821–826.

Gould, J. L. (1982). *Ethology: The Mechanisms and Evolution of Behaviour*. New York: Norton.

Gray, P. H. (1964). Interactions of temporal and releasing factors in familial recognition of own and ancestral species. *Perceptual and Motor Skills*, 18, 445–448.

Green, P. R. (1982). Problems in animal perception and learning and their implications for models of imprinting. In *Perspectives in Ethology*, Vol. 5, ed. P. P. G. Bateson & P. H. Klopfer, pp. 243–273. New York: Plenum Press.

Hoffman, H. S. & Kozma, F.Jr. (1967). Behavioral control by an imprinted stimulus: long term effects. *Journal of the Experimental Analysis of Behavior*, 10, 495–501.

Hoffman, H. S. & Ratner, A. M. (1973). A reinforcement model of imprinting: Implications for socialization in monkeys and men. *Psychological Review*, 80, 527–544.

Hoffman, H. S. & Segal, M. S. (1983). Biological factors in social attachments: A new view of a basic phenomenon. In *Advances in the Analysis of Behaviour*, Vol. 3, ed. M. D. Zeiler & P. Harzem, pp. 41–61. London: John Wiley.

Hogan, J. A. (1988). Cause and function in the development of behavior systems. In *Handbook of Behavioral Neurobiology*, Vol. 9, ed. E. M. Blass, pp. 63–106. New York: Plenum Press.

Hollis, K. L., ten Cate, C. & Bateson, P. (1991). Stimulus representation: a subprocess of imprinting and conditioning. *Journal of Comparative Psychology*, 105, 307–317.

Horn, G. (1985). *Memory, Imprinting and the Brain*. Oxford: Clarendon Press.

Immelmann, K. (1969a). Song development in the zebra finch and other Estrildine finches. In *Bird Vocalizations*, ed. R. A. Hinde, pp. 61–74. Cambridge: Cambridge University Press.

Immelmann, K. (1969b). Ueber den Einfluss frühkindlicher Erfahrungen auf die geschlechtliche Objektfixierung bei Estrildiden. *Zeitschrift für Tierpsychologie*, 26, 677–691.

Immelmann, K. (1972a). The influence of early experience upon the development of social behavior in estrildine finches. *Proceedings XV International Ornithological Congress, Den Haag*, 1970, pp. 316–338. Leiden: Brill.

Immelmann, K. (1972b). Sexual and other long-term aspects of imprinting in birds and other species. *Advances in the Study of Behavior*, 4, 147–174.

Johnson, M. H. & Bolhuis, J. J. (1991). Imprinting, predispositions and filial preferences in chicks. In *Neural and Behavioural Plasticity*, ed. R. J. Andrew, pp. 133–156. Oxford: Oxford University Press.

Johnson, M. H., Davies, D. C. & Horn, G. (1989). A sensitive period for the development of a predisposition in dark-reared chicks. *Animal Behaviour*, 37, 1044–1046.

Johnson, M. H. & Horn, G. (1988). Development of filial preferences in dark-reared chicks. *Animal Behaviour*, 36, 675–683.

Johnston, T. D. (1988). Developmental explanation and the ontogeny of birdsong: nature/nurture redux. *Behavioral and Brain Sciences*, 11, 617–663.

Johnston, T. D. & Gottlieb, G. (1985). Effects of social experience on visually imprinted maternal preferences in Peking ducklings. *Developmental Psychobiology*, 18, 261–271.

Kent, J. P. (1987). Experiments on the relationship between the hen and chick. *Behaviour*, 102, 1–14.

Klopfer, P. H. (1965). Imprinting: A reassessment. *Science*, 147, 302–303.

Klopfer, P. H. (1967). Stimulus preferences and imprinting. *Science*, 156, 1394–1396.

Klopfer, P. H. & Hailman, J. P. (1964). Perceptual preferences and imprinting in chicks. *Science*, 145, 1333–1334.

Kovach, J. K. (1971). Interaction of innate and acquired: Color preferences and early exposure learning in chicks. *Journal of Comparative and Physiological Psychology*, 75, 386–398.

Kovach, J. K. (1979). Genetic influences and genotype-environment interactions in perceptual imprinting. *Behaviour*, 68, 31–60.

Kovach, J. K. (1983a). Constitutional biases in early perceptual learning: I. Preferences between colors, patterns, and composite stimuli of colors and patterns in genetically manipulated and imprinted quail chicks (*C. Coturnix japonica*). *Journal of Comparative Psychology*, 97, 226–239.

Kovach, J. K. (1983b). Perceptual imprinting: Genetically variable response tendencies, selective learning, and the phenotypic expression of colour and pattern preferences in quail chicks (*C. coturnix japonica*). *Behaviour*, 86, 72–88.

Kovach, J. K. (1983c). Constitutional biases in early perceptual learning: II. Visual preferences in artificially selected, visually naive and imprinted quail chicks (*C. coturnix japonica*). *Journal of Comparative Psychology*, 97, 240–248.

Kruijt, J. P. (1985). On the development of social attachments in birds. *Netherlands Journal of Zoology*, 35, 45–62.

Kruijt, J. P., ten Cate, C. J. & Meeuwissen, G. B. (1983). The influence of siblings on the development of sexual preferences in zebra finches. *Developmental Psychobiology*, 16, 233–239.

Lickliter, R. (1990). Premature visual stimulation accelerates intersensory functioning in bobwhite quail neonates. *Developmental Psychobiology*, 23, 15–27.

Lickliter, R & Gottlieb, G. (1985). Social interaction with siblings is necessary for the visual imprinting of species-specific maternal preference in ducklings (*Anas platyrhynchos*). *Journal of Comparative Psychology*, 99, 371–379.

Lickliter, R. & Gottlieb, G. (1986). Training ducklings in broods interferes with maternal imprinting. *Developmental Psychobiology*, 19, 555–566.

Lickliter, R. & Gottlieb, G. (1988). Social specificity: interaction with own species is necessary to foster species-specific maternal preference in ducklings. *Developmental Psychobiology*, 21, 311–321.

Lickliter, R. & Hellewell, T. B. (1992). Contextual determinants of auditory learning in bobwhite quail embryos and hatchlings. *Developmental Psychobiology*, 25, 17–31.

Lickliter, R. & Stoumbos, J. (1991). Enhanced prenatal auditory experience facilitates postnatal visual responsiveness in bobwhite quail chicks. *Journal of Comparative Psychology*, 105, 89–94.

Lickliter, R. & Virkar, P. (1989). Intersensory functioning in bobwhite quail chicks: Early sensory dominance. *Developmental Psychobiology*, 22, 651–667.

Lorenz, K. (1935). Der Kumpan in der Umwelt des Vogels. *Journal für Ornithologie*, 83, 137–213, 289–413.

Marler, P. (1970). A comparative approach to vocal learning: song development in the white crowned sparrows. *Journal of Comparative and Physiological Psychology Monographs*, 71, 1–25.

Marler, P. (1987). Sensitive periods and the roles of specific and general sensory stimulation in birdsong learning. In *Imprinting and Cortical Plasticity*, ed. J. P. Rauschecker & P. Marler, pp. 99–136. New York: John Wiley & Sons.

Marler, P. (1990). Innate learning preferences: Signals for communication. *Developmental Psychobiology*, 23, 557–568.

Matthews, W. A. & Hemmings, G. (1963). A theory concerning imprinting. *Nature*, 198, 1183–1184.

Moltz, H., Rosenblum, L. & Stettner, L. J. (1960). Some parameters of imprinting effectiveness. *Journal of Comparative and Physiological Psychology*, 53, 297–301.

Pepperberg, I. M. (1985). Social modeling theory: a possible framework for understanding avian vocal learning. *The Auk*, 102, 854–864.

Pepperberg, I. M. (1991). Learning to communicate: the effects of social interaction. In *Perspectives in Ethology*, Vol. 9, ed. P. P. G. Bateson & P. H. Klopfer, pp. 119–164. New York: Plenum Press.

Petrinovich, L. (1988). The role of social factors in white-crowned sparrow song development. In *Social Learning: Psychological and Biological Perspectives*, ed. T. R. Zentall & B. G. Galef, Jr., pp. 255–278. Hillsdale, NJ: Lawrence Erlbaum.

Petrinovich, L. & Baptista, L. (1987). Song development in the white-crowned sparrow: Modification of learned song. *Animal Behaviour*, 35, 961–974.

Petrinovich, L. & Patterson, T. L. (1982). Field studies of habituation: V. Evidence for a two-factor, dual process system. *Journal of Comparative and Physiological Psychology*, 96, 284–296.

Reese, E. P., Schotte, C. S., Bechtold, R. E. & Cowley, V. L. (1972). Initial preference of chicks from five rearing conditions for a hen or a rotating light. *Journal of Comparative and Physiological Psychology*, 81, 76–83.

Salzen, E. A. (1969). Contact and social attachment in domestic chicks. *Behaviour*, 33, 36–51.

Salzen, E. A. & Sluckin, W. (1959). The incidence of the following response and the duration of responsiveness in domestic fowl. *Animal Behaviour*, 7, 172–179.

Schulman, A. H., Hale, A. H. & Graves, H. B. (1970). Visual stimulus characteristics for initial approach response in chicks. *Animal Behaviour*, 18, 461–466.

Shapiro, L. J. & Agnew, R. L. (1975). The development of preferences for live models in White Peking ducklings. *Bulletin of the Psychonomic Society*, 5, 140–142.

Sluckin, W. (1972). *Imprinting and Early Learning*, 2nd edn. London: Methuen.

Smith, F. V. & Bird, M. W. (1963). The relative attraction for the domestic chick of combinations of stimuli in different sensory modalities. *Animal Behaviour*, 11, 300–305.

Smith, F. V. & Hoyes, P. A. (1961). Properties of the visual stimuli for the approach response in the domestic chick. *Animal Behaviour*, 9, 159–166.

Sonnemann, P. & Sjölander, S. (1977). Effects of crossfostering on the sexual imprinting of the female zebra-finch *Taeniopygia guttata*. *Zeitschrift für Tierpsychologie*, 45, 337–348.

Staddon, J. E. R. (1983). *Adaptive Behaviour and Learning*. Cambridge: Cambridge University Press.

Stevenson-Hinde, J. (1973). Constraints on reinforcement. In *Constraints on Learning: Limitations and Predispositions*, ed. R. A. Hinde & J. Stevenson-Hinde, pp. 285–299. London: Academic Press.

Storey, A. E. & Shapiro, L. J. (1979). Development of preferences in white peking ducklings for stimuli in the natural post-hatch environment. *Animal Behaviour*, 27, 411–416.

Suboski, M. D. (1989). Recognition learning in birds. In *Perspectives in Ethology*, Vol. 8, ed. P. P. G. Bateson & P. H. Klopfer, pp. 137–171. New York: Plenum Press.

ten Cate, C. (1982). Behavioural differences between zebra finch and Bengalese finch (foster)parents raising zebra finch offspring. *Behaviour*, 81, 152–172.

ten Cate, C. (1984). The influence of social relations on the development of species recognition in zebra finch males. *Behaviour*, 91, 263–285.

ten Cate, C. (1986a). Does behaviour contingent stimulus movement enhance filial imprinting in Japanese quail? *Developmental Psychobiology*, 19, 263–285.

ten Cate, C. (1986b). Listening behaviour and song learning in zebra finches. *Animal Behaviour*, 34, 1267–1268.

ten Cate, C. (1989a). Behavioral development: toward understanding processes. In *Perspectives in Ethology*, Vol. 8, ed. P. P. G. Bateson & P. H. Klopfer, pp. 243–269. New York: Plenum Press.

ten Cate, C. (1989b). Stimulus movement, hen behaviour and filial imprinting in Japanese quail (*Coturnix coturnix japonica*). *Ethology*, 82, 287–306.

ten Cate, C. (1991). Behaviour-contingent exposure to song and zebra finch song learning. *Animal Behaviour*, 42, 857–859.

ten Cate, C. & Bateson, P. (1989). Sexual imprinting and a preference for 'supernormal' partners in Japanese quail. *Animal Behaviour*, 38, 356–358.

ten Cate, C., Los, L. & Schilperoord, L. (1984). The influence of differences in social experience on the development of species recognition in zebra finch males. *Animal Behaviour*, 32, 852–860.

ten Cate, C. & Mug, G. (1984). The development of mate choice in zebra finch females. *Behaviour*, 90, 125–150.

Todt, D. (1975). Social learning of vocal patterns and models of their applications in Grey Parrots. *Zeitschrift für Tierpsychologie*, 39, 178–188.

Todt, D., Hultsch, H. & Heike, D. (1979). Conditions affecting song acquisition in nightingales (*Luscinia megarhynchos*). *Zeitschrift für Tierpsychologie*, 51, 23–35.

Turner, E. R. A. (1964). Social feeding in birds. *Behaviour*, 24, 1–46.

van Kampen, H. S. & Bolhuis, J. J. (1991). Auditory learning and filial imprinting in the chick. *Behaviour*, 117, 303–319.

van Kampen, H. S. & Bolhuis, J. J. (1993). Interaction between auditory and visual learning during filial imprinting. *Animal Behaviour*, 45, 623–625.

van Kampen, H. S. & De Vos, G. J. (1991). Learning about the shape of an imprinting object varies with its colour. *Animal Behaviour*, 42, 328–329.

West, M. J., King, A. P. & Duff, M. A. (1990). Communicating about communicating: when innate is not enough. *Developmental Psychobiology*, 23, 585–598.

West, M. J., Stroud, A. N. & King, A. P. (1983). Mimicry of the human voice by European starlings: the role of social interaction. *Wilson Bulletin*, 95, 635–640.

Williams, H. (1990). Song learning in the zebra finch: do fathers supply the models for their sons' songs? *Animal Behaviour*, 39, 745–757.

7

The development of action patterns
KENT C. BERRIDGE

How do action patterns develop in early life? In this chapter, I examine the early development of patterns of action such as walking, hatching, feeding, and grooming. Such behavior patterns I consider to be instinctive or innate. Use of the terms instinct or innate has been strongly criticized by ethologists and comparative psychologists, for essentially two reasons, so I will begin with a few words to justify my use here.

The first criticism traditionally attacked the role of innate as a causal explanation for behavior (e.g. Dewey, 1918; Kuo, 1921; Beach, 1955; Lehrman, 1953, 1970; Hinde, 1968). Kruijt has condensed this criticism succinctly: ' ... for ontogenetic purposes, the term instinctive or innate can be said to completely empty. No light at all is thrown on the nature of the factors underlying the development of instinctive behavior except for the fact that certain specific factors are not necessary' (Kruijt, 1971, p. 10); and 'the term innate does not invite one to ask causal questions about the ontogeny to its final consequences' (Kruijt, 1964, p. 5).

The second criticism has attacked innate as a classification for behavior. If innate, or instinctive, is meant to refer to behavior that is absolutely predetermined by genotype, independent of experience, and unmodifiable by feedback, then no behavior can be found that qualifies. Instead, concerning this absolute sense of innate, the conclusion has prevailed that there might be said to exist only innate behavioral *differences* between individuals, if their developmental histories were identical yet their behavior differed (e.g. Jensen, 1961; Hinde, 1968, 1970; Bateson, 1991).

Innate was rejected as a concept because it presumes a dichotomy that has no correspondence in reality. One may ask, however, why the term learned was not also generally rejected on identical grounds (with few exceptions: for example, Kuo, 1929, 1967, rejected both concepts). After all, no example of 'pure' learned behavior can be found, in the sense of being untouched by inherited influences, any more than one can find 'pure' innate behavior. Learning itself operates by rules that are set in advance. As the early learning theorist E. L. Thorndike wrote: 'The learning of an animal is an instinct of its neurones' (1911; reprinted 1966, p. 186; see also Lorenz, 1965). This argument has recently been extended in modern terms by Marler (1991). It may be that psychologists and ethologists accepted learned, as a concept and term, but not innate, partly because it is difficult to discuss behavioral development without reference to either polarity. But a more principled reason for retaining learning was that it can be identified by certain positive features, while instinct was defined solely in terms of what it lacked (Beach, 1955).

Based on the objections that have been raised to instinctive or innate as a pure concept, many authors choose to avoid the use of these terms (e.g. Bateson, 1991). Instead, when dealing with instinctive aspects of behavior, euphemisms are often used for classification. Terms such as species-typical, species-specific, modal, or stereotyped are usually chosen. Ostensibly of different meaning, these words nonetheless are often used simply in place of instinctive or innate. Although, as descriptive terms, they are more literally true than innate is, these euphemisms are too inclusive. They apply to many types of behavior that are not meant by their authors. They are even less precise than the original terms about the important distinguishing features of the behavior they refer to. Finally, if such words are used simply in place of the word instinctive, they do not avoid the concept – fuzzy as that concept may be – they merely decline to name it (cf. Lorenz, 1965).

I would prefer to reconsider the rejection of innate as a concept for ethology and comparative psychology. The constrained status that has allowed learned to be retained can be applied also to innate. If defined by certain limited features, as learned is, the concept of innate might be usefully employed to highlight important aspects of behavior, just as the concept of learned does (Marler, 1991). If so, the concept of innate behavior would have to be understood to be subject to constraints just as learned behavior is. To begin with, the criticisms mentioned above must be acknowledged to constrain the meaning of the term.

This chapter focuses on the development of innate behavior in a constrained sense. Instinctive action will be taken to refer to distinctive patterns of behavior that share certain features. This use of innate does not deny an experiential influence on such behavior. Innate patterns are not pure, preformed, and inflexible. But they are predictable in appearance, and remain predictable in the face of a wide variety of environmental rearing conditions. The instinctive patterns considered in this chapter are highly organized, characteristic of the species, robust in their ability to appear stable across individuals raised in environments that may vary within the entire natural range (and even beyond this range to quite unnatural settings), and stereotyped in their form and elicitation. Instinctive features, in this sense, do not constitute a complete definition or an exclusive classification of any behavior, much less an explanation for it. But they do help to identify patterns of behavior that have a special use for the study of behavioral development. The interplay between maturation and experience during the earliest days of life may be most visible in such patterns. As I will show in this chapter, instinctive behavior, defined by these features, paradoxically provides a window into the acquisition of behavioral competence (see also Marler, 1991).

The development of instinctive action patterns
General principles of development

Throughout this chapter I will refer to two issues that have general developmental significance in that they allow one to compare phenomena from different individuals and species. The first involves the concepts of differentiation and integration as ways of describing changes in the form of a behavioral pattern over time.

Differentiation is a concept originally emphasized by Coghill (e.g. 1929; see also Fentress & McLeod, 1986; Haverkamp & Oppenheim, 1986), who stressed that early behavior events were crude, but not uncoordinated. Lawful coordination could be seen even in the earliest embryonic movements. Later patterns seemed to be drawn out of those early coordinations as specialized derivations.

Integration refers to the building of new central connections between existing preformed parts (Carmichael, 1934). For example, the development of complex social aggression in young chicks was described by Kruijt (1964) as a gradual connection of separate behavioral components (locomotion, attraction to other chicks, and pecking) into coordinated aggressive attack (see also Hogan, 1988). Stated as hypotheses, progressive differentiation and integration are antitheses of each other. This contradiction has more to do with the way we phrase the concepts, however, than with the phenomena themselves. Both differentiation and integration can occur simultaneously within the same behavior, as will be discussed later.

The second general issue concerns the problem of continuity in development. The appearance and disappearance of many behavior patterns seems to happen abruptly, without any obvious gradual process of formation. For example, many an observer who watches young rat pups or kittens change day by day has been struck by the dramatic metamorphosis on the day on which the young infant's eyes open. All at once the infant begins to scramble over its world, probing every nook and cranny, investigating each new thing. Other behavior patterns disappear just as new ones emerge. Suckling by young mammals, hatching movements by amphibians or birds, and newborn stepping by human infants are examples of action patterns that are generated by the young nervous system, and that may be crucial to early development, but which disappear later in life. This theme is related to the notion of 'development as encephalization' (Teitelbaum, 1971), which posits that development involves a gradual maturation of hierarchical neural controls that usurp the autonomy of existing systems. Other ways of explaining such abrupt changes will be discussed later.

The development of locomotor patterns in birds

Early leg movements and walking

Within a week of the laying of the egg that contains it, the central nervous system of a chick inside can generate patterns of motor output (Hamburger, 1963; Hamburger & Oppenheim, 1967). These patterns appear to the observer first as twitching pulses of the legs, wings, and body (called Type I movements). In the earliest days, the patterned motor outputs can be detected best by recording muscle activity directly through EMG electrodes. EMG recordings show a series of rhythmic cycles of activation in leg muscles, each cycle lasting 2 to 3 seconds, as early as the fourth day of embryonic development. Behavioral manifestation of these patterns of muscle activation, as actual movements of the leg with respect to the body, generally appears as random jerks, more fragmented in form than the underlying EMG patterns might have led one to expect. Coordinated patterns of movement are not clearly visible until around day 14 (coordinated leg movements; Bekoff, 1976) to day 17 (coordinated postural shift prior to hatching; Hamburger & Oppenheim, 1967).

Why do the leg movements appear relatively incoherent to the eye in the first 2 weeks of embryonic life, despite the presence of EMG regularities by day 4 (Bekoff, Stein & Hamburger, 1975; Bekoff, 1976)? One reason is that there may be a great deal of noise in the signals delivered by the central pattern generators in the brain and spinal cord to muscles that participate in the movement. These generators appear to be capable of producing only a degenerate or partial pattern that contains frequent errors during the first days of activation. Bradley & Bekoff (1990) report for chick embryos that in a set of four leg muscles participating in a rhythmic pattern, one or another muscle would often miss a cycle. Although it was difficult to predict which muscle would fail, some failure could be expected. Random gaps in muscle activity would certainly disrupt the regularity of the resulting movement. Randomness might also arise from 'mechanical' sources inherent in immature muscles and joints. Chick muscles, for example, lack adult-like myotube contractile properties until the third week of embryonic life. Finally, it should be recognized that even constant features of a neural motor signal, which might activate particular muscles at particular times, would produce different patterns of movement if the limb began the movement from different initial positions. All of these features might combine to obscure behavioral manifestation of the patterned regularity produced by the developing nervous system.

In spite of these obstacles, Watson and Bekoff (1990) conducted a meticulous behavioral study of the slow-motion kinematics of videotaped leg movements of chick embryos, and found that certain quantitative aspects of the movements were quite stable and patterned. Even though the duration of the movements and the participation roles of particular joints were highly variable, other aspects of the movement, such as the coordination of joints into a flexion or extension phase of movement, the degree of symmetry between these two phases, and the synchronous timing of the joints in a given movement, were quite predictable.

This description of behavioral kinematics fits well the *integrative* notion of development. At least some of the features that characterize the full pattern appear to be produced with good precision very early by the nervous system. The 10-day-old embryo appears to specialize in certain features of the task (e.g. joint coordination). With more time, it extends its competence to encompass other features of the movement as well (e.g. duration).

Differentiation or modification of existing components must also take place during embryonic development. Even among muscles that participate in embryonic leg movements on the 10th day of incubation within the egg, there are important differences in embryonic patterns from the pattern the same muscles will show later. In embryonic movements, the bursts of activity in antagonistic pairs of muscles (which move the leg in opposite directions) are highly symmetrical. Extensor muscles alternate with flexor muscles, and the activation period for each group is similar (Bradley & Bekoff, 1990). In later walking, however, although alternation between extensor and flexor groups is preserved, the excitation periods differ from those seen earlier. Extensor muscles typically have longer periods of excitation than do flexor muscles during mature walking, unlike embryonic movements.

In general, a striking feature of embryonic movement patterns is that they fail to correspond precisely to *any* particular pattern that the chick will show later in life. Embryonic movements seem to form a pattern that is unique, being both simpler and different from such functional movements as walking, hatching, etc. Bradley & Bekoff (1990) suggest that embryonic movements may be a type of *primal movement pattern*, which gives rise later to a host of different patterns, each derived from this original basic unit (e.g. Coghill, 1929). This process of differentiation must involve specialized modifications to the basic pattern, as well as the incorporation of new components, in order to produce the individualized patterns of hatching, walking, scratching, or swimming that are seen later. Each of these mature patterns shares some features, such as alternation between flexor/extensor groups, and synergies among joint movements, with the original embryonic leg movements but they also differ from the precursor, as well as from each other, in a variety of ways that can be seen in EMG activation patterns.

Function of embryonic movements

Do the movements made by a developing embryo serve a purpose? It is difficult to imagine any immediate benefits that might be gained by the embryo itself from behavior within the egg or uterus (with the exception of hatching). It is conceivable, however, that benefits might accrue later to an individual that performed embryonic movements. These benefits might take the form of 'practice effects' , which improve the performance of patterning circuits within the brain and spinal cord, physical exercise and facilitation of muscle or joint growth, or of stimulation of neural development by sensory feedback, etc. Hypotheses of this sort are extraordinarily difficult to test in mammals. A test would require the ability to manipulate or prevent embryonic movements, so as to be able to see later the effect of this manipulation or prevention on later behavior. The same question has been more amenable to investigation in amphibian species, however, and investigations along this line have been pursued for nearly 100 years.

Early studies (e.g. Harrison, 1904; Carmichael, 1926; Matthews & Detwiler, 1926) used a paralytic agent to immobilize developing amphibian embryos, and prevent them from performing normal embryonic patterns of movement. The question asked was whether this immobilization would impair the coordination of swimming movements after hatching. The conclusion from these studies was generally that it did not. When the larval amphibian emerged, it appeared quite capable of patterned action, despite having been denied the opportunity to engage in movement during embryonic development. The results of more recent careful studies on *Xenopus* frog and *Ambystoma* axolotl embryos, which were designed to extend the range of manipulations and to close interpretive loopholes existing from earlier studies, have tended to support this conclusion (e.g. Haverkamp & Oppenheim, 1986). Although the formation of neuromotor synapses and the differentiation of 'fast twitch' versus 'slow twitch' muscle types may be altered by embryonic paralysis, only transient effects on subsequent patterns of behavior have been found by these studies. Behavioral activity may be suppressed during the first few hours after release from immobilization, but later behavior appears normal (Haverkamp & Oppenheim, 1986).

A study that asked the same question of birds has found similar results. After chick embryos were immobilized with d-turbocurarine from embryonic day 6 to day 12, their later spinal motor neuron activity followed normal patterns of flexor/extensor alternation (Landmesser & Szente, 1986). So far, the evidence from amphibians and birds suggests that the emission of early embryonic movements does not carry important consequences for final behavioral competence. Whether this is true also for more complex behavioral patterns or for the behavior of mammals remains unexplored.

Neural basis of embryonic movements in birds and mammals

The adult spinal cord is capable of generating by itself the alternating pattern of left-right leg movements used in mature walking (e.g. Grillner, 1985). This is true for both mammals and birds. Also within the avian spinal cord are the neural circuits responsible for generating the basic pattern of rhythmic embryonic leg movements, as well as of the later egg-escape movements seen during hatching on embryonic day 21, and of walking by the chick after hatching. If a high (cervical) transection of the spinal cord is made in anesthetized chicks on the first or second day after hatching, eliminating the control of movement by the brain above the transection, the spinalized chicks still show stepping movements when they are suspended so that their feet touch a moving treadmill belt (Bekoff, Kauer, Fulstone & Summers, 1989).

Coordinated stepping appears within 2 hours of the time of transection, although the step cycle is slow compared to normal. The opposite effect is seen in young mammals: the rate of stepping actually increases after spinal transection. This has been interpreted to mean that the mammalian brain exerts a constant inhibition over spinal stepping circuits (see below). Although one might be tempted to suppose that the interpretation ought to be reversed for chicks, and that the slowing of the chick's step cycle after spinal transection should imply a tonic excitatory influence by the brain over spinal circuits, this is not the case. A slow stepping cycle is also produced in the chick by mere hindlimb deafferentation, which simply removes the flow of sensory feedback from the legs to the spinal cord, and does not disrupt descending projections from the brain (Bekoff, Nusbaum, Sabichi & Clifford, 1987). In other words, a slowed cycle appears to be the characteristic response of the chick's nervous system to quite diverse neural manipulations. It may be a 'default' output produced by basic circuitry in the spinal cord whenever total function of the integrated nervous system involved in walking is impaired in any serious way.

Spinal chicks show good coordination of stepping muscles within a single leg. Spinal chicks also coordinate steps between the left and right legs so that the normal pattern of alternation is seen. At times, however, the two legs drift out of phase, so that they are no longer in strict alternation. This 'phase drift' must occur whenever the step cycle duration of the left and right legs is not identical. If one leg is slightly slower than the other, then it will fail to maintain proper timing with its pair. Even if it begins 180° out of phase with the other, it will enter the step cycle increasingly later and later with successive steps. Before long, it may be activated in each cycle nearly simultaneously with its partner, rather than in opposite phase. Later still, it will drift further back to an alternating timing, and so on. Phase drift between legs, in other words, appears to reflect an uncoupling of the two cycles (Gallistel, 1980). They become uncoupled both in the sense that they are no longer held together in strict alternation and in the sense that the intrinsic oscillation of each leg's own cycle appears to become independent of the other and to run by different clocks (Bekoff *et al.*, 1989).

Spinal control of hatching

A chick embryo begins to show its first functional behavior, hatching, on its 21st day in the egg (although it must manoeuver itself into position several days earlier via a postural shift known as 'tucking'). Hatching involves patterned leg movements, as well as head and neck movements. In hatching, the legs move synchronously together, rather than in alternation, in a slow rhythmic push that is repeated approximately every 20 seconds. The stimulus that elicits hatching behavior is present well before the actual behavior appears. The folded position of the chick within the egg causes its head to be pushed down upon its right shoulder. It is this postural displacement of the head by the egg that appears crucial to elicit hatching in a 21-day-old embryo. Although hatching normally would be performed only once in a lifetime, the patterned movements can be re-elicited in a chick that has hatched already by gently bending its head again to the right (the position in which most embryos develop), and placing the chick inside an artificial glass 'egg' (Bekoff *et al.*, 1987). If the spinal cord of a newly hatched chick is transected at a thoracic level, the sensory signals produced by head displacement are isolated from the spinal cord, and no leg movements for hatching can be elicited in the glass egg (e.g. Oppenheim, 1975). If the transection is placed higher at the cervical level, however, so that sensory signals from the neck during head displacement can reach the spinal cord, the slow, synchronous pattern of leg movements is seen within the glass egg and can be recorded by EMG (Bekoff *et al.*, 1989). This pattern appears essentially normal in a spinal chick, although a nonspecific stimulus such as a mild pinch to the wing may be needed to cause the movement to persist beyond a few seconds. The effect of a mild pinch on spinal hatching is an instance of more general principles relating nonspecific arousal to behavior (e.g. Fentress, 1977). Similar effects are seen for walking, grooming, etc. in both normal and spinal-transected individuals from many species, both in infants and adults, (e.g. Bekoff *et al.*, 1989; Fentress, 1984).

Development of locomotor patterns in mammals

Walking in newborn mammals

Compared to chicks, the mammals used for developmental studies of action patterns are altricial: relatively immature and incompetent at birth. Kittens and rat pups do not normally begin to walk independently until a number of days after birth, and human children not until nearly 12 months. In spite of this, the basic neural circuits responsible for generating walking-like patterns have been shown to exist at birth or even before.

Human newborns show a form of reflexive stepping in the first weeks after birth, if held upright and slightly forward so that the feet touch a firm surface, which disappears during the second postnatal month (e.g. Thelen & Cooke, 1987). During these steps, the two legs move in alternation and, for each leg, the hip, knee, and ankle joints flex together simultaneously.

Until recently, studies of the development of walking in kittens have placed the earliest time of appearance of coordinated stepping of hindlimbs at approximately 14 days after birth. Prior to recent years, studies that examined walking in kittens did so by supporting the kittens above a moving treadmill belt so that their paws just touched. Those studies did not find coordinated stepping until the end of the second postnatal week. In contrast, using a more 'naturalistic' procedure, Bradley & Smith (1988) have found that patterned hindlimb stepping could be elicited even in newborns, with good rhythmic and interlimb coordination. If the kitten was gently removed a very short distance away from its mother, 'the newborn kitten initiated weight-supported steps to return to its mother's side' (Bradley & Smith, 1988, p.48). (Good interlimb coordination of stepping has also been reported in rat pups on the first day after birth; Bekoff & Trainer, 1979.) If merely supported over a treadmill, on the other hand, kittens at this age simply cried and, as a component of their struggle, either flexed their legs up against the trunk or extended them back and away from the body. Only many days later did kittens cease to be distressed by the treadmill test, with its inherent handling and separation from the mother, and begin to show coordinated treadmill stepping. Bradley & Smith's (1988) observations stand as compelling testimony to the importance of attention to the natural behavioral requirements of the infant even for the study of the simplest action patterns. Clearly, no action pattern is so simple or automatic that it is immune to contextual influences.

A similar lesson may be drawn from Bradley & Smith's account of the development of 'airstepping'. Airstepping is a pattern of stepping movements made when the kitten is suspended above the ground, by fingers held underneath its trunk or by the scruff of the neck. Rhythmic, alternating strokes by the forelimbs (and under some conditions by the hindlimbs) are made by kittens thus suspended in the first week of life. These movements do not occur in a contextual vacuum: 'Typically, these movements were associated with a high level of arousal' (p. 41), and may well be an ineffectual form of escape elicited by the distress of being handled and separated from the litter. Airstepping disappears by the second week of life for normal kittens, but can be seen in adults after spinal transection (Giuliani & Smith, 1985). It might be that the 'disappearance' of this pattern is due actually to its replacement by a more effective and coordinated strategy of escape as the kitten grows older, complemented by increased assurance as the kitten becomes accustomed to handling. Such encephalized influences over the motor patterns produced by the spinal cord would be eliminated by spinal transection, leading to the potential re-emergence of airstepping. The general lesson may be that environmental context exerts a powerful influence over the expression of early behavior. The role of this environmental information appears to be best described as either facilitatory or inhibitory rather than as generative. Experiential factors can allow, prevent, or distort the appearance of the pattern, but appear relatively unimportant for generation of the details of the pattern itself.

Newborn stepping: the precursor of walking?

Are early stepping movements the precursor of later walking? Or is newborn stepping a separate 'reflex' that is replaced by 'real' walking? The issue of continuity between newborn stepping and later walking has been addressed most explicitly in studies of human infants, who show a period between 3 months and 11 months when coordinated stepping cannot be elicited easily (e.g. Thelen & Cooke, 1987; Kamm, Thelen & Jensen, 1990). In early stepping by human newborns, the hip, knee, and ankle joints all flex or extend together, each leg in alternation with the other. This simultaneous activation of joints changes with maturity. During the second month, the ankle drops out of the group and is no longer flexed in synchrony with the others. Stepping becomes difficult to elicit by the end of the second month, and remains so until independent walking appears around 1 year of age. Gradual changes in the step pattern continue to occur even in the second year. Striking the ground with the heel to initiate a step at the end of a swing cycle appears at around 18 months. At the same time, a knee-flexion wave begins to accompany the strike to cushion the blow. Thelen and her colleagues have argued that the development of the mature walking pattern in humans is continuous and gradual (e.g. Thelen & Cooke, 1987; Kamm et al., 1990). Change continues to occur throughout the early years, owing in part perhaps to slow maturation of neuronal pattern generators but also, they argue, due to changes in the way physical torque, gravity, and inertia interact with the child's movement as growth alters body mass, proportion, posture, and movement topology.

Spinal control of walking in kittens and rat pups

Walking patterns of leg movements still can be elicited from young mammals after spinal transection. By comparison to the rapid onset in spinal chicks, stepping movements take at least 24 hours to appear after spinal transection in altricial young mammals. The step cycle of the kitten's transected spinal cord is faster than a normal kitten's. The speeding up of the rate of stepping in a kitten by removing the descending control of the brain over spinal function has been interpreted to mean that the young mammalian brain exerts a tonic inhibition over spinal patterning circuits (e.g. Robinson & Goldberger, 1985). Consistent with the hypothesis that the brain normally inhibits an otherwise competent spinal generator is the remarkable report that stepping in the conventional treadmill task appears days earlier in kittens that have undergone spinal transections compared to normal kittens (Robinson & Goldberger, 1985). Thus the control of stepping by newborn kittens seems to be governed by a strong degree of hierarchical organization (Gallistel, 1980).

Fetal action patterns by mammals prior to birth

The study of behavior of a mammalian fetus presents a formidable challenge. Although bioimaging techniques are increasingly able to visualize fetal movement *in utero*, it is difficult to deliver stimuli or other forms of manipulation to the fetus. Unlike the chick embryo, where a literal window through which to observe and manipulate developing patterns can be provided by removal of a piece of eggshell, the mammalian fetus is not easily accessible to behavioral study. Still, a number of clever and sophisticated techniques that allow one to elicit, observe, and manipulate action patterns have been applied successfully in the last 15 years by resourceful investigators.

It is not possible to observe complex behavioral patterns *in utero* for species such as rats, mice, or cats, because multiple embryos develop crowded together as a single litter at the same time. It is possible, however, to observe individual fetal behavior outside the uterus while preserving the physiological and environmental conditions needed by the fetus (e.g. Bekoff & Lau, 1980; Smotherman & Robinson, 1987, 1988, 1990). In a typical experiment of this sort, the mother is anesthetized with a general anesthetic that has a short duration. The equivalent of a Caesarean section is performed: an abdominal incision is made, the wall of the uterus is opened, and the fetuses in their amniotic sac are gently lifted out and placed next to the mother in a warmed bath of saline, which duplicates the temperature and tonicity of the intrauterine environment. The umbilical connections are left intact, preserving the normal physiological exchange of oxygen, nutrients, and waste metabolites between the blood supply of the mother and her fetuses. Because this exchange also may allow transfer to a fetus of the general anesthetic, which protects the mother from pain during delivery of the fetuses, in some experiments a surgical transection of the maternal spinal cord is made after the fetuses have been removed, and the maternal anesthesia is afterwards discontinued for the period of observation. The spinal transection produces a complete spinal block for the mother, which prevents any pain or movement when the anesthesia wears off. The behavior of a fetus can then be videotaped while it is in a drug-free state.

Fetal walking patterns

Rats have a gestation period of 21 days. On embryological day 20, most of the spontaneous movements of exteriorized fetuses appear random to the eye. Occasionally, however, bursts of patterned 'stepping' or 'swimming' actions can be seen as the fetus floats in its warm bath (Bekoff & Lau, 1980). These bursts of coordinated movement may last up to 10 seconds. During the bursts, synchronized alternation is seen between the left and right sides, both for the forelimbs and hindlimbs, indicating that interlimb coordination is already present on fetal day 20. Coordination between the front and back pairs of limbs, however, has not been observed in fetuses, unlike in the swimming and walking movements that will be produced after birth (Bekoff & Lau, 1980).

Development of feeding patterns

In the normal course of life, a young rat pup will encounter only one foodstuff in the first few weeks after birth: its mother's milk. All that is needed until weaning in the third week after birth is the infantile ability to attach to a maternal nipple and suckle. But in spite of the lack of necessity for altricial mammals to recognize and ingest external foods until weaning occurs, the developing pup is able to do so at birth and even before (Schwartz & Grill, 1985). Rat pups can discriminate the taste of sucrose and quinine from water by days 2–3 of postnatal life (Grill & Norgren, 1978; Hall & Bryan, 1981; Ganchrow, Steiner & Canetto, 1986). Infusions of sweet sucrose directly into the mouth on postnatal day 3 elicit more hedonic tongue protrusions, rhythmic mouth movements of the jaw, and general movements of the head, limbs, and body, than infusions of water. Infusions of milk into the mouth elicit even higher levels of these behaviors. Oral infusions of milk also elicit a characteristic action, the stretch response (in which the limbs are pushed out and the body extended), which usually occurs only during suckling (Hall & Bryan, 1981). Infusions of bitter quinine, on the other hand, elicit gapes (wide openings of the mouth that are larger and last longer than ingestive mouth movements) at this age (Ganchrow, Steiner & Canetto, 1984).

Human newborns too can recognize sweet and bitter tastes and respond appropriately to them within the first days of life. Sweet tastes elicit a 'smile-like' hedonic relaxation of the face, sour elicits pursing of the lips, and bitter elicits an arching of the mouth into a 'disgust' expression on day 1 of postnatal life (e.g. Steiner, 1979). The same expressions are seen in premature newborns (7 months gestation) as in normal 9-month term infants (Steiner, 1979). The early capacity to discriminate between sweet and bitter tastes, and to respond with appropriate reactions, is the provenance of neural circuits within the brainstem for both rats and humans. The basic discriminations among tastes can be made both by decerebrate rats that have lost their forebrain (above the midbrain superior colliculus) by transection, and by anencephalic human infants born with a malformation or absence of forebrain structures (Grill & Norgren, 1978; Steiner, 1979).

Feeding patterns of the mammalian fetus

Even before birth, the brainstem's capability for ingestive reaction is ready. Rat fetuses, which have been removed from the uterus as described above, respond discriminatively on embryonic day 17 (4 days before birth) to flavor stimuli that are infused into the mouth (Smotherman & Robinson, 1990). By embryonic day 19, infusions of milk elicit the 'suckling stretch' response mentioned above, whereas infusions of lemon extract elicit a characteristic forepaw wiping movement of the face (Smotherman & Robinson, 1987, 1988). A similar aversive pattern of face wiping is shown by adult rats to sour and bitter tastes (Grill & Norgren, 1978). Fascinatingly, although rat fetuses show face wiping in response to lemon infusions on the last few days before birth, wiping disappears as a response within a day after birth, and does not reappear again for nearly 2 weeks (Robinson & Smotherman, 1992). One possible explanation for this remarkable temporary 'submersion' of behavior is that competition develops for forelimb motor control soon after birth from other action systems that are involved in locomotion, postural adjustment, and escape movements (Smotherman & Robinson, 1990).

Development of rodent grooming

Grooming by mouse and rat pups

So far, only simple action patterns have been considered: short, discrete actions or simple rhythmic alternations. What about more complex patterns of action? An excellent source of complex patterns for developmental study is grooming. Spontaneous bouts of face grooming by rat pups have been reported to occur as early as the day after birth (Bolles & Woods, 1964). Grooming employs many components involving the forelimbs, hindlimbs, body, and face. These components appear over the first weeks in an order that is roughly cephalocaudal, beginning with actions directed at the head and moving downward to the tail (Richmond & Sachs, 1978). Facial strokes by the forelimbs are the first to appear. Paw licking and scratching appear at the end of the first week, body licking a few days later, while tail licking is delayed until approximately the third week.

In general, however, spontaneous grooming by newborn infants is quite rare. Facial grooming requires free use of the forepaws. That, in turn, requires postural control over the body, to lift the upper half of the body so that the infant does not lie upon its limbs. Young rat or mouse pups are not able to exercise such postural control until well into the second week after birth. As a consequence, such grooming bouts as a pup shows before this time must largely depend upon lucky accidents in which it happens to find itself 'propped up' against a littermate or other object in a way that keeps its forelimbs free (Golani & Fentress, 1985). When fortuitous propping of this sort occurs, the pup may well groom its face.

An investigator can take advantage of this latent capacity to groom by providing the necessary circumstances directly, and propping up the pup with artificial support. Golani & Fentress (1985) did just this in order to conduct an exquisitely detailed study of the development of grooming movements that were emitted by newborn mice. Each mouse pup was supported by inserting it halfway up through a snug hole in the chamber floor, which held the pup around the abdomen and kept it upright, like a circular life-preserver.

The detailed analysis of Golani & Fentress (1985) focused upon changes that occur in the form of forelimb strokes of mice during the first 2 weeks of life. They noted that these changes appeared to reflect three separate phases of development, each lasting roughly 100 hours. The first phase, occupying the first 4 days after birth (0 to 100 hours), contained grooming strokes with many different trajectories. These strokes were poorly controlled in the sense that they often 'overshot' their target as they ascended to the face and sometimes missed the face entirely, instead passing through the air above the surface of the face. During the second phase, lasting roughly throughout postnatal days 5–9 (100–200 hours), facial strokes became more highly controlled. The forelimbs ceased to miss the face, the two forepaws became linked so that each moved in synchrony with the other, and the trajectories of the forepaws wandered less variably over the surface of the face, instead following simpler, more predictable paths. The third and final phase described by Golani & Fentress (1985) began roughly on day 9 (200–300 hours). From this point, the trajectory of facial strokes began to diversify again, the two forepaws more commonly moved independent of each other, and successive strokes again became connected into continuous bouts. For the first time, in this third phase, the grooming actions of the pups began to resemble closely adult grooming patterns.

Recently, several undergraduate students in my laboratory have extended Golani & Fentress' (1985) documentation of early changes in the first weeks of grooming to rats and other species. By using a notation system that converts movement into readily comprehended symbols, the sequential structure of grooming patterns can be made apparent. We have provided mouse and rat pups with postural support to enable grooming in a number of ways: by the Golani & Fentress 'life-preserver' method; by wrapping the lower body in a cotton sheath; by floating the pup in a warm saline bath; by lifting the upper body with support threads; or by simply holding the pup gently in one's fingers. In our experience, at least for rats, an experimenter's fingers provide the best postural support for grooming because they can be relaxed and adjusted so as not to provoke the pup into escape attempts. Mouse pups are somewhat more tolerant of restraint.

On the first day of postnatal life, supported rat pups emit strokes in short bursts as described for mice by Golani & Fentress (1985). Strokes are performed equally often by a single paw or by both paws together, and generally ascend above the eye to the upper part of the face. After a supported pup has emitted several of these bouts, however, facial strokes become replaced by 'pushing' movements into the air, giving an appearance of swimming. Under normal circumstances in the nest, such movements would effectively push the pup away from the situation that held it.

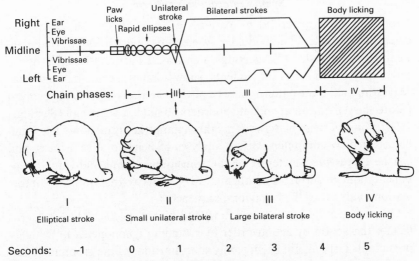

Figure 7.1 A syntactic grooming chain. The pattern is depicted by a choreographic notation system for grooming behavior, which displays the trajectories and sequence of facial strokes and the sequential occurrence of other grooming components. Drawings depict action types. Time proceeds from left to right. The horizontal axis represents the center of the nose. The line above the horizontal axis denotes movement of the right forepaw along the face; the line below the axis denotes movement of the left forepaw. Small rectangles denote paw licks. Large rectangle denotes body licks. Chain phases are: (I) 5 to 8 rapid elliptical strokes around the nose (6.5 Hz); (II) unilateral strokes of small amplitude; (III) symmetrically bilateral strokes of large amplitude; (IV) licking of the torso. Syntactic chains are typically completed (by Phase IV onset) within 5 seconds of the first Phase I ellipse. Adapted from Aldridge, Berridge & Herman (1990).

From day 2 to 7, the accuracy of facial strokes increases in the sense that more successful contacts with the face are made (cf. Golani & Fentress, 1985). The percentage of strokes that touch the face rises from 40 % on day 1 to better than 70 % on day 7.

Spontaneous grooming by infant rats, without postural support from the experimenter, begins to be commonly observed about day 8. Grooming is still highly dependent upon context and is aided by extra postural support at this time. Pups do not yet groom when by themselves in the home cage, perhaps because they still lack sufficient postural control to hold themselves upright without leaning against littermates.

Syntax of grooming patterns

Grooming is characterized by patterns of serial order. These patterns determine the order of individual movements during a grooming bout (e.g. Fentress & Stilwell, 1973; Berridge, Fentress & Parr, 1987). Each of the three major suborders of rodents show similar sequential patterns, suggesting that these patterns may be phylogenetically old, having evolved before the separation of rodent suborders 60 million years ago (Berridge, 1990). Some of these patterns apply throughout the entire bout. These can be detected by computer-assisted analyses of behavioral sequences, and can be represented in terms of the transitions among individual components and in terms of mathematical stereotypy (measured by information analysis) of the behavioral sequence.

Other sequential patterns are short and highly specific. These patterns dictate the action-by-action order of grooming components in a highly predictable fashion, but occupy only short portions of the grooming bout. A remarkably stereotyped pattern of this sort was described for rats by Berridge & Fentress (1986), and called a 'syntactic chain' because of the lawful nature of its serial order or syntax. This syntactic chain pattern connects 15 to 25 discrete grooming actions into an easily recognized sequence (actions can be predicted with over 90% accuracy based upon a syntax rule), which comprises four phases (Figure 7.1). The pattern begins with Phase I: a series of tight elliptical strokes over the nose, made by the two forepaws simultaneously. This is followed by Phase II: one or several small strokes with a single paw; then Phase III: a series of large strokes over the entire face synchronously by both paws; and finally Phase IV: in which the animal bends downwards and to the side and begins to lick its body. This highly stereotyped pattern is also displayed, with minor variations, by other rodents ranging from mice, to hamsters, gerbils, squirrels and guinea pigs.

A number of developmental questions have been posed in my laboratory concerning this syntactic chain pattern. Does the elaborate pattern appear from the first day that grooming strokes are emitted? Or does the pattern require time to develop? If the pattern appears only after some delay, then does it come in fully formed the first time it is seen? Or does it develop gradually?

To some extent, the answers to these questions appear to depend upon which rodent species we study. A crucial factor is the relative maturity of the species at birth. Rodents such as rats and mice are born in an altricial state: blind, hairless, grub-like creatures (Figure 7.2). A strong contrast to

Figure 7.2 Relative physical maturity of newborn rat and guinea pig. An altricial rat pup and a precocial infant guinea pig on postnatal day 1.

this picture is provided by the guinea pig. Highly precocial at birth, after a long gestation period of 65 days (compared to 22 days for rats), the newborn guinea pig appears rather like a small adult (Figure 7.2). Eyes open, able to feed immediately on solid food, with a developed coat, excellent postural control, and fully able to walk and run about, the infant guinea pig needs no artificial support in order to groom, even on the first day of postnatal life; and on day 1 it shows complete syntactic grooming chains.

Altricial development of grooming syntax

Rat pups, during the first week of postnatal life, emit grooming actions as described above. Nothing can be seen that resembles the syntactic chain pattern before postnatal day 7 (Figure 7.3).

On day 7 or 8, the first 'ellipses' are seen: rapid strokes made simultaneously by both paws, with small, symmetrical and elliptical trajectories. The ellipses almost invariably occur singly during the first days in which they appear. This contrasts strongly to their organization in adult grooming. In adults, single ellipses are virtually never seen, but instead combine in series to form Phase I of syntactic chains. Although

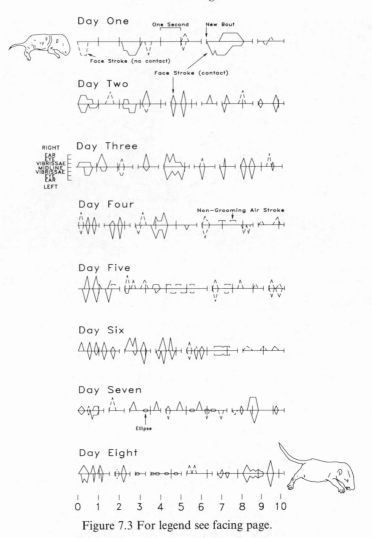

Figure 7.3 For legend see facing page.

large facial strokes often congregate into continuous bouts on this day, there is no hint of the syntax or phase structure between stroke types that will characterize syntactic chains later.

A dramatic transformation of grooming patterns occurs over postnatal days 9 to 15. This period marks a transition from a lack of any recognizable chain syntax to syntactic chains that are nearly perfect in form (i.e. matching the modal adult pattern). The development of syntactic chains between days 9 and 15 proceeds by the addition of successive phases (Figure 7.3). First, single ellipses become grouped together into bouts to

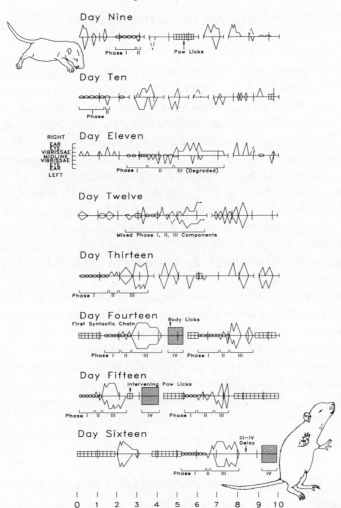

Figure 7.3 The development of syntactic grooming patterns of rat pups. Characteristic grooming sequences for each day of life, from birth to postnatal day 16. Each line shows the most 'syntactic chain-like' grooming sequence displayed by pups from day 7 onwards; no chain-like patterns are seen on earlier days. Drawings show the physical appearance of rat pups at postnatal days 1, 8 to 9, and 16 (size is proportional). Note the dramatic improvement of syntactic chain structure between day 7 (the first occurrence of a Phase I-type elliptical stroke) and day 14 (the first occurrence of a chain that was syntactically complete through Phase IV).

form Phase I around day 9. Next, Phase I bouts tend to be followed within a day by small-amplitude strokes made by one paw. These unilateral strokes become Phase II in the context of syntactic chains. Occasionally at first, but with increasing regularity over the next days, Phase II strokes are followed by Phase III: a series of large strokes made symmetrically by both paws, which reach up to the ear. Fascinatingly, the boundaries between components of different phases are blurred in these early chains on days 11 and 12. These chains often blend together aspects of Phase II and Phase III into an amalgamation of both. Soon, however, the strokes become more distinct and the boundaries between the phases sharpen (Figure 7.3).

Early syntactic chains typically lack the final Phase IV of body licking until at least day 14. Although Phases I, II, and III, are arranged in serial order, they are degraded in several ways. The actions within each phase tend to be slow compared to adults. The form of component actions also is degraded. Bilateral strokes, for example, are often strikingly asymmetrical and 'lopsided'.

Chains do not develop, in other words, by first perfecting the details of individual strokes, next connecting strokes of a similar type together to form separate phases, then perfecting those phases, and only then connecting the phases together to form a global syntactic chain. Instead, syntactic aspects of the 'whole' chain are present as early as day 10 or 11. The components within this global structure, however, are impoverished. Over days 12 and 13, the form of chain components within each phase gradually becomes more adult like (Figure 7.3).

Programmed allometric timing of syntactic movement

The timing of adult movement patterns differs among different rodent species, even for shared syntactic patterns that are similar in form and sequence. The single most important factor that determines the timing of syntactic chain movements for adults appears to be very simple: body mass or average weight (Berridge, 1990). The relationship can be seen most clearly by comparing the duration of elliptical Phase I strokes, the chain component that is most highly stereotyped in form and duration. The relationship is allometric: a doubling of body size does not double the duration of the behavioral ellipse, but increases it instead in a logarithmic fashion (Figure 7.4). The increase in duration can be described by an allometric equation of the type, $y = a \cdot x^b$, where a and b are constants, y is movement duration and x is body mass. Calculation of the allometric

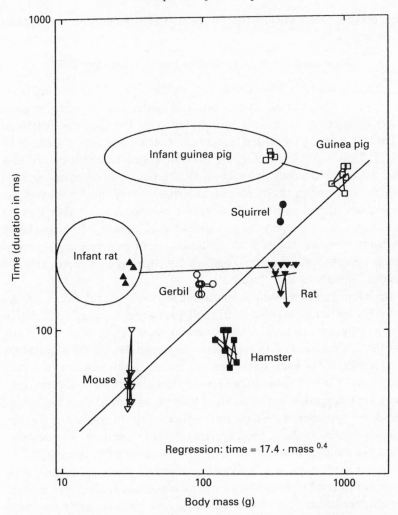

Figure 7.4 Programmed allometry of action timing. Ellipse stroke duration is shown for rat pups and young guinea pigs, and for full-grown adults of six species (Phase I components of syntactic grooming chains). The duration of elliptical strokes for adults can be seen to be influenced allometrically by body size. The regression line shows the timing predicted for adults by the allometric equation ($y = ax$). Infant rats (mouse-sized on postnatal day 15) and infant guinea pigs (rat-sized on day 18) have timing appropriate to their adult conspecifics even though the infants weigh much less. During their development, infants will 'slide' horizontally into their adult positions on the graph, indicating pre-programmed timing of the movement pattern. They do not instead 'climb' the diagonal regression line, as they would if movement timing were controlled directly by body mass. Adult data from Berridge (1990).

equation for adult rodents has shown that ellipse duration is related to body size as,

$$\text{ellipse duration} = 17.44 \cdot \text{Body Mass}^{0.39} \text{ (Berridge, 1990).}$$

Is this allometric relation due directly to the physics of movement? That is, is it a passive consequence of physical inertia and the force required from muscles to move limbs of varying size? Or does the allometric relationship arise from within the nervous system? The answer seems to be the latter. Some evidence for this assertion comes from the observation that allometry depends as much upon the pauses between movements (which are exempt from inertia influence) as upon the movements themselves (Berridge, 1990). Pauses must be programmed, since they do not depend directly upon the physics of the movement. More conclusive evidence, however, comes from the development of movement timing in infant rodents. Infants do not conform directly to the allometric rule but instead 'anticipate' it (Figure 7.4).

An infant guinea pig weighs only as much as an adult rat. But its ellipses resemble the timing of those of an adult guinea pig. A young rat at day 20 weighs only as much as an adult mouse; but the rat pups' ellipses are more like those of rat adults than like those of mouse adults. Young rodents time their strokes to the body weight they will have as adults rather than to their actual size. They conform to the allometric relation based on their parents' mass (or, put another way, their own future mass). Rather than have ellipse durations appropriate to their own weight, they are slower by just the amount needed to make them species-typical in duration. Possession of adult like timing parameters by an infant cannot possibly be explained by a 'physics of movement' allometry. Neither body mass nor body proportions are fully adult-like early in life. Instead, allometry must be programmed in advance into the pattern generators that control the timing of ellipse movements.

Brain mechanisms of grooming syntax

What events in the brain are crucial to the behavioral changes that we see in development? What enables the precocial brain of a newborn guinea pig to generate perfect syntactic chains of grooming on its first day in the world? Why do altricial rats and mice need many days of postnatal development to achieve the same competency? What happens within the brain of a young rat between day 12 and day 16 that makes it suddenly able to coordinate syntactic chains of grooming actions?

Tides of change sweep through the developing brain both before and after birth. New neurons are born; many existing neurons die. Fibers of connection grow, synaptic connections form, glial cells ensheathe neuronal axons in myelin, the synthesis and transport of neurotransmitters progress. Birth provides no absolute divide for these changes. Even though neural division and cell death may occur largely prenatally for most species, leaving other changes such as myelinization to proceed largely postnatally, many exceptions are found on both sides.

For adult rats, loss of either the cerebellum or motor cortex does not seriously disrupt the performance of syntactic chains of grooming actions (Berridge & Whishaw, 1992). It is therefore unlikely that the development of these structures would be crucial to the appearance of grooming syntax in infants. Instead the evidence from neurobehavioral studies of adults points to the *brainstem* and to the subcortical forebrain *striatum* as substrates for the development of action syntax (Berridge & Fentress, 1987; Berridge, 1989a, 1989b; Cromwell & Berridge, 1990; Berridge & Whishaw, 1992).

Anatomical connections to the striatum form during the time of dramatic improvement in syntactic grooming chains, between days 7 and 17 (Iñiguez, De Juan, Al-Majdalawi & Gayoso, 1990). Iñiguez *et al.* studied the growth in these connections by injecting a retrograde tracer (horseradish peroxidase) into the striatum of rat pups at different times after birth and tracing the transport of the label along connecting axons to other brain regions. When injections are successfully transported to another site in the brain, connections can be inferred to have formed. Retrograde transport of label from the striatum to the cortex is first observed around postnatal day 7. Between day 7 and day 14, the density of transported label increases markedly, indicating that new connections are formed during that time. The density of label in midbrain sites that connect to the striatum, such as the substantia nigra, tegmentum, and red nucleus, also increases between postnatal days 7 and 14. Within the striatum itself, organization into patch and matrix compartments becomes more adult-like from day 7 to day 20, as the striatal matrix grows and patches shrink in size (Lanca, Boyd, Kolb & Van der Kooy, 1986). Finally, efferent projections from the striatum to the globus pallidus, which mediate striatal output, begin to appear during the second postnatal week. In other words, dramatic changes in the anatomical circuits that link the striatum to other brain structures, and in the internal organization of the striatum itself, occur simultaneously with the behavioral maturation of syntactic grooming sequences during the second week after birth.

General indices of neuronal maturation provide a clue as to the source of differences in behavioral competence between altricial and precocial species. The physical development of neurons in rat and guinea pig brains proceeds by quite different time courses relative to birth. Nucleation, the creation of new neurons by the division of precursors, and neuronal elongation, especially the growth of axons, are the chief processes of structural change in a developing brain. These processes draw upon a replenishing reservoir of tubulin, a protein that forms the basic building block for microtubule assembly. Tubulin forms the structural skeleton for the axons and dendrites of a developing neuron, and also participates in the nucleation reaction (Lennon, Francon, Fellous & Nunez, 1980). Because tubulin provides a basic building block for growing neurons, its concentration within the brain, and the rate at which it is synthesized *in vitro*, provide indicators of the rate of structural growth of the brain at that time (although it is not a perfect neuronal indicator: tubulin is also used for glial cell formation).

Infant rats and mice have slow rates of brain microtubule assembly, and do not reach adult-like levels until at least 25 days after birth. Structural changes, in other words, occur within their brains on a large scale for nearly a month. Newborn guinea pigs, by contrast, are already adult like in brain tubulin concentration, and in microtubule assembly rates, at birth (Lennon *et al.*, 1980). At 10 days of postnatal age, rats are far below guinea pigs in rates of microtubule assembly (and mice are even further below rats). Thus there is reason to believe that structural development of neurons relative to birth is delayed in rats and mice compared to guinea pigs.

Neurochemical maturation also is important to behavioral development, and the maturation of neurotransmitter systems shows intriguing correlations with the development of grooming syntax. For syntactic grooming patterns, a crucial neurotransmitter is nigrostriatal dopamine (Berridge, 1989b). Levels of the monoamine neurotransmitters, dopamine, norepinephrine, and serotonin are quite low in the rat fetus (e.g. Altar, Bogar, Oei & Wood, 1987; Kalsbeek *et al.*, 1989). These neurotransmitters and their metabolites begin to rise at birth, and reach plateaus at postnatal day 26 (dopamine and norepinephrine) and day 70 (serotonin) that are at least 10 times higher than their starting points (30 times for dopamine and 45 times higher for serotonin).

The metabolites for these neurotransmitters also increase gradually from a few days before birth until 2 to 3 weeks after, but not in direct parallel with the neurotransmitters themselves. This may have interesting consequences. The *ratio* between the level of metabolites for a neuro-transmitter and the neurotransmitter itself is thought to provide a measure of the rate of use or turnover of that neurotransmitter (Altar *et al.*, 1987; Kalsbeek *et al.*, 1989; but see Commissiong, 1985). Since metabolites are produced when a neurotransmitter is released into a synapse and then broken down, higher ratios of metabolites for a given level of neuro-transmitter suggest that the neurotransmitter is being released, broken down, and replaced at a fast rate. Thus high metabolite:neurotransmitter ratios are presumed to reflect high activity among the neurons that use that neurotransmitter.

A peak in 'whole brain' DOPAC:dopamine ratios can be seen around postnatal day 16 in data published for the developing brains of young rats (Herregodts *et al.*, 1990). This peak was not discussed explicitly by the authors, but its timing is intriguing from the point of view of grooming syntax development. From postnatal day 8 to day 16, the ratio of DOPAC:dopamine rises by 25%. After day 16, this ratio declines dramatically over the next few days. By day 20, the ratio is at a level considerably lower than that of day 8. Does the sharp rise in the ratio leading up to day 16 reflect a surge in brain dopamine activity that occurs when syntactic grooming chains are being assembled for the first time? It is tempting to speculate that a dopamine surge between days 12 and 16 might cause the improvement in sequential coordination, given the importance of striatal dopamine to adult grooming syntax (Berridge, 1989b). It is not possible to tell from the data of Herregodts *et al.* (1990) the precise day-by-day timing of this peak in DOPAC:dopamine ratios, but their results are compatible with the hypothesis that the peak in dopamine function coincides with the improvement of grooming syntax. Further support for a dopaminergic hypothesis of grooming development is added by reports that dopamine neurons to the striatum increase their firing rate and 'pacemaker' activity between days 5 and 13 (Tepper, Trent & Nakamura, 1990), and that striatal D_1 and D_2 receptors for dopamine more than double in density between days 7 and 16 (Murrin & Zeng, 1990; Rao, Molinoff & Joyce, 1991).

Within other brain structures relevant to grooming syntax, specific changes in dopamine levels and receptors also occur that are correlated with the development of syntactic chains. The midbrain tegmentum and substantia nigra region together send massive dopamine projections to the striatum and form a primary determinant of striatal function. This midbrain region undergoes a dramatic increase in dopamine content and in dopamine receptor binding between days 5 and 15 (Noisin & Thomas,

1988). This may reflect a sudden boost in positive feedback from sources outside of the tegmentum, such as the striatum. If so, the change might signify that these neuroanatomical loops come 'on line' at this time, becoming functional systems that can influence the pattern of action.

Of course at this point such speculations are highly tenuous. They rest on correlations between events that have been observed to occur at a certain time in the development of behavioral structure and events that occur in the development of brain systems. Although justifiable as an approach, for now any attempt to construct a developmental 'one-to-one mapping' of correlated brain to behavioral events must be recognized to rest on an extraordinarily small data base. It is a major task to chart the development of even a single behavioral or brain feature over the course of early life. With so many possible features to choose from in both brain and behavior, and so few that have been studied so far, finding the right causal connection at present could only be attained by a massive stroke of luck.

The task of mapping developmental brain-behavior relations is far more difficult than the proverbial search for a single needle in a haystack. This search is instead for two needles, one neural and one behavioral, each in a different haystack. Somewhere in the two haystacks are two needles that match. Even this analogy oversimplifies the true state of affairs, for it assumes a simple one-to-one correspondence between brain system and behavioral function, rather than the multidetermined causal relationship that is likely to exist (Fentress, 1990, 1991). But an analogy constructed on that premise becomes almost too awful to contemplate. Perhaps it is wiser to focus instead on the progress that has been made in spite of the obstacles.

Conclusions

I began by framing the themes of 'integration of originally separate parts' and 'differentiation of crude elements' as a useful bipolar framework for thinking about individual instances of development (see Fentress, 1972). Each example of instinctive behavior discussed in this chapter displays aspects of progressive integration. The walking patterns of embryos and neonates begin as movements by separate joints, and become reliably integrated in a gradual fashion to produce the final coordinated pattern. The feeding patterns of the mammalian fetus and infant become integrated later into adult feeding behavior and into affective displays. The grooming syntax of the rat pup emerges slowly as actions are linked in integrative series to form syntactic chains.

Differentiation also can be identified in these examples. Not so much undoing the work of integration as marching in parallel with it, differentiation operates on slightly separate aspects of early action. Embryo movements, for instance, are at first hardly recognizable as patterned events. Later the pattern of these embryonic movements becomes sharper and distinct.

But clearly these polarized conceptual trends do not exhaust the range of developmental generalizations. Differentiation is not the only conceptual opposite to integrative 'bottom-up' organization. 'Top-down' emergence of structure was encountered in this chapter for the development of grooming syntax. The global structure of the syntactic chain pattern appears before its elements are fully perfected. Instead of being formed by a combination of existing elements, the pattern takes the elements at their crudest and gradually perfects them. Top-down global structure, like the programming of adult-like movement, may imply an imbedded pattern organizer within the nervous system, which exists independently of the elements it uses for expression.

Finally, as presaged by the notion of appearance and disappearance, developmental events may occur in stepwise fashion, not gradually at all. Sudden changes in behavioral pattern may reflect equally sudden events in the underlying nervous system. Grooming syntax and DOPAC:dopamine peak ratios may be an instance of such connections.

Whatever precise aspect of action development we wish to understand, our investigation can be guided in a general way by the advice of Kruijt, to whom this book is dedicated:

How does ... behaviour develop in the life of the individual? The problem can be approached in much the same way as is done by the embryologist. What is needed is, first, a description of the changes that take place during the life of the individual under natural circumstances, and second, an analysis of the factors that underly the observed changes. (Kruijt, 1971, p. 11).

Acknowledgements

I thank Jill Becker, Ann Bekoff, Johan Bolhuis, Casey Cromwell, Jerry Hogan, Theresa Lee and Elliot Valenstein for helpful comments on an earlier draft of the manuscript. Most of this chapter was written while I was on leave at the University of Birmingham in England. I am very grateful to the faculty and students of the School of Psychology at the University of Birmingham for their warm hospitality, and to the James McKeen Cattell Foundation for sabbatical support. The studies of grooming syntax development described here were supported by a grant from the NIH (NS

23952) and conducted by a large group of undergraduate students. I thank them all for their patient efforts, especially Matthew Colonnese, who constructed Figure 7.3. I also thank Darlene Tansil, who drew the adult and infant rat portraits.

References

Aldridge, J. W., Berridge, K. C. & Herman, M. (1990). Single unit activity in the rat caudate-putamen during stereotypical grooming sequences. *Society for Neuroscience Abstracts*, 16, 233.

Altar, C. A., Bogar, W. C., Oei, E. & Wood, P. H. (1987). Dopamine autoreceptors modulate the in vivo release of dopamine in the frontal, cingulate, and entorhinal cortices. *Journal of Pharmacology and Experimental Therapeutics*, 242, 115–120.

Bateson, P. P. G. (1991). Are there principles of behavioural development? In *The Development and Integration of Behaviour*, ed. P. Bateson, pp. 19–39. Cambridge: Cambridge University Press.

Beach, F. A. (1955). The descent of instinct. *Psychological Review*, 62, 401–410.

Bekoff, A. (1976). Ontogeny of leg motor output in the chick embryo: a neural analysis. *Brain Research*, 106, 271–291.

Bekoff, A., Kauer, J. A., Fulstone, A. & Summers, T. R. (1989). Neural control of limb coordination. *Experimental Brain Research*, 74, 609–617.

Bekoff, A. & Lau, B. (1980). Interlimb coordination in 20-day-old rat fetuses. *Journal of Experimental Zoology*, 214, 173–178.

Bekoff, A., Nusbaum, M. P., Sabichi, A. L. & Clifford, M. (1987). Neural control of limb coordination. I. Comparison of hatching and walking motor output patterns in normal and deafferented chicks. *Journal of Neuroscience*, 7, 2320–2330.

Bekoff, A., Stein, P. S. & Hamburger, V. (1975). Coordinated motor output in the hindlimb of the 7-day chick embryo. *Proceedings of the National Academy of Science (USA)*, 72, 1245–1248.

Bekoff, A. & Trainer, W. (1979). The development of interlimb coordination during swimming in postnatal rats. *Journal of Experimental Biology*, 83, 1–11.

Berridge, K. C. (1989a). Progressive degradation of serial grooming chains by descending decerebration. *Behavioural Brain Research*, 33, 241–253.

Berridge, K. C. (1989b). Substantia nigra 6-OHDA lesions mimic striatopallidal disruption of syntactic grooming chains: A neural systems analysis of sequence control. *Psychobiology*, 17, 377–385.

Berridge, K. C. (1990). Comparative fine structure of action: rules of form and sequence in the grooming patterns of six rodent species. *Behaviour*, 113, 21–56.

Berridge, K. C. & Fentress, J. C. (1986). Contextual control of trigminal sensorimotor function. *Journal of Neuroscience*, 6, 325–330.

Berridge, K. C. & Fentress, J. C. (1987). Disruption of natural grooming chains after striatopallidal lesions. *Psychobiology*, 15, 336–342.

Berridge, K. C., Fentress, J. C. & Parr, H. (1987). Natural syntax rules control action sequence of rats. *Behavioural Brain Research*, 23, 59–68.

Berridge, K. C. & Whishaw, I. Q. (1992). Cortex, striatum, and cerebellum: control of syntactic grooming sequences. *Experimental Brain Research*, 90, 275–290.

Bolles, R. C. & Woods, P. J. (1964). The ontogeny of behaviour in the albino rat. *Animal Behaviour*, 12, 427–441.

Bradley, N. S. & Bekoff, A. (1990). Development of coordinated movement in chicks: I. Temporal analysis of hindlimb muscle synergies at embryonic days 9 and 10. *Developmental Psychobiology*, 23, 763–782.

Bradley, N. S. & Smith, J. L. (1988) Neuromuscular patterns of stereotypic hindlimb behaviors in the first two postnatal months. I. Stepping in normal kittens. *Developmental Brain Research*, 38, 37–52.

Carmichael, L. (1926). The development of behavior in vertebrates removed from the influence of external stimulation. *Psychological Review*, 34, 34–47.

Carmichael, L. (1934). An experimental study in the pre-natal guinea pig of the origin and development of reflexes and patterns of behavior in relation to the stimulation of specific receptor areas during the period of active fetal life. *Genetical Psychology Monographs*, 16, 337–491.

Coghill, G. E. (1929). *Anatomy and the Problem of Behavior*. Cambridge: Cambridge University Press.

Commissiong, J. W. (1985). Monoamine metabolites: their relationship and lack of relationship to monoaminergic neuronal activity. *Biochemistry and Pharmacology*, 34, 1127–1131.

Cromwell, H. C. & Berridge, K. C. (1990). Anterior lesions of the corpus striatum produce a disruption of stereotyped grooming sequences in the rat. *Society for Neuroscience Abstracts*, 16, 233.

Dewey, J. (1918). *Human Nature and Conduct*. New York: Holt.

Fentress, J. C. (1972). Development and patterning of movement sequences in inbred mice. In *The Biology of Behavior*, ed. J. Kiger, pp. 83–132, Corvallis, Oregon: Oregon State University Press.

Fentress, J. C. (1977). The tonic hypothesis and the patterning of behavior. *Annals of the New York Academy of Sciences*, 290, 370–395.

Fentress, J. C. (1984). The development of coordination. *Journal of Motor Behavior*, 2, 99–134.

Fentress, J. C. (1990). Organizational patterns in action: Local and global issues in action pattern formation. In *Signal and Sense: Local and Global Order in Perceptual Maps*, ed. G. M. Edelman, W. E. Gall & W. M. Cowan, pp. 357–382. New York: Wiley.

Fentress, J. C. (1991). Analytical ethology and synthetic neuroscience. In *The Development and Integration of Behaviour*, ed. P. Bateson, pp. 77–120. Cambridge: Cambridge University Press.

Fentress, J. C. & McLeod, P. (1986). Motor patterns in development. In *Handbook of Behavioral Neurobiology*, Vol. 8. *Developmental Processes in Psychobiology and Neurobiology*, ed. E. M. Blass, pp. 35–97. New York: Plenum Press.

Fentress, J. C. & Stilwell, F. P. (1973). Grammar of a movement sequence in inbred mice. *Nature*, 224, 52–53.

Gallistel, C. R. (1980). *The Organization of Action: A New Synthesis*. Hillsdale, NJ: Lawrence Erlbaum.

Ganchrow, J. R., Steiner, J. E. & Canetto, S. (1986). Behavioral displays to gustatory stimuli in newborn rat pups. *Developmental Psychobiology*, 19, 163–174.

Giuliani, C. A. & Smith, J. L. (1985). Development and characteristics of airstepping in chronic spinal cats. *Journal of Neuroscience*, 5, 1276–1282.

Golani, I. & Fentress, J. C. (1985). Early ontogeny of face grooming in mice. *Developmental Psychobiology*, 18, 529–544.

Grill, H. J. & Norgren, R. (1978). The taste reactivity test I and II. *Brain Research*, 143, 263–297.

Grillner, S. (1985). Neurobiological bases of rhythmic motor acts in vertebrates. *Science*, 228, 143–149.

Hall, W. G. & Bryan, T. E. (1981). The ontogeny of feeding in rats: IV. Taste development as measured by intake and behavioral responses to oral infusions of sucrose and quinine. *Journal of Comparative and Physiological Psychology*, 95, 240–251.

Hamburger, V. (1963). Some aspects of the embryology of behavior. *Quarterly Review of Biology*, 38, 342–365.

Hamburger, V. & Oppenheim, R. (1967). Prehatching motility and hatching behavior in the chick. *Journal of Experimental Zoology*, 166, 171–203.

Harrison, R. G. (1904). An experimental study of the relation of the nervous system to the developing musculature in the embryo of the frog. *American Journal of Anatomy*, 3, 197–220.

Haverkamp, L. J. & Oppenheim, R. W. (1986). Behavioral development in the absence of neural activity: effects of chronic immobilization on amphibian embryos. *Journal of Neuroscience*, 6, 1332–1337.

Herregodts, P., Velkeniers, B., Ebinger, G., Michotte, Y., Vanhaelst, L. & Hooghe-Peters, E. (1990). Development of monoaminergic neurotransmitters in fetal and postnatal rat brain: analysis by HPLC with electrochemical detection. *Journal of Neurochemistry*, 55, 774–779.

Hinde, R. A. (1968). Dichotomies in the study of development. In *Genetic and Environmental Influences on Behavior*, ed. J. M. Thoday & A. S. Parkes, pp. 3–14. Edinburgh: Oliver and Boyd.

Hinde, R. A. (1970). *Animal Behaviour: A Synthesis of Ethology and Comparative Psychology* (2nd edn). Tokyo: McGraw-Hill Kogakusha.

Hogan, J. A. (1988). Cause and function in the development of behavior systems. In *Handbook of Behavioral Neurobiology*, Vol. 9, ed. E. M. Blass, pp. 63–106. New York: Plenum Press.

Iñiguez, C., De Juan, J., Al-Majdalawi, A. & Gayoso, M. J. (1990). Postnatal development of striatal connections in the rat: a transport study with wheat germ agglutinin-horseradish peroxidase. *Developmental Brain Research*, 57, 43–53.

Jensen, D. D. (1961). Operationism and the question 'Is this behavior learned or innate?' *Behaviour*, 17, 1–8.

Kalsbeek, A., Buijs, R. M., Hofman, M. A., Matthijssen, M. A., Pool, C. W. & Uylings, H. B. (1989). Monoamine and metabolite levels in the prefrontal cortex and the mesolimbic forebrain following neonatal lesions of the ventral tegmental area. *Brain Research*, 479, 339–343.

Kamm, K., Thelen, E. & Jensen, J. L. (1990). A dynamical systems approach to motor development. *Physical Therapy*, 70, 763–775.

Kruijt, J. P. (1964). Ontogeny of social behaviour in Burmese red jungle fowl. *Behaviour*, Supplement 12.

Kruijt, J. P. (1971). Early experience and the development of social behavior in junglefowl. *Psychiatria, Neurologia, Neurochirurgia* 74, 7–20.

Kuo, Z.-Y. (1921). Giving up instincts in psychology. *Journal of Philosophy*, 17, 645–664.

Kuo, Z.-Y. (1929). The net result of the anti-heredity movement in psychology. *Psychological Review*, 36, 181–199.

Kuo, Z.-Y. (1967). *The Dynamics of Behavior Development*. New York: Random House.

Lanca, A. J., Boyd, S., Kolb, B. E. & Van der Kooy, D. (1986). The development of a patchy organizaton of the rat striatum. *Developmental Brain Research*, 27, 1–10.

Landmesser, L. T. & Szente, M. (1986). Activation patterns of embryonic chick hindlimb muscles following blockade of activity and motoneurone cell death. *Journal of Physiology*, 380, 157–174.

Lehrman, D. S. (1953). A critique of Konrad Lorenz' theory of instinctive behavior. *Quarterly Review of Biology*, 28, 337–363.

Lehrman, D. S. (1970). Semantic and conceptual issues in the nature–nurture problem. In *Development and Evolution of Behavior*, ed. L. R. Aronson, E. Tobach, D. S. Lehrman & J. S. Rosenblatt, pp. 17–52. San Francisco: Freeman.

Lennon, A. M., Francon, J., Fellous, A. & Nunez, J. (1980). Rat, mouse, and guinea pig brain development and microtubule assembly. *Journal of Neurochemistry*, 35, 804–813.

Lorenz, K. (1965). *Evolution and Modification of Behavior*. Chicago: University of Chicago Press.

Marler, P. (1991). The instinct to learn. In *The Epigenesis of Mind*, ed. S. Carey & R. Gelman, pp. 37–66. Hillsdale, NJ: Lawrence Erlbaum.

Matthews, S. A. & Detwiler, S. R. (1926). The reaction of *Amblystoma* embryos following prolonged treatment with chloretone. *Journal of Experimental Zoology*, 45, 279–292.

Murrin, L. C. & Zeng, W. (1990). Ontogeny of dopamine D1 receptors in rat forebrain. *Developmental Brain Research*, 57, 7–13.

Noisin, E. L. & Thomas, W. E. (1988). Ontogeny of dopaminergic function in the rat midbrain tegmentum, corpus striatum and frontal cortex. *Brain Research*, 469, 241–252.

Oppenheim, R. W. (1975). The role of supraspinal input in embryonic motility: a re-examination in the chick. *Journal of Comparative Neurology*, 160, 37–50.

Rao, P. A., Molinoff, 9P. B. & Joyce, J. N. (1991). Ontogeny of dopamine D1 and D2 receptor subtypes in rat basal ganglia. *Developmental Brain Research*, 60, 161–177.

Richmond, G. & Sachs, B. D. (1978). Grooming in Norway rats: the development and adult expression of a complex motor pattern. *Behaviour*, 75, 82–96.

Robinson, G. A. & Goldberger, M. E. (1985) Interfering with inhibition may improve motor function. *Brain Research*, 346, 400–403.

Robinson, S. R. & Smotherman, W. P. (1992). Motor competition in the prenatal ontogeny of species-typical behaviour. *Animal Behavior*, 44, 89–99.

Schwartz, G. J. & Grill, H. J. (1985). Comparing taste-elicited behaviors in adult and neonatal rats. *Appetite*, 6, 373–386.

Smotherman, W. P. & Robinson, S. R. (1987). Prenatal expression of species-typical action patterns in the rat fetus (*Rattus norvegicus*). *Journal of Comparative Psychology*, 2, 190–196.

Smotherman, W. P. & Robinson, S. R. (1988). Behavior of rat fetuses following chemical or tactile stimulation. *Behavioral Neuroscience*, 102, 24–34.

Smotherman, W. P. & Robinson, S. R. (1990). The prenatal origins of behavioral organization. *Psychological Science*, 2, 97–106.

Steiner, J. E. (1979). Human facial expressions in response to taste and smell stimulation. *Advances in Child Development and Behavior*, 13, 257–295.

Teitelbaum, P. (1971) The encephalization of hunger. In *Progress in*

Physiological Psychology, *Vol.* 4, ed. E. Stellar & J. M. Sprague, pp. 319–350. New York: Academic Press.

Tepper, J. M., Trent, F. & Nakamura, S. (1990). Postnatal development of the electrical activity of rat nigrostriatal dopaminergic neurons. *Developmental Brain Research*, 54, 21–33.

Thelen, E. & Cooke, D. W. (1987). Relationship between newborn stepping and later walking: a new interpretation. *Developmental Medicine and Child Neurology*, 29, 380–393.

Thorndike, E. L. (1911). The law of effect. In *Animal Intelligence*, New York: MacMillian, Reprinted in *Motivation*, ed. D. Bindra & J. Stewart, pp. 184–187. Baltimore: Penguin, 1966.

Watson, J. & Bekoff, A. (1990). A kinematic analysis of hindlimb motility in 9- and 10-day-old chick embryos. *Journal of Neurobiology*, 4, 651–660.

Part three

Development of behaviour systems

8

The ontogeny of social displays: interplay between motor development, development of motivational systems and social experience

TON G. G. GROOTHUIS

Intraspecific communication is of crucial importance for survival and reproduction in the majority of animal species. Much of this communication takes place by means of 'displays', conspicuous, stereotyped and species-specific postures, movements and vocalizations that are specifically adapted to serve as a signal to another member of the species (cf. Tinbergen, 1952). Because of these characteristics, social displays provide interesting material for the study of behavioural development. First, displays are especially suitable for the study of the development of complex stereotyped motor patterns that may be influenced by social experience. Second, the study of display development can provide a better understanding of the ontogeny of social behaviour in general and of the immediate causation of adult social behaviour (Kruijt, 1964). Third, displays are believed to be derived in the course of evolution from intention movements (such as to flee, to attack, to preen) as a result of ritualization and emancipation (Tinbergen, 1952; see below). Because phylogeny is modified ontogeny, changes in ontogeny may reflect changes that have occurred in evolution. Consequently, the study of the ontogeny of displays may also provide insight into the evolution of displays.

Despite these interesting properties of displays, their ontogeny has hardly been analysed quantitatively. A notable exception is the development of bird song. Because of its relation to learning, the discovery of the neural systems involved in song, and the possibility of manipulating feedback by deafening the bird, the study of bird song has become one of the most flourishing fields in ethology. This topic is reviewed by DeVoogd (this volume) and will be dealt with only briefly in this chapter. For an extensive comparison between song development and development of other displays see Groothuis (1993b).

This chapter deals with the relation between motor development,

development of motivation and social experience, and covers the field between development of non-social motor patterns, discussed by Berridge (this volume), and motivational systems, discussed by Hogan (this volume). The main part of this chapter deals with the ontogeny of the form of displays. I will first review studies describing form development of postural and vocal display in different species. The possible mechanisms underlying this development will be discussed next. Finally, the phenomenon of form fixation and changes in context and motivation of display will be discussed briefly.

Display development in normal ontogeny: the descriptive level
Development in frequency of display

The widespread notion that complex display patterns only gradually emerge in young animals is not always true. Young animals are not only incomplete adults, they also show special adaptations to their own needs in each stage of development (Oppenheim, 1981). For example, in several bird species, distinct call types can already be heard from the egg (see below) and these may function in controlling the breeding behaviour of the parents. In humans, young babies are able to control the behaviour of their caretakers substantially by crying or smiling. Young of many species perform begging displays soon after hatching, in order to obtain food from their parents. Some of these displays may disappear from the repertoire later in ontogeny, illustrating that ontogeny deals not only with progression, but with regression as well (i.e. the general problem of continuity in development; see also Berridge, this volume).

In some species, soon after hatching, the young may perform agonistic displays, e.g. in lizards (Roggenbruck & Jenssen, 1986; Stamps, 1978), or in several gull species (Groothuis, 1989a). Where data from the field are available, this behaviour is clearly related to the need of the young to defend their own territory. In the lizard species, hatchlings do not receive parental care while they have to defend food resources. In gulls, which breed in dense colonies, the parents are frequently absent from the territory in order to collect food for their young, leaving the latter to defend the nest site.

However, in species in which the young are able to feed themselves and do not defend territories, but aggregate in groups, social displays often emerge later in life. This has been documented quantitatively in many fish species (salmon, Noakes, 1978; blue gourami, Tooker & Miller, 1980; some cichlid fish species, Wyman & Ward, 1973, and Groothuis & Ros, in preparation; four species of centrarchid fish, Brown & Colgan, 1985) and in several bird species (junglefowl, Kruijt, 1964; turkeys, Schleidt, 1970; gulls, Groothuis, 1989a). In almost all of these cases a gradual increase in frequency of threat and courtship displays is seen, in concordance with an increase of overt aggressive and sexual behaviour. This suggests that development of motivational factors for agonistic and sexual behaviour is involved in the emergence of display.

Development in form of display
Postural displays

In his classic study on the ontogeny of social behaviour in junglefowl, Kruijt (1964) found that male displays develop by differentiation of and/or integration of components that had previously emerged independently. For example, frontal threat seems to develop out of the erect alert posture that chickens may adopt when they accidentally bump against each other during undirected, early hopping. Adult wing flapping may be seen as a stationary modification of the latter. Head zigzagging, performed during frontal threat, can be interpreted as the integration of head shaking (explaining the sideward movement of the head) and ground pecking (explaining the downward component). The sexual display tidbitting can be seen as the integration of head zigzagging, and ground pecking and ground scratching (more or less modified feeding behaviours), and the utterance of a special call that has its origin in the food run-call.

Modification, differentiation and integration have been found to occur in display development of many other animal species. On the basis of similarities in form, Wyman & Ward (1973) concluded that in the cichlid fish *Etroplus maculatus*, all agonistic displays develop from either the 'glance' or the 'micronip', two feeding behaviours. The glance consists of a swimming movement to and away from the body of the parent in a circular path with the body curved, resulting in physical contact between the body walls. This is thought to give rise to the lateral displays (carouselling, lateral display and tailbeating). The authors believe that other displays with a frontal orientation (frontal display, pendulum and mouth fighting), arise from the 'micronip', by which the young approach another fish or an object, nipping it with their mouths.

Similarly, Platz (1974) interpreted the ontogeny of some agonistic and sexual displays in ducks as modifications of an early greeting or appeasement movement performed by young chicks to parent and siblings. By changes in the orientation, the accompanying call, and head and neck movements, greeting seems to give rise to the displays chin-lifting, neck-stretching, sneezing and head-raising. The inciting display of females (a movement in which the neck is stretched sidewards towards a male intruder followed by a more upward movement away from him and in the direction of her partner), seems to emerge as an integration of chin-lifting, facing away, and a threat movement. Platz also noticed the similarity in form of chin-lifting and appeasement displays with pre-flight movements, and of the bobbing display with head movements during locomotion.

Groothuis (1989a) described display development in the black-headed gull quantitatively, by analysing changes in duration, combinations and position of separate elements of the displays such as bill, neck, body, and wing. The results suggest that the adult displays emerge gradually by differentiation of two fear postures, and by integration of these with the alert posture. First, crouching in a hiding place, which is performed silently by chicks when disturbed, changes gradually into the adult 'choking' posture, in part by addition of elements characteristic for aggressive behaviour. In adult choking, the bird stands or crouches openly in front of the intruder, with the body tilted to the ground and the carpal joints bent away from the body (as if the bird is preparing itself to jump on its legs to attack). Furthermore, typical head movements are shown with the bill pointing downward (which may be interpreted as redirected aggressive pecking to the ground) while the characteristic rhythmic choking call is uttered. Second, the erect adult oblique display emerges from an intermediate form of choking, by addition of an erect neck position, walking or running towards the intruder and higher-pitched notes in the call, all elements that chicks perform during overt attack.

Third, the forward display, in which the neck is extended and the head is held in front of the body, seems to arise from the first escape movements, by which chicks run away with the head between the shoulders. Here again, elements seen in overt attack, such as the raising of the carpal joints and the extended neck, become incorporated in the adult display. Finally, the begging display, in which the head is moved up and down in a pumping fashion, can be interpreted as the integration of the alert posture (in which the chick scans the surroundings with the neck extended vertically), and juvenile forms of the forward display. With increasing age, the latter component becomes more pronounced, possibly because the young become increasingly fearful due to the many agonistic interactions they experience in the colony.

Düttmann (1992) measured the development of the adult whistle–shake display in the shell duck in even more detail. By analysing form and duration of the different components, (tail wagging, body shaking, neck, head, and bill movements, and the accompanying vocalization), he showed that the display emerges from the early comfort movement, body shake. In the course of ontogeny, neck and head movements become more conspicuous in horizontal and vertical directions (showing in its forward direction possibly some similarity with the threat display). Somewhat later in ontogeny, the bill-tossing component and the call are added to the vertical movement of the neck, by which this part of the display shows similarity with pre-flight or escape movements.

In human babies, the species-specific display of pointing to something with stretched arm and stretched finger develops gradually (e.g. Butterworth, 1993). First, extension of the index finger is performed during face-to-face interactions at the age of 2 months. The whole hand becomes involved at the age of 6 months. Intentional extension of both arm and index finger together with gazing in the proper direction in a communicative context are first observed in the beginning of the second year. Also in humans, the full-blown social smile is frequently performed after the age of 3 months; before that age, asymmetrical smiles are often performed, with a less sudden onset and end (Papoušek & Papoušek, 1993).

Thus, in the course of ontogeny, the form of many social displays emerges gradually. Some authors report a clear increase in stereotypy (Platz, 1974, and Dane & van der Kloot, 1964, for duck display; Groothuis, 1989a, for gull displays), but others find this to be true for minor details only. Wiley (1973) analysed in detail temporal features of the complex strut display in the sage grouse. One-year-old juveniles displayed in a quicker tempo, but variation in temporal characteristics was not larger than in adults. However, juveniles showed a larger variation in the upwards heaving of the chest sac. Düttmann (1992, personal communication) found an increase in stereotypy only with respect to the incorporation of the bill-tossing component in the whistle–shake display of the shell duck, becoming more smooth with increasing age. The other components, although changing in form, showed a persistent, small amount of variation.

In contrast to these examples of gradual development over a substantial time period, display has been found to be present in the adult form early in ontogeny. One of the classic examples is bill-pecking of young gull chicks, which stimulates regurgitation of food by the parents. However, even in this case, Hailman (1967) found a change in form in the course of the first days after hatching: an increasing accuracy of pecking and the addition of the rotation component of the head. Both aspects only developed completely after experience. Furthermore, Groothuis (1989a) showed that the form of this display changes considerably both in amplitude and neck position after the second week of age. Schulman (1970) reported the occurrence of the adult strutting display in 1-week-old turkey chicks. Unfortunately, no detailed quantitative measurements of the developmental course of this display are available. Stamps (1978) analysed the temporal characteristics of several bobbing displays in the lizard, *Anolis aneus*. All displays, except the courtship display, were found to be present in hatchlings and juveniles. Adult display was somewhat longer in duration, and juveniles showed larger form variation. However, the latter was mainly due to interindividual variation, which stresses the danger of relying on cross-sectional data only. Although the data point to the presence of adult display very early in ontogeny, it is still possible that changes in form do occur during the first days after hatching. In fact, this was found by Roggenbruck & Jenssen (1986) for the development of bobbing displays in *Sceloporus undulatus*. Although hatchlings of this species, as in *Anolus*, perform the displays in the adult form already in the first week after hatching, some changes in duration and in details of the bobbing pattern do occur soon after hatching.

Vocal displays

Distinct call types, such as begging, contentment and distress calls, may be present in young birds even before hatching. However, the variability within each call type is relatively large, and calls may change considerably in form after hatching (ten Thoren & Bergmann, 1987; Groothuis, 1989a). Almost all authors studying non-passerines report a decrease in pitch of the calls with increasing age (Würdinger, 1970; Platz, 1974; Schleidt & Shalter, 1973; Meinert & Bergmann, 1983; ten Thoren & Bergmann, 1987). Some of these authors also report an increase in duration and amplitude of the calls.

As in the case of postures, vocal displays may differentiate from other vocal displays. Würdinger (1970) reports that two different calls in some species of goose, the lament call and the distance call, develop from one and the same precursor, the distress call. Similarly, Groothuis (1989a) argued that all agonistic calls in the black-headed gull emerge from the harsh call, a distress call. The most complex adult call in this species, the long-call, develops as follows (see also Groothuis, 1993b). First, note duration and note interval duration of the distress call become smaller and more stereotyped, while the sound quality becomes increasingly more harsh and bout duration increases (during the first 2 weeks after hatching). Then higher-pitched notes become interspersed with these notes and increase in duration (weeks 3–8). Next, the notes become much clearer and higher in pitch, while the stereotyped temporal structure of the call emerges: a sequence of higher-pitched notes of long duration followed by lower-pitched notes of shorter duration (around week 10). Thereafter, the call becomes much harsher in quality again (around week 40), and such a change is often called 'the breaking of the voice'. This is followed by an increase in amplitude, by which finally the adult long call emerges (around 1 year of age).

Concerning stereotypy, several authors report a relatively high frequency of occurrence of intermediate call forms in the beginning of ontogeny (Meinert & Bergmann, 1983; ten Thoren & Bergmann, 1987; Groothuis, 1989a). In contrast, Schleidt & Shalter (1973) reported a high amount of stereotypy in form and duration of the 'wet-my-lip' call of the European quail early in ontogeny, although the call changed in frequency with age. However, these results were obtained from birds that had received testosterone injections, a point to which I will return. Kroodsma (1984) studied vocal development in the sub-oscine species, the alder and willow flycatcher. Their song consists of relatively simple two-part calls. The basic form of these calls, although with somewhat more noise and smaller amplitude, is already present in the third week after hatching.

The ontogeny of vocal displays in primates has been reviewed recently by Symmes & Biben (1992). It is sufficient to note here that in several species of non-human primates, as in birds, infantile calls may be present early in ontogeny and may disappear from the repertoire afterwards; that many other calls show a gradual change in duration, pitch and sometimes even frequency modulation; that adult calls may differentiate from other calls; and that in some cases pitch variability decreases with age.

In conclusion, most of the postural and vocal displays gradually emerge in ontogeny over a relatively long time span. Form changes may be described as differentiation, modification and integration. In many, but not all, cases an increase in stereotypy occurs. Other displays are present early in ontogeny, but still show changes in form before they emerge completely.

Mechanisms underlying display development

The general idea among behavioural biologists about display development seems to be that, although aspects of the occurrence of motor patterns might be under the influence of experience, the development of motor patterns themselves is not. This is mainly based on the finding that animals, reared in isolation from conspecifics, do not show clear deviations in the form of display when adult. However, detailed quantitative evidence supporting these findings is lacking almost entirely. Furthermore, as we shall see, there is evidence that social experience may affect form development of display. Moreover, results of isolation experiments alone are not very informative. If animals develop aberrant displays in such experiments, this may be caused by several mechanisms, such as the lack of imitation or practising, or the development of aberrant motivational states. If animals reared in isolation show normal display, this result tells us nothing about factors influencing behavioural development. In the different subsections of this section, I will discuss different possible mechanisms underlying aspects of display development and their inter-relationship.

Maturation

Limitations on the motor side or in the central nervous system (CNS) may be the reason why the form of displays is different in young animals, compared to adults. One often concludes in favour of this possibility when experience seems not to be involved in the developmental process. This is, however, negative evidence and therefore not very reliable. Furthermore, experience other than that which has been manipulated could be involved. Moreover, maturation has been shown to be dependent on experience early in ontogeny. Muscle activity may influence differentiation of muscle types, the number of neural connections with them, and the development of joints (see, e.g. Baerends, 1991). Sex steroids (the production of which is dependent on maturation of the gonads) may influence maturation of both the CNS (e.g. DeVoogd, this volume) and the periphery, and the production of these hormones is often influenced by social interactions (see below). In many cases, an explanation in terms of maturation only attributes developmental changes to (often assumed) structural changes but does not explain why these changes take place as they do.

Postural displays

Except for form deviations very early in ontogeny, such as the case of bill-pecking in gull chicks and head-bobbing in hatchlings of the lizard (see above), limitations on the motor side of young animals do not seem to be a likely explanation for the form of juvenile postural display. For example, in many cases of display development in birds, the different positions of bill, head, neck, body, and legs, characteristic of the complete adult display, are often performed early in ontogeny during other activities such as locomotion or feeding (Groothuis, 1993a). In humans it has been shown that all but one of the many different action patterns that constitute facial expressions are already present in the neonate (Oster, 1978).

Vocal displays

Maturation of the periphery may well explain aspects of vocal development. Würdinger (1970) concluded, based on both correlational and experimental evidence, that during ontogeny in geese: (1) the pitch of calls decreases by the increase in size and thickness of the syringeal membranes; (2) increase in amplitude, duration or bout-length of calls is related to increase in the pressure of the air sac (*Saccus clavicularis*); (3) changes in pitch and amplitude of the calls are related to growth in length and size of the trachea. Platz (1974) suggested that the decrease in variability of calls in ontogeny is due to a decrease in elasticity of the syrinx. Finally, Groothuis & Meeuwissen (1992) found that testosterone treatment in gull chicks resulted in enlargement of syringeal structures and suggested that this may be one of the causes of enhanced call development in these birds. Decrease in pitch, and increase in amplitude and duration of calls have been related to physical growth in the literature on primate call development as well (see Symmes & Biben, 1992).

However, both in primates (see Symmes & Biben, 1992) and in birds (ten Thoren & Bergmann, 1987), juvenile calls have been observed to occur, although rarely, in adult animals again. Although maturational effects of sex hormones may be reversible when hormone production is low, these observations do restrict the explanatory value of maturation of peripheral structures for the development of vocal displays. Maturation of central mechanisms will be discussed below.

Imitation

One of the ways in which social experience may influence display development is by providing models of display to copy. Most authors agree that imitation can be seen as a two-step process, although especially the involvement of the second step can be questioned (Groothuis, 1993b). First, information about the form of display, as performed by the other animal, is acquired. This process may be stimulated by social interactions, as has been found in the case of song development (see Clayton, this volume). Next, this information is translated into motor output. The translation is likely to take place on the basis of some sort of motor matching in which the form of the motor output is compared with and adjusted to the acquired information. The frequency of motor matching depends, of course, on the frequency with which the display patterns are performed, and this in turn is influenced by motivational factors and social stimulation.

Postural displays

In postural development, imitation seems unlikely to take place. Most animals cannot see their own visual output. Therefore they have to compare visual input, seeing the display of the tutor, with proprioceptive output, the feedback from their own motor performance. Such cross-modal comparison may require considerable cognitive abilities that may not be present in many animal species. Indeed, in a recent review, Whiten & Ham (1992) came to the conclusion that there is no conclusive evidence for the occurrence of postural imitation in animal species, other than chimpanzees and humans. As far as humans are concerned, imitation of facial expression has been claimed to occur in babies less than 3 weeks old (Meltzoff & Moore, 1977; Field *et al.*, 1982). These babies perform several facial expressions, e.g. tongue protrusion, mouth opening (or a surprised expression), lip protrusion (disgust), and smiling, at greater than chance level when these expressions were performed by a tutor. However, it is possible that these expressions are not the result of imitation, but are already developed, fixed action patterns that are triggered by specific stimuli, such as similar facial expressions of a conspecific (cf. Eibl-Eibesfeldt, 1984). However, the possibility of bimodal perceptual integration in young infants has been indicated by other research. MacKain *et al.* (1982) showed that prelinguistic human infants recognise structural correspondence between auditory aspects of speech and the accompanying mouth movements of the tutor.

Experiments to test the influence of imitation on the development of postural display have nearly always been done by rearing animals in social isolation, thereby depriving them of the possibility to imitate. Many authors report normal displays in such animals when adult (see, e.g. Huntingford & Turner, 1987). Likewise, normal facial expressions representing the basic emotions, smiling, crying and disgust, are present in children born blind (e.g. Eibl-Eibesfeldt, 1984; Papoušek & Papoušek, 1993). However, detailed results are often lacking. Furthermore, although animals may be able to develop normal display in isolation, imitation may occur when models to copy are present. Thus, Groothuis (1992) reared black-headed gull chicks (almost) from hatching on, individually with foster parents or peers of a related species, either the common gull or the little gull. None of these black-headed gulls copied elements of the postural displays of the other species. However, the black-headed gulls were not acoustically isolated from each other and in this species there is an intimate connection between development of calls and postures. Consequently, hearing conspecifics may have guided the development of postural displays. Nonetheless, the results indicate that imitation is not likely to be part of the mechanism underlying development of postural display in birds. The occurrence of local dialects in smiling in humans (Fridlund & Seaford, 1976, cited in Fridlund, Ekman & Oster, 1987) suggests that imitation does occur in development of display in humans.

Vocal displays

The complication of cross-modal comparison is not present in imitation of vocal display. Here, an animal may compare auditory input (hearing the tutor) with auditory output (hearing its own call production). Vocal imitation has indeed been shown to occur in many species of songbirds, hummingbirds, and parrots. However, there is no conclusive evidence for the occurrence of vocal imitation in display development in non-passerines. Adult chickens (Konishi, 1963), turkeys (Schleidt, 1964), and doves (Nottebohm & Nottebohm, 1971) show normal species-specific calls when deafened early in life. But, here again, detailed data are lacking and these are necessary because imitation may shape only details of calls. There is a detailed analysis for call development in the male sub-oscine eastern phoebe (a new world flycatcher), deafened early in life (Kroodsma & Konishi, 1991). These birds did show normal calls when adult. The authors relate the lack of imitation in this species to the absence of some forebrain nuclei present in the brain of songbirds – and also lacking in a non-passerine, the ring dove (Cheng, 1992). Finally, a deafening experiment was carried out on a 5-day-old squirrel monkey which did not result in deviating call development (Winter *et al.*, 1973).

An even better way to investigate the influence of imitation is to rear young animals with tutors performing slightly different calls. This was, in a way, done by Marler, Kreith & Willis (1961). They reared young cockerels in mixed groups of three strains in two separate rooms and implanted them with pellets of testosterone. They found no consistent differences in seven characteristics of the calls between the groups of the two rooms. However, the effect of testosterone, as well as the very large interindividual differences, could have masked effects of vocal learning. Kroodsma (1984) reared some species of the new world flycatcher with tutor tapes containing the calls of a related species or with conspecific song that was slightly modified artificially (Kroodsma, 1989). He did not find indications of imitation. However, as is well known from the literature on song learning (see Ten Cate, Clayton, this volume), the lack of live tutoring may have masked the ability to imitate. To overcome this, Groothuis (1993b) reared young black-headed gull chicks with two related gull species (see above). He found one gull performing long-calls that were different from the normal adult long-call of its own species, but rather similar to the long-call of the foster species, the little gull. Similarly, Matsataka & Fujita (1989) found in cross-fostering experiments with Japanese and rhesus macaques that the yearling's food-calls show a resemblance to that of the foster mother.

Conclusion

Imitation of postural displays is not likely to occur, based on both theoretical considerations and experimental results. There is ample evidence that vocal displays other than those of songbirds, hummingbirds and parrots, can develop normally, at least with regard to gross co-ordination, without the opportunity to imitate. Interestingly, there is also some support for the idea that imitation of such vocal displays does occur if models to copy are provided together with the opportunity to interact with the tutor.

Social interactions

If opponents react consistently differently to different forms of displays, these reactions may guide form development of displays on the basis of operant conditioning. The influence of interactions with conspecifics on the development of the form of displays has been made very likely for song development in a song bird species, the cow bird (West & King, 1988). Here, the female performs a wing-stroke when the male is singing her preferred song. This stroke shows some similarity with her copulation posture and seems to shape song development in the male.

Radesäter (1974) found that goslings often meet each other frontally before the turning-away component becomes integrated in the greeting display. This new orientation inhibits aggression of the opponent and this may reinforce the development of facing-away. In line with his suggestion, he showed that in one of the two species he studied, goslings separated by wire from other goslings, and therefore not experiencing aggressive reactions, did not develop the head-flagging component.

Although rare, there have been reports of abnormal display in animals reared with abnormal interactions. De Lannoy (1967) found abnormal display in some ducks reared in small groups. Blurton Jones (1968) found an aberrant display in a hand-reared great tit. Groothuis (1992) found stereotyped deviations in form of postural and vocal display and in their combinations in gulls reared in small groups. Hopkins & Savage-Rumbaugh (1986, cited in Symmes & Biben, 1992) reported abnormal vocalizations in a pygmy chimpanzee reared with humans. These findings show that display development shows plasticity, and it seems very likely that these abnormalities are all due to abnormal social interactions. The abnormalities that Groothuis reported were not similar in form among birds of the same group, indicating that they were not due to imitation. Because gulls reared in isolation did not show such deviations, abnormal social reactions in the abnormally small groups must have been the primary cause. This is in line with the finding that the isolates showed deviations in their begging calls only. These calls were performed to their caretakers in the only interactions they had. Groothuis (1992) suggested that the abnormal calls and threat postures developed through a process of operant conditioning. In the unusually small groups of gulls, birds must have been very familiar with each other, and therefore no pressure might have been present to develop the most pronounced form of display, the adult one, showing the highest tendency of the bird to attack.

The fact that abnormal forms of display have been found to develop in abnormally small groups seems to contradict the finding that animals reared in social isolation may develop normal display. Two explanations are possible: (1) social experience is not indispensable for normal display development, but it guides this development as a safeguard; (2) abnormal experience distorts normal development. In either case the results point to an important aspect of behavioural development: more than one mechanism may influence the same developmental process or even lead to the same behavioural outcome.

Motor matching

Animals reared in isolation that perform normal adult display must have developed this ability by mechanisms other than imitation or guidance by social reactions. One possibility is that these animals did so by matching their motor output against some sort of template containing information about the species-specific form of display. (Although it remains un-answered how in this case the template is acquired by the animal.) There are at least four ways to investigate this hypothesis. First, since motor matching would involve adjusting motor output by trial and error by comparing the output with a template, we would expect development from more or less random form variation to a more stereotyped form. Second, one may block the feedback resulting from the performance of motor patterns upon which motor matching must be based. Although this is relatively easy to do in the case of vocal development by deafening the animal, it is difficult in the case of postural development. However, animals may be reared in very small cages, or the adequate stimulus for display to occur may be withheld. Third, one may try to enhance development by stimulating the practising of displays, e.g. by social stimulation. Finally, one may look for the performance of complete adult display before the animal has had the opportunity to practise its motor output.

Postural displays

Groothuis (1992) did find form retardation in isolation-reared gulls that had hardly practised their display. However, this retardation was correlated with retardation of testosterone production, suggesting that not motor matching, but sex hormones influence display development. Furthermore, results of normally reared gulls indicate that display development is not so much an increase in co-ordination of the different form elements, as well as an increase in likelihood for these elements to occur simultaneously: if an element occurred during display in early ontogeny, it already did so in the proper combination with other elements. Moreover, when suddenly confronted with a stuffed adult black-headed gull, the gull chicks sometimes performed, although rarely, the complete adult displays (Groothuis, 1989a).

The latter finding has been reported for other species as well. Klopman (1961, cited in Andrew, 1966) observed the adult greeting display in 3-day-old Canada goslings when they returned to their brood after a period of separation. Similarly, Kruijt (1964) observed aggressive interactions, including frontal threat and irrelevant head movements, in 1-week-old junglefowl chicks when suddenly confronted with other chicks after a week of separation. Likewise, Groothuis & Ros (in preparation) found that young cichlid fish only 4 weeks of age will perform adult display when confronted with each other after 2 days of social isolation. Andrew (1966) could induce tidbitting (and copulation) in young chicks by special hand movements of the experimenter. Schulman (1970) saw strutting display in turkey chicks and Groothuis (unpublished observations) observed the adult crow posture in young fowl chicks when they were suddenly confronted with a stimulus object. These findings indicate that the complete adult display is already present early in ontogeny. Consequently, display development is not so much a change in co-ordination of the different form elements, but a lowering of the threshold for these elements to occur. This may explain why early in ontogeny, only a relatively strong stimulus will trigger the adult display. In the course of ontogeny these thresholds lower, and consequently the frequency of complete adult display increases.

In the case of courtship and threat displays, this change in threshold is under the influence of sex hormones. For example, early treatment with testosterone induced the precocious performance of adult-like strutting in turkey chicks (Schein & Hale, 1960), and of the adult crowing posture in domestic fowl chicks (Andrew, 1966). Theoretically, in these cases testosterone might have enhanced motor development by stimulation of motor matching. Apart from the examples of precocious display in untreated young, mentioned above, this has been made unlikely by Groothuis & Meeuwissen (1992). They provided full-grown (to anticipate maturational effects) juvenile gulls with pellets of testosterone subcutaneously, and measured changes in the frequency and form of their displays in the course of 2 weeks after the start of treatment. These birds had been reared in social deprivation, and observations confirmed that they had not practised display before the start of hormonal treatment. Nonetheless, many birds consistently performed complete adult display shortly after this treatment. Moreover, some birds showed no juvenile display at all during the observation periods. Furthermore, even young chicks were able to perform complete adult display when treated with testosterone, without the opportunity to practise the display patterns.

Vocal displays

Testosterone enhances the expression of the adult vocalization in domestic chicks (Andrew, 1969) and black-headed gulls (Groothuis & Meeuwissen, 1992). In the latter case this involves a change in calls towards longer duration, a decrease in pitch, and the emergence of the adult temporal structure. However, although highly speeded up, this development took longer (1 week or more) than in the case of postural displays, which might be due to the fact that maturation of the syrinx may be important for calls to develop (see above). Even in the case of song development in songbirds, where learning plays such an important role, testosterone has been found to induce precociously species-specific adult features (e.g. Heid, Güttinger & Pröve, 1985; cf. Groothuis, 1993b). Furthermore, the results of the deafening experiments discussed above (by which feedback, resulting from motor matching, was blocked) also indicate that the development of at least the cross-co-ordination of vocal displays in non-songbird species does not depend on motor matching.

Conclusion

Motor matching does not seem very important for display development. The ontogeny of display, at least the development after birth or hatching in non-altricial species, is not so much a development in co-ordination of the motor patterns themselves, but a gradual change in expression of a neural co-ordination mechanism that is present early in ontogeny. This development takes place by changes in internal states, affecting the expression of the frequency and form of species-specific displays.

Motivation

The behaviour of an animal is controlled by internal factors in interaction with external factors. Furthermore, different, but often functionally related, motor patterns share the same internal causal, or motivational, factors. The entity of these factors together with the group of motor patterns they control, and from which they may receive feedback, are referred to as a 'behaviour system' (see Baerends, 1976; Hogan, 1988, Chapter 1, this volume). These systems not only 'drive' the animal to perform certain behaviour, but switch behaviour on and off, in relation to specific external or internal factors. Different behaviour systems may be activated simultaneously. In the majority of cases, the system that is activated most strongly will be expressed in the final motor output. However, in other cases, more than one system may be expressed in one and the same motor pattern, resulting in ambivalent or conflict behaviour (see Hinde, 1970, for a review).

Displays are thought to be derived in the course of evolution from various intention movements such as those of the locomotion, comfort, escape, or aggression systems. Such intention movements may occur because the complete expression of one behaviour system is blocked by the activation of another. Therefore, many displays would be derived from conflict between competing systems such as aggression and escape. This was formulated explicitly in the 'conflict hypothesis' (Tinbergen, 1952; for a review see Baerends, 1975; see also Baerends, this volume). By a process called 'ritualization', the motor patterns changed gradually into a conspicuous display that served a communication function. Concurrently, by a process called 'emancipation' the display changed in its underlying causation; that is, the display became more or less independent of the original causal factors.

Evidence in support of these ideas about the evolution of displays comes in part from a comparison of the form of display behaviour with motor patterns from which they are throught to be derived, between and within species. However, this type of evidence is not always convincing, and additional evidence has been marshalled from the analysis of the immediate causation of displays, deduced from the analysis of the context of display and the temporal sequences of these motor patterns with attack and escape behaviour. This type of analysis has evoked considerable debate on functional grounds, mainly from game model theorists who have suggested that an animal should not reveal information about how likely it is to attack during agonistic interactions (see Groothuis, 1989b, for a discussion). However, recent game theoretical models have shown that under natural conditions, which are more complex than the conditions originally assumed, animals could reveal their tendencies to attack and to flee, and this has been supported by detailed analysis of real animal conflicts (for a brief review, see Wilson, 1992).

Postural displays

It has often been suggested that threat and courtship displays develop in concordance with the development of agonistic and/or sexual behaviour. Kruijt (1964) explicitly addressed the question of how the development of behavioural systems for escape, attack and sex is related to display development. On the basis of analysis of form, context and temporal structure of the behaviour of junglefowl in various stages of ontogeny, he came to the conclusion that: (1) these systems develop, or come to expression, in that order; (2) the displays, including courtship display, develop under the influence of an increasing interaction between the systems for escape and aggression; (3) the sex system influences the occurrence of courtship display via its stabilizing influence on the interaction between the two agonistic systems.

Groothuis (1989b) undertook a similar approach. He found crouching (hiding), withdrawal (escape), aggression and sexual behaviour to emerge in this order. Initially, erratic transitions between overt aggression and escape occurred frequently, but became increasingly interspersed with sudden breaks in the movement. These breaks became longer in duration and were accompanied by display behaviour. This indicates that the behavioural systems for aggression and fear become increasingly activated simultaneously, leading to more 'balanced' behaviour and the performance of (more complete) displays.

These ideas about causation of display development have been tested in three ways. The first method is indirect: by rearing animals in social isolation, which is known to result in development of erratic agonistic behaviour. Kruijt (1964), Tooker & Miller (1980) and Groothuis & Ros (in preparation) found erratic agonistic behaviour and the performance of display in a somewhat 'sloppy' fashion (e.g. short duration, erratic transitions) in, respectively, junglefowl, blue gourami fish and cichlid fish. The last authors also observed aberrant display in some fish showing erratic transitions between attack and escape when tested with a con-specific. This display appeared to be a mixture of the most aggressive display, frontal gill display, and the most fearful display, the inferior display: it shows the erect gill covers and open mouth of the former, but the folded fins and lifted tail of the latter. Black-headed gulls reared in isolation showed, when tested with a conspecific, an extremely large variation in overt aggression which was correlated with form and frequency of the three agonistic displays (Groothuis & van Mulekom, 1991). Thus in animals reared in isolation, aberrant agonistic behaviour correlates to a certain extent with form of agonistic display.

The second method consisted of manipulation of external factors (Groothuis & Ros, unpublished data). Young hand-reared gull chicks performed the begging display when the caretaker entered their cage, but performed choking display when confronted with a stuffed adult con-specific. Confronting the chicks with both stimuli simultaneously reliably induced a mixture of both displays in which the speed of the bobbing movement was slowed down and the downward phase of it ended in the choking display. This confirmed the difference and to some extent the nature of the causation of these displays, and the idea that activation of different behavioural systems may come to expression as ambivalent behaviour in one and the same motor pattern.

The third method concerned hormonal treatment. Groothuis & Meeu-wissen (1992) tested two specific predictions on the basis of the causal analysis of the form development of gull display. The first concerns the relation between aggression and display development. Testosterone is known to induce aggressive behaviour in many animal species. If, in the young gulls, display and aggression are causally related, we would therefore expect that early treatment with testosterone: (1) increases display frequency and induces adult-like display in proportion to the increase of aggression; (2) increases the frequency of the three agonistic displays only (choking, oblique and forward), but not the frequency of the begging display, because it was postulated that the latter is not under the control of aggression. The second prediction concerns the relation between the choking display and the tendency to stay put. In young gulls before fledgling, this tendency is relatively high and as a consequence chicks perform predominantly choking-like display. After fledging, the tendency to defend the territory and thereby to stay put decreases considerably. Consequently, in chicks testosterone would mainly induce choking. In fledged juveniles testosterone would mainly induce the oblique and forward display. These expectations were confirmed.

Discussion

The neural co-ordinating mechanisms for adult display are present early in ontogeny. In this respect, displays are no exception, because this has been found for many other motor patterns as well, such as for walking, flying, grooming, and feeding, both in birds and mammals, including humans, as well as for some features of song in songbirds (Groothuis 1993a, 1993b; see also Berridge, this volume). The expression of these neural mechanisms or central pattern generators changes in the course of ontogeny, due to changes in motivational factors. These changes are influenced by sex hormones, (other) age-dependent factors yet unknown, and social experience. These factors may lower the threshold for the different form elements of display to occur. They may also raise this threshold, as in the case of juvenile displays that gradually disappear from the repertoire. In this case, the co-ordination mechanisms for these motor patterns are still present later in ontogeny, as is indicated by the finding that incidently infantile behaviours may suddenly return in the repertoire. Delius (1973) and Groothuis (1989a) reported crouching in adult gulls whose wings were clipped and could therefore not escape. The latter author also reported the performance of the juvenile begging display in an adult gull.

Thresholds for display to occur may also rise due to habituation to conspecifics. This may explain why young of several species perform display precociously after a period of social isolation or during the presentation of a new object, as was discussed earlier.

The idea that changes in the expression of display behaviour are not due to changes in the relevant motor mechanisms themselves, but in factors regulating the activation of these mechanisms (including differences in receptor systems for, and metabolism of, sex hormones), may even be applicable to explain behavioural differences between sexes and species. There is growing evidence that, at least in monomorphic species, both sexes may perform, although rarely, motor patterns typical for the other sex. Similarly, Lorenz (1941) found in ducks a hybrid male to show display never seen in its parents species and typical for another species of ducks.

This suggests that the display has been present in these parent species, but with too high a threshold to occur, and that this threshold was lowered in the hybrid. How the motor mechanisms for display themselves develop might be studied in altricial species, showing a slower development than most of the species discussed so far.

The interpretation of the gradual development of display as ritualization in ontogeny (see above) is in fact not correct. Ritualization was originally coined for the process by which, in evolution, new motor patterns developed. What we often see in ontogeny is the gradual expression of motor systems, already present early in ontogeny, not the formation of new patterns. However, the finding that animals may develop aberrant display (see above) indicates that formation of new motor mechanisms is still possible in the course of ontogeny, and this provides interesting opportunities for further studies on ritualization of communication behaviour.

How testosterone affects the behavioural systems controlling display behaviour is not yet clear. Although it may affect the motor mechanisms themselves, for example at the level of the spinal cord, it certainly does more than that: it changes the sensitivity for and interpretation of specific external stimuli, such as conspecifics, and the co-ordination of displays with each other and with other behaviours such as aggression.

Finally, although the incomplete juvenile forms of display may be interpreted as merely epiphenomena of a gradual rise in activation of the proper motivational state, one has to realize that these incomplete forms may still fulfil important functions for the young animal. They may signal the infantile status of the animal; they may be part of a special strategy to solve aggressive interactions with stronger adults (Groothuis, 1989a); they may be a tool for the young to gain experience in agonistic interactions; and they may bring the young into a situation which stimulates the production of testosterone, either by social stimulation (Groothuis, 1992), or by behavioural self-feedback (Cheng, 1992).

Form fixation and emancipation of displays

In many species, displays emerge gradually in form, and may show a relatively large form variation in the young, while in the adult animal the complete form, including its individual characteristics, is stable. This form fixation creates reliable species- and individual-specific communication signals. Although form fixation or crystallization is an important item in the study of bird song development, it has hardly received attention in other ontogenetic studies (but see Hogan, 1988). Groothuis (1993b) suggested that form fixation and change in causation of the displays occurred in his gulls on the basis of the following findings. (1) Although some form variation still exists in adult display, juvenile forms of display are rarely performed by the adult. (2) The stereotyped form deviations of display, developed by birds reared in small groups, were still performed by these birds after they were replaced for months in large groups of unknown conspecifics. Thus these deviations persisted outside the social situation in which they developed originally. (3) Social isolation affected form development of display more severely when it occurred before the emergence of complete display than when it occurred thereafter. Thus, once emerged completely, displays are relatively resistant to effects of isolation. (4) Once emerged under the influence of testosterone, displays are expressed more easily after a second treatment with testosterone and may also become (temporarily) independent of this hormone. This latter finding indicates that display may become independent of the original causal factor, as has been postulated to have occurred in the evolution of displays.

Form fixation may in part be caused by irreversible growth of the motor apparatus. It may also be caused by a stabilization of the motivational state underlying display performance. However, the primary cause of form fixation seems to be a change in causation of displays. This is clearly the case in song learning, where blocking of auditory feedback and lesions of the brain areas X and MAN only severely affect song quality before, but not after, crystallization (Bottjer, Miesner & Arnold, 1985; see DeVoogd, Bolhuis, this volume). One has the impression that after form fixation the motor patterns, both song and other displays, become a behavioural routine, easy to recall in different situations.

The factor inducing form fixation is not yet clear. It is not age because gulls reared in isolation and retarded in their behavioural development were still able to develop stereotyped display after their replacement in large groups when adult. Nor is it dependent on whether the gull has reached the normal species-specific form, because aberrant displays became stable too. Three other factors may induce form fixation. First, a sufficiently high production of testosterone may induce structural changes in the nervous system. Interestingly, testosterone has been shown to induce song crystallization in a number of songbirds (Marler et al., 1988; see also Groothuis, 1993b). Second, repeated performance of the same motor pattern (possibly as an indirect effect of testosterone) may sensitize neural networks controlling these patterns. Such a mechanism has been proposed for the fixation of stereotypies in the behaviour of farm animals (for a review see Mason, 1991). These stereotypies show interesting similarities with display development. Like display, they develop gradually within certain species-specific limits, from intention movements, as the result of long-lasting motivational conflicts, after which these patterns become 'routine' and emancipate from the original causal factors. Third, form fixation and emancipation may be under the influence of social experience, if an animal learns to use its display as an instrument, more or less independently of its exact motivational state.

Changes in the context in which display occurs

It is generally thought that the social behaviour of young animals is less predictable and more chaotic than that of adults. Although hardly quantified in this respect, play in young mammals is a classical example of this idea. Furthermore, in several species of fish, birds and mammals, including humans, it has been documented that young animals solve their agonistic interactions predominantly with overt aggression. Later in life, these interactions become more complex, longer lasting, and more balanced, with a decrease in overt aggression and escape and an increase in display performance. Moreover, adults may use the display in other contexts than young. Such changes in the context of social behaviour are discussed here.

It has often been suggested that these changes in social behaviour depend on social experience. Indeed, as Kruijt (1964) was one of the first to show, animals reared in isolation often show agonistic behaviour, characteristic of inexperienced young ones. Isolates may show excessive aggression or fear, erratic transitions between both, lack of adequate reaction to the displays performed by the opponent, and courtship displays directed to males instead of females. This has now been quantified in detail for several species (e.g. Tooker & Miller, 1980, for the blue gourami; Groothuis & Ros (in preparation) for a cichlid fish; Groothuis & van Mulekom, 1991, for the black-headed gull; see also Huntingford & Turner, 1987, pp. 211–216). Groothuis & van Mulekom suggested that both classical and operant conditioning of the displays are involved in the learning process. The former may take place because the animal starts to associate agonistic interactions or opponents with harmful attacks, leading to an increase of fear and consequently the inhibition of overt aggression.

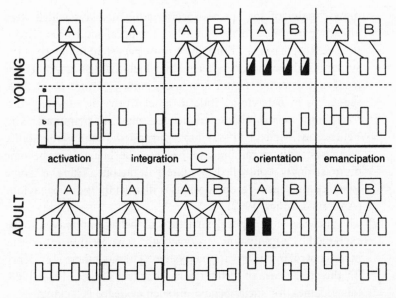

Figure 8.1 Alternative models for the ontogenetic changes in context and temporal structure of display behaviour, in terms of changes in motivational control of display. Large squares represent motivational systems; small rectangles represent motor patterns. Upper part: situation in young animals; lower part: situation in adult animals. Above the dotted line: postulated motivational control; below the dotted line: corresponding observed temporal structure in behaviour; linked rectangles represent a clear structure, unlinked rectangles depict a more or less random sequence. Black and white rectangles: motor patterns occurring in two stimulus situations.

This creates the motivational state for display behaviour, which may act as an operant during instrumental conditioning on the basis of the reactions of opponents to that display.

All these changes in the context of displays are, of course, reflected in changes in the temporal sequencing of these displays with each other and with other motor patterns such as those involved in overt aggression or copulation. Such changes must be due to changes in motivational organization of social behaviour, and I conclude this chapter with a discussion of five, not mutually exclusive, possibilities that are depicted in Figure 8.1.

(1) An increase in *activation* of the motivational system, controlling the motor patterns, may lead to two changes. (a) A lowering of the threshold for the different motor patterns to occur, and thereby an increase in predictable sequences of these motor patterns. Indeed, treatment of young cichlid fish with testosterone has shown to

induce many of the species-specific adult displays which were
already performed in the adult-like temporal sequences (Groothuis & Ros, in preparation). (b) A more persistent influence of the
system on these motor patterns, by which the distractability for
irrelevant external stimuli decreases, inducing more predictable
sequences in behaviour. Indeed, a relatively high level of distractability seems characteristic for play in mammals, and
Fentress & Mcleod (1987) have suggested that distractability
decreases with increasing age because the behaviours become
more strongly controlled by internal factors. There is some
evidence that testosterone decreases distractability in behaviour
(Andrew & Jones, 1992).

(2) The increase in temporal structure of behaviour may also be
explained by a gradual *integration* of the separate motor patterns
in the control by a motivational system. Although there is evidence
for this in the case of feeding behaviour (see Hogan, 1984, 1988,
this volume), for social behaviour such evidence is lacking.

(3) An *integration* may also occur at a *higher level*. Kruijt (1964)
proposed that a system controlling sexual behaviour (square C in
Figure 8.1) 'stabilizes' the interaction between the systems for
aggression and escape (A and B). By this he could explain why the
overt expression of the two lower order systems, attack and
escape, decreased in frequency. These systems increasingly inhibit
each other's overt expression, leading to an increase of conflict
behaviour, the displays, in the course of ontogeny. Furthermore,
the involvement of the sexual system in this way explains why
agonistic display, controlled by the two lower order systems, starts
to occur in a sexual context, courtship, in older animals. However,
the hypothesis postulates that even courtship displays are under
the control of the two agonistic systems as well. This seems
unlikely, at least for display in experienced adults of several
species. Courtship displays are very frequently performed without
any sign of activated aggression.

(4) The *context* of a display may change because the orientation to
external stimuli changes, while the motivational background stays
the same. A shift in occurrence of display from a clearly agonistic
context to a sexual one has been found in the waltzing display of
junglefowl (Kruijt, 1964) and the quivering display of cichlid fish
(Groothuis & Ros, in preparation). On the basis of experimental
evidence, the latter authors suggest that young male fish mix their

agonistic and sexual behaviour because they are not yet able to recognise the proper sex. Only after experience with both sexes do they direct sexual behaviour to females, and aggressive behaviour mainly to males. Likewise, in black-headed gulls, specific external stimuli, such as the performance of specific displays by a specific opponent, become increasingly important in the regulation of display (Groothuis, 1989b).

(5) Finally, behaviour may change in context and temporal structure because it *emancipates* from its original motivational system and becomes hooked up to another system. There is increasing evidence for this in the literature. Groothuis (1989b) found that whereas in young gulls the oblique, choking and forward displays are performed exclusively in a clearly agonistic context, these displays are often performed in other contexts in older birds, such as during courtship, or when calling their mate or young. Hulscher-Emeis (1992) studied the colour patterns of a cichlid fish species, which may change rapidly during agonistic interactions in which the fish perform several threat and courtship displays. These colour patterns indicate different motivational states. Hulscher-Emeis found that the intensity of these colour patterns decreased in experienced fish, although the fish still performed the same displays. This suggests that the motivational states controlling the displays become activated less in the course of ontogeny and that the displays become (partially) independent of these. Similarly, in the literature on the ontogeny of human signal behaviour, a change in emotional states underlying the motor patterns is more than once explicitly reported. For example, both the social smile and the expression of disgust are thought to be coupled to specific emotions early in life. However, already at the end of the first year, babies may use these patterns instrumentally in a wide array of social contexts, even in a deceitful way (Eibl-Eibesfeldt, 1984; Fridlund et al., 1987).

Acknowledgements

I thank Gerard Baerends, Johan van Rhijn, Johan Bolhuis and Jerry Hogan for comments on the manuscript.

References

Andrew, R. J. (1966). Precocious adult behaviour in the young chick. *Animal Behaviour*, 14, 485–500.

Andrew, R. J. (1969). The effect of testosterone on avian vocalizations. In *Bird Vocalizations*, ed. R. A. Hinde, pp. 97–131. Cambridge: Cambridge University Press.

Andrew, R. J. & Jones, R. B. (1992). Increased distractability in capons: an adult parallel to androgen-induced effects in the domestic chick. *Behavioural Processes*, 26, 201–210.

Baerends, G. P. (1975). An evaluation of the conflict hypothesis as an explanatory principle for the evolution of displays. In *Function and Evolution of Behaviour*, ed. G. P. Baerends, C. Beer & A. Manning, pp. 187–227. Oxford: Oxford University Press.

Baerends, G. P. (1976). The functional organisation of behaviour. *Animal Behaviour*, 24, 726–738.

Baerends, G. P. (1991). On spontaneity in behaviour, the model (fixed) action pattern and play. *Netherlands Journal of Zoology*, 40, 565–584.

Blurton Jones, N. E. (1968). Observations and experiments on causation of threat displays of the great tit (*Parus major*). *Animal Behaviour Monographs*, 1, 75–158.

Bottjer, S. W. , Miesner, E. A. & Arnold, A. P. (1985) Forebrain lesions disrupt development but not maintenance of song in passerine birds. *Science*, 224, 901–903.

Brown, J. A. & Colgan, P. W. (1985). The ontogeny of social behaviour in four species of centrarchid fish. *Behaviour*, 92 , 254–276.

Butterworth, G. & Franco, F. (1993). Motor development: communication and cognition. In *Motor Development in Early and Later Childhood: Longitudinal Approaches*, ed. A. F. Kalverboer, B. Hopkins & R. H. Geuze, pp. 153–165. Cambridge: Cambridge University Press.

Cheng, M.-F. (1992). For whom does the female dove coo? A case for the role of vocal self-stimulation. *Animal Behaviour*, 43, 1035–1044.

Dane, B. & van der Kloot, W. G. (1964). An analysis of the display of the goldeneye duck (*Bucephala clangula*). *Behaviour*, 14, 265–281.

de Lannoy, J. (1967). Zur prägung von Instinkthandlungen. *Zeitschrift für Tierpsychologie*, 30, 162–200.

Delius, J. D. (1973). Agonistic behaviour in juvenile gulls: a neuroethological study. *Animal Behaviour*, 21, 236–246.

Düttmann, H. (1992). *Veränderungen im Individual- und Sozialverhalten während der Jugendentwicklung der Brandente (Tadorna tadorna)*. PhD thesis, Universität Osnabrück.

Eibl-Eibesfeldt, I. (1984). *Die Biologie des Menschlichen Verhaltens*. München: Piper.

Fentress, J. C. & Mcleod, P. J. (1987). Motor patterns in development. In *Handbook of Neurobiology*, Vol. 8, ed. E. M. Blass, pp. 38–97. New York: Plenum Press.

Field, T. M., Woodson, R., Greenberg, R. & Cohen, D. (1982). Discrimination and imitation of facial expressions by neonates. *Science*, 218, 179–181.

Fridlund, A. J., Ekman, P. & Oster, H. (1987). Facial expressions of emotions: review of the literature, 1970–1983, In *Nonverbal Behavior and*

Communication, ed. A. Siegman & S. Feldstein, pp. 143–223. Hillsdale, NJ: Lawrence Erlbaum.

Groothuis, T. (1989a). On the ontogeny of display behaviour in the black-headed gull: I. The gradual emergence of the adult forms. *Behaviour*, 109, 76–124.

Groothuis, T. (1989b). On the ontogeny of display behaviour in the black-headed gull: II. Causal links in the development of aggression, fear and display behaviour: emancipation reconsidered. *Behaviour*, 110, 161–204.

Groothuis, T. (1992). The influence of social experience on the development and fixation of the form of displays in the black-headed gull. *Animal Behaviour*, 43, 1–14.

Groothuis, T. G. G. (1993a). The ontogeny of social displays: form development, form fixation and change in context. *Advances in the Study of Behavior*, 22, 269–322.

Groothuis, T. G. G. (1993b). Song development in songbirds and display development in other species: different mechanisms or similar principles? *Netherlands Journal of Zoology*, 43, 172–192.

Groothuis, T. & Meeuwissen, G. (1992). The influence of testosterone on the development and fixation of the form of displays in two age classes of young black-headed gulls. *Animal Behaviour*, 43, 189–208.

Groothuis, T. & van Mulekom, L. (1991). The influence of social experience on the ontogenetic change in the relation between aggression, fear and display behaviour in black-headed gulls. *Animal Behaviour*, 42, 873–881.

Hailman, J. P. (1967). The ontogeny of an instinct. *Behaviour* Supplement, 15, 1–159.

Heid, P., Güttinger, H. R. & Pröve, E. (1985). The influence of castration and testosterone replacement on the song architecture of canaries (*Serinus canaria*). *Zeitschrift für Tierpsychologie*, 69, 224–236.

Hinde, R. A. (1970). *Animal Behaviour: A Synthesis of Ethology and Comparative Psychology*. New York: McGraw-Hill.

Hogan, J. A. (1984). Pecking and feeding in chicks. *Learning and Motivation*, 15, 360–376.

Hogan, J. A. (1988). Cause and function in the development of behavior systems. In *Handbook of Behavioral Neurobiology*, Vol. 9, ed. E. M. Blass, pp. 63–106. New York: Plenum Press.

Hulscher-Emeis, T. M. (1992). The variable colour patterns of *Tilapia zilii* (*Cichlidae*): integrating ethology, chromatofore regulation and the physiology of stress. *Netherlands Journal of Zoology*, 42, 1–36.

Huntingford, F. & Turner, A. (1987). *Animal Conflict*. London: Chapman & Hall.

Konishi, M. (1963). The role of auditory feedback in the vocal behaviour of the domestic fowl. *Zeitschrift für Tierpsychologie*, 20, 349–367.

Kroodsma, D. E. (1984). Songs of the alder flycatcher (*Empidonaxalnorum*) and willow flycatcher (*Empidonax trailii*) are innate. *Auk*, 101, 13–24.

Kroodsma, D. E. (1989). Male eastern phoebes (*Sayornis phoebe*; *tyrannidae*, *passeriformes*) fail to imitate songs. *Journal of Comparative Psychology*, 103, 227–232.

Kroodsma, D. E. & Konishi, M. (1991). A suboscine bird (eastern phoebe, *Sayornis phoebe*) develops normal song without auditory feedback. *Animal Behaviour*, 42, 477–487.

Kruijt, J. P. (1964). Ontogeny of social behaviour in Burmese red junglefowl (*Gallus gallus spadiceus*). *Behaviour* Supplement 12, 1–201.

Lorenz, K. Z. (1941). Vergleichende Bewegungsstudien an Anatiden. *Journal für Ornithologie*, 89, 194–293.

MacKain, K., Studdert-Kennedy, M., Spieker, S. & Stern, D. (1982). The bimodal perception of speech in infancy. *Science*, 219, 1138–1148.

Marler, P., Kreith, M. & Willis, E. (1961). An analysis of testosterone-induced crowing in young domestic cockerels. *Animal Behaviour*, 10, 48–54.

Marler, P., Peters, S., Ball, G. F., Dufty, A. M. & Wingfield, J. C. (1988). The role of sex steroids in the acquisition and production of birdsong. *Nature*, 336, 770–772.

Mason, G. (1991). Stereotypies: a critical review. *Animal Behaviour*, 41, 1015–1037.

Matsataka, N. & Fujita, K. (1989). Vocal learning of Japanese and rhesus monkeys. *Behaviour*, 109, 191–199.

Meinert, U. & Bergmann, H. (1983). Jugendentwicklung der Lautäuserungen beim Birkhun, *Tetrix tetrix*. *Behaviour*, 85, 242–260.

Meltzoff, A. N. & Moore, M. K. (1977). Imitation of facial and manual gestures by human neonates. *Science*, 198, 75–78.

Noakes, D. L. G. (1978). Ontogeny of behaviour in fishes: a survey and suggestions. In *The Development of Behaviour*, ed. G. M. Burghardt & M. Bekoff, pp. 103–125. New York: Garland.

Nottebohm, F. & Nottebohm, M. E. (1971). Vocalizations and breeding behaviour of surgically deafened ring doves (*Steptopelia risoria*). *Animal Behaviour*, 29, 313–327.

Oppenheim, R. W. (1981). Ontogenetic adaptations and retrogressive processes in the development of the nervous system and behaviour: a neuroembryological perspective. In *Maturation and Development: Biological and Psychological Perspective*, ed. K. J. Conolly & H. F. R. Prechtl, pp. 73–109. Philadelphia: Lippincott.

Oster, H. (1978). Facial expressions and affect development. In *The Development of Affect*, ed. M. Lewis & L. Rosenblum, pp. 43–76. New York: Plenum Press.

Papoušek, H. & Papoušek, M. (1993). Early interactional signalling: the role of facial movements. In *Motor Development in Early and Later Childhood: Longitudinal Approaches*, ed. A. F. Kalverboer, B. Hopkins & R. H. Geuze, pp. 136–152. Cambridge: Cambridge University Press.

Platz, F. (1974). Untersuchungen über der Ontogenie der Ausdruckbewegungen und Lautäuserungen der Kolbenenten (*Netta rufina pallas*) mit einem Beitrag zur Anatomie des Stimmapparatus. *Zeitschrift fur Tierpsychologie*, 36, 293–428.

Radesäter, T. (1974). On the ontogeny of the orienting response in the triumph ceremony in two species of geese (*Anser anser L.* and *Branta canadensis*) *Behaviour*, 50, 1–15.

Roggenbruck, M. E. & Jenssen, T. A. (1986). The ontogeny of display behaviour in *Sceloporus undulatus* (*Sauria, Inguanidae*). *Ethology*, 71, 153–165.

Schein, M. W. & Hale, E. B. (1960). The effect of early social experience on male sexual behaviour of androgen injected turkeys. *Animal Behaviour*, 7, 189–200.

Schleidt, W. M. (1964). Uber the spontaneität von Erbkoordinationen. *Zeitschrift für Tierpsychologie*, 21, 235–256.

Schleidt, W. M. (1970). Precocial sexual behaviour in turkeys. *Animal Behaviour*, 18, 760–761.

Schleidt, W. M. & Shalter, M. D. (1973). Stereotypy of a fixed action pattern during ontogeny in *Coturnix coturnix coturnix*. *Zeitschrift für Tierpsychologie*, 33, 35–37.

Schulman, A. H. (1970). Precocial sexual behaviour in imprinted male turkeys. *Animal Behaviour*, 18, 758–759.

Stamps, J. A. (1978). A field study of the ontogeny of social behaviour in the lizard *Anolis aeneus*. *Behaviour*, 66, 1–32.

Symmes, D. & Biben, M. (1992). Vocal development in nonhuman primates. In *Nonverbal Vocal Communication: Comparative & Developmental Approaches*, ed. H. Papoesek, U. Jurgens & M. Papoesek, pp. 123–140. Cambridge: Cambridge University Press.

ten Thoren, B. & Bergmann, H. (1987). Veränderung und Konstanz von merkmalen in der jugendlichen Stimmentwicklung der Nonnengans (*Branta leucopsis*). *Behaviour*, 100, 61–91.

Tinbergen, N. (1952). 'Derived activities': their causation, biological significance, origin and emancipation during evolution. *Quarterly Review of Biology*, 27, 1–37.

Tooker, C. P. & Miller, R. J. (1980). The ontogeny of agonistic behaviour in the blue gourami, *Trichogaster trichopterus* (*pisces, anabantoidei*). *Animal Behaviour*, 28, 973–988.

West, M. J. & King, A. P. (1988). Female visual display affects the development of male song in the cowbird. *Nature*, 334, 244–246.

Whiten, A. & Ham, R. (1992). On the nature and evolution of imitation in the animal kingdom: Reappraisal of a century of research. *Advances in the Study of Behavior*, 21, 239–284.

Wiley, R. H. (1973). The strut display of male sage grouse: a fixed action pattern. *Behaviour*, 47, 129–152.

Wilson, J. D. (1992). A functional analysis of the agonistic display of great tits, (Parus major). *Behaviour*, 121, 168–214

Winter, P., Handley, P., Ploog, D. & Schott, D. (1973). Ontogeny of squirrel monkey calls under normal conditions and under acoustic isolation. *Behaviour*, 47, 230–239.

Würdinger, I. (1970). Erzeugung, Ontogenie, und Funktion der Lautäuserungen bei vier Gänse-Arten (*Anser indicus*, *A. caerulescens*, *A. albifrons*, und *Branta canadensis*). *Zeitschrift für Tierpsychologie*, 27, 257–302.

Wyman, R. L. & Ward, J. A. (1973). The development of behaviour in the cichlid fish *Etroplus maculatus* (Bloch). *Zeitschrift für Tierpsychologie*, 33, 461–491.

9

Psychobiology of the early mother–young relationship

ALISON S. FLEMING AND ELLIOTT M. BLASS

This chapter is concerned with the development of interactions between mammalian mothers and their infants, especially rat and human mothers and their young. We first consider the hormonal and experiential factors that determine the mother's earliest responses to her newborn. We then focus on the infant and the events that control and guide its changing responses to its early social and physical environment. Throughout, we emphasize the ways in which the behavior of the mother and infant are complementary and mutually dependent, and are based on experiences acquired during the interaction.

As an example of complementarity between mother and infant, consider maternal retrieving behavior. When leaving the nest to retrieve her young, the female moves slowly in a low body position along a seemingly random path until she reaches her hidden pup. At that time, she lifts the pup by the nape of the neck, turns directly to the nest, and with head held high, she returns directly to the nest in a canter-like motion keeping the pup well above the substrate. For their part, the pups become limp and offer no resistance to movement when picked up by the scruff of the neck by dams (or by sharp, tooth-like tweezers by experimenters, Brewster & Leon, 1980). Lifting pups by other parts of the body causes them to twist and squirm. On such occasions that we have seen mothers hastily retrieve pups by parts other than the nape of the neck, the pups were dragged along the substrate. Thus, complementarity exists in the apparent service of preventing damage during retrieval and transport.

The mother's perspective

Mother rats are maternally responsive to their newborn pups as they emerge from the birth canal. At parturition, dams strip the amniotic sac, eat the placenta and clean their pups. Within 30 minutes after parturition, pups are retrieved to the nest site, and gathered in a huddle. Mothers mouth and lick them and adopt a crouched nursing posture over them. This scenario occurs immediately even in primiparous females. Virgin animals, by contrast, are not maternally responsive when first presented with newborn foster pups. In fact, they initially move away from the pups and actively avoid them. However, within 1 to 2 days of continuous daily pup stimulation, virgins habituate to and lie down with pups; after 5 to 10 days of continuous contact with foster pups, virgins eventually respond maternally to them. This process is known as pup-induction (see Rosenblatt, 1967; Fleming & Luebke, 1981).

Unlike the apparent homogeneity of rat maternal behavior at parturition, there are large individual and cultural differences in human maternal feelings and behavior. Although many women experience heightened maternal feelings at about 20 weeks of gestation, the time of first fetal movements, and again at the parturition, for others, feelings of nurturance are not experienced until after birth. In primipara, these feelings often develop gradually over the first few weeks postpartum; in fact, mothers often admit retrospectively that their first feelings of attachment did not occur until the first eye-to-eye contact or first infant smiles. There is also considerable variability in the quality of maternal behavior shown by new mothers. Skin-to-skin infant contact often enhances maternal stroking and touching, but many mothers do not even touch their infants on first presentation (for review, see Corter & Fleming, 1990).

Effects of parturitional hormones

Maternal responsiveness

What accounts for the substantial differences in maternal responsiveness between newly parturient and nulliparous female rats? We can first eliminate a number of factors, including mating experience, physical changes associated with gestation, and parturition itself. The hormonal changes that precede parturition, however, are critical and we discuss them now.

Hormones associated with late pregnancy and parturition account for the rapid onset of maternal responsiveness in rat dams at parturition. Gradual changes in responsiveness during late pregnancy may be mediated by increased concentrations of circulating placental lactogens, possibly in combination with elevated densities of hypothalamic estradiol receptors. The immediate expression of maternal behavior at parturition is activated by high titres of estradiol, prolactin and, possibly, oxytocin against a backdrop of declining progesterone (see Bridges, 1990; Insel, 1990; Rosenblatt, 1990). The neurochemical systems through which these hormones act are not known, although likely candidates include the catecholamines, particularly norepinephrine, and the opioids. These neurotransmitters undergo substantial change at the time of parturition, especially within maternally relevant medial preoptic area (MPOA) structures, and are influenced by the parturitional hormones. Noradrenergic antagonists block the expression of maternal behavior in some mammals, while opioid antagonists facilitate it (see Heritage, Grant & Stumpf, 1977; Bridges, 1990; Keverne & Kendrick, 1990).

The role of hormones in facilitating maternal responsiveness and in accounting for its large variability in women has only recently been explored. During pregnancy, many of the same hormones change as in rats; however, the large shift in the ratio of estradiol to progesterone occurring at the end of pregnancy in the rat is not as dramatic in women. Although maternal feelings rise at the time of parturition, we found no significant relation between any of the steroid and protein hormones or ratio of plasma hormones and any maternal attitude at any point during pregnancy or the postpartum period. There is, however, a relationship between postpartum maternal behavior and the adrenal hormone, cortisol, but the hormonal effects interact with psychological factors such as maternal attitudes (see Fleming & Corter, 1988, for review).

Emotional behavior and affect

Although changing hormonal profiles facilitate maternal behavior both directly and indirectly, different hormones and neurochemicals have different behavioral effects and any one hormone or neurochemical probably exerts multiple effects. Our task is to sort out this interconnecting web of behavioral–hormonal interdependencies.

Progesterone and estradiol may facilitate maternal behavior onset in rat dams by reducing the incidence of competing non-maternal behaviors. Specifically, these hormones induce an attraction to pup odors, reduce fear of the pups, and facilitate learning about pup characteristics, possibly by augmenting pups' reinforcing value.

Relieving fear through hormonal change unmasks maternal propensities which continue even after the period of hormonal priming. Evidence for this conceptualization is broadly based. First, at the time of birth, females are less timid and neophobic than when virgin: parturient females approach an unfamiliar intruder, and enter and explore novel environments more readily. Further, the regimen of progesterone and estradiol that facilitates maternal behavior in virgin rats also reduces the animals' timidity in an open-field. Finally, manipulations that reduce the animals' timidity or anxiety, such as benzodiazepines, lesions of the amygdala, or early handling, also facilitate maternal responding, further suggesting that reduced timidity contributes to elevated maternal responsiveness (see Mayer, 1983; Hansen, Ferreira & Selart, 1985; Fleming, 1987; Fleming & Corter, 1988).

Human mothers seem to undergo mood fluctuations and intensifications during the early postpartum period, reflected in postpartum 'euphorias', 'blues' and 'lability'. Although it is widely assumed that puerperal depression is hormonally mediated, in fact the evidence is mixed and to date no single hormone, neurotransmitter, or combination thereof, has been consistently implicated in its etiology. Instead, a variety of situational and experiential factors may well influence postpartum mood, most notably prior experience with children: women with more childcare experience tend to be less depressed postpartum (see Fleming & Corter, 1988; Fleming, 1990; Steiner, 1992).

Regardless of the etiology of mood, affective state strongly influences the quality of a mother's interactions with her offspring. At birth the happier mothers show more instrumental responding; at 1 and 3 months postpartum they respond more contingently to their infants (e.g. vocalizing when their infant vocalizes) and are more affectionate altogether (Fleming, Ruble, Flett & Shaul, 1988). The long-term, adaptive consequences of heightened affection and sensitivity are not known either from the perspectives of physical growth and development, emotional stability and independence, or what the infant learns from its mother. Cultural factors have to be assessed in this complex matrix as well.

Olfactory responses

Parturitional hormones in rats also influence responsiveness to pup-related cues, especially odors. For instance, new mothers with no direct experience with pups prefer bedding from the nest of a new mother and her pups to material taken from a diestrous female's nest or clean material, whereas virgins show no such preference. Moreover, virgins treated with progesterone and estradiol also exhibit a preference for pup-related odors. Experience with those odors can facilitate maternal responding. Adult virgins show more rapid maternal inductions after pre-exposure to pup odors and vocalizations during early development, and pre-exposure to distal pup cues (primarily odors and vocalizations) in adulthood also facilitates maternal responding during induction tests, at least among females whose baseline responsiveness is high to begin with. Finally, virgins rendered anosmic or hyposmic do not avoid donor pups. They exhibit rapid onset of maternal behavior, suggesting that pups' odors in the context of other pup cues are aversive (Bauer, 1983; Fleming & Rosenblatt, 1974a, 1974b; Fleming, Vaccarino, Chee & Tambosso, 1979; Orpen & Fleming, 1987).

Neuroanatomical evidence supports these behavioral effects of hormones. We know, for instance, that the elevated hedonic and affective responses experienced by parturient females are probably mediated by olfactory–limbic pathways known to contain estradiol receptors and mediate olfactory and emotional behaviors within other functional contexts (Fleming, 1987). Interestingly, these pathways project onto the MPOA, a nucleus whose activation is essential for the full expression of maternal behavior (Numan, 1988).

Although human mothers can recognize their offspring based on their odors (Schaal & Porter, 1991), the extent of olfactory influence on emotional state and maternal responsiveness has only recently been addressed. In a recent study, groups of day 2 postpartum mothers, 1 month postpartum mothers, and female and male non-parent controls were asked to rate the pleasantness of a variety of infant- and non-infant related odorants, using magnitude estimation procedures, (Fleming *et al.*, 1993). The results show that new mothers give higher hedonic ratings to infant t-shirts than do non-mothers, while not differing in their responses to other stimuli. However, new mothers were also more extreme in their ratings of the infant t-shirts, which suggests the importance of early postpartum experience. Mothers who gave positive ratings experienced shorter postpartum intervals to their first extended contact and nursing of their infants and heightened maternal responsiveness, measured both behaviorally and by self-report.

Thus, new mothers are more attracted to the general body odors of infants. This attraction varies with early postpartum contact and experiences. Whether postpartum hormones contribute to this effect is not known. Cortisol may facilitate mothers' recognition of their infants' odors (Fleming, unpublished) and, in general, may sensitize women at parturition to their infants and the circumstances of maternal care. According to this formulation within the context of good health, a benign pregnancy and parturition, and a positive developmental history, this sensitization generates elevated positive nurturant attitudes. Negative valences of any or all of these factors will push maternal attitudes and dispositions in the other direction. The interactions among all of these factors and their ensuing tensions, habituations and dishabituations predict considerable affective fluctuation that is exaggerated by high cortisol titers.

Effects of maternal experience

Hormonal effects, although powerful, are relatively short lived. Within 4 to 5 days after parturition, the continued expression of maternal behavior until weaning seems to be sustained by experiences acquired during the period of hormonal priming and beyond. If rat dams are prevented from interacting with their pups at birth, then maternal responsiveness declines over the next 3–5 days, dropping to virgin levels by day 10. However, mothers that interact with pups within 24 to 36 hours of delivery for as little as 30 minutes, remain maternal even after 10 days of separation (Orpen & Fleming, 1987). The retention of maternal behavior is related to the length of the early exposure period (Cohen & Bridges, 1981; Fleming & Sarker, 1990). We ask here what aspects of experience are important for the continued expression of maternal behavior.

Behavioral changes postpartum

The initial interactions between mothers and their newborn infants are intense and sustained. After the nest has been constructed and the pups retrieved into it, dams actively mouth and lick their pups, especially their anogenital regions. They lick male pups more often than they lick female pups, apparently in response to testosterone (Moore, 1990). This behavior induces urination and defecation by the newborns, which lack these eliminatory reflexes, and help maintain the dam's fluid balance, another instance of complementarity. As will be shown below, this licking is vital to engaging the infant's central nervous system too. The dam then gathers her infants underneath her ventrum, permitting pups to gain access to her teats and suckle. Once suckling begins, the dam usually adopts a high arch crouch posture over them and becomes immobile for a period. In fact, both mother and pups are in slow wave sleep during milk exchange. The arched back allows pups room to shift from nipple to nipple after extracting milk from a nipple. This maximizes milk gain and may also facilitate the next milk letdown (Stern, 1989, 1990).

During their early interactions with pups, dams can receive tactile, chemosensory and thermal stimulation, each of which contributes to behavioral maintenance. Maternal behavior in females that were prevented from crouching over their young during the postpartum period declined before weaning completion, even though they still received distal stimulation (Jakubowski & Terkel, 1986; Stern, 1983). Moreover, Orpen & Fleming (1987) found that females separated from their litters by a wire mesh floor during the 1-hour postpartum exposure phase, showed no long-term benefit of limited maternal experience that lacked tactile or proximal chemosensory stimulations. They responded to pups on day 10 as did virgins. Reciprocally, pups gently placed in contact with the dam are very inactive without the vigorous stimulation that she normally provides. Ventral stimulation is probably essential to the maternal experience effect. The additional findings that dams need to receive somatosensory perioral input to exhibit normal maternal licking and crouching (Stern, 1989), and that licking during exposure is correlated with responsiveness during test (Morgan, Fleming & Stern, 1992), point to the importance of chemosensory and perioral stimulation (i.e. different classes of contact and motor pattern expression).

A role for somatosensory input was demonstrated by Morgan *et al.* (1992), who found that either somatosensory perioral or trunk stimulation at birth can produce long-term changes. When both sources of stimulation are precluded (by perioral anesthetization or the application of a ventral–trunk spandex jacket), the remaining modalities, i.e. distal ones that do not elicit contact, are not sufficient for maintaining maternal state. Dams also learn about specific olfactory and chemosensory features of the pups during interactions. Pups scented with an artificial odorant during the exposure phase are responded to more rapidly by the familiarized dam tested 10 days later. This effect is dependent on the association of the odors with pups. Pre-exposure to the odor alone, in the absence of pups, does not produce the phenomenon (Malenfant, Barry & Fleming, 1991). Learned recognition or maternal 'imprinting' to infant odors is even more striking in species which develop a selective and specific attachment to a single offspring (as opposed to litter recognition), as seen for ewes with their lambs (Poindron & Levy, 1990).

Moreover, maternal response quality to pup-associated scents at day 10 reflects the quality of her interaction with pups during the initial pup-odor pairings. Mothers that had interacted proximally with pups during the pairing, by sniffing and licking them, developed a strong long-term preference for that scent over a novel scent. Mothers that were not interactive during the pairing, however, did not develop a preference for the paired scent. It seems, therefore, that early, active olfactory contact mutually labels mother and infants, inducing response selectivity. The odors, therefore, allow more proximal aspects of mother and infant to gain control over suckling and caretaking behaviors for infant (see below, p. 230) and mother respectively. Whether this strong olfactory presence reflects the specific interaction with young or whether other interactions, e.g. with food, would suffice, is not yet known.

Little is known in human mothers about which aspects of early mother–infant interactions influence subsequent responsiveness. Mothers receive a complex of stimuli when they hold, cradle, nurse, look at and talk to their infants. They soon come to recognize their infants individually, based on particular infant features. For instance, new mothers when 'rooming-in' discriminate their own infants' cries from those of other infants; indeed, they selectively awaken to their own infant's cries (Formby, 1967). Mothers are also able to recognize their infants based on the tactile characteristics of the dorsal surface of their hands, an effect that occurs after only a few hours of experience of interacting with the infant (Kaitz, Lapidot, Bronner & Eidelman, 1992). Finally, new mothers can discriminate their own infants' soiled t-shirts from those of same-aged infants, requiring very little interaction with their infants to do so (Schaal & Porter, 1991). Thus, during their early interactions, mothers are becoming increasingly familiar with their infants, a process which may well contribute to the development of attachment. Reciprocally, newborns can learn about tactile stimulation (Blass, Ganchrow & Steiner, 1984) and sounds that predict feeding.

The human early postpartum period may constitute a time of heightened sensitivity when new mothers can most easily feel nurturant and become attached to their infants (Klaus & Kennell, 1982). According to this sensitivity hypothesis, if interactions are prevented during this time, maternal responsiveness should decline and the process of attachment become delayed. Conversely, for mothers who have contact with their infants during this sensitive period, subsequent responsiveness and attachment should be strengthened (Goldberg, 1983; Fleming, Steiner & Anderson, 1987). How responsiveness and attachment are temporally related to each other, and whether they constitute the same or different processes are not known.

Short-term experience effects on early maternal responding have been suggested by studies that vary the timing of mother–infant contact during the first postpartum days. Grossman, Thane & Grossman (1981) observed mothers who received 30 minutes of skin-to-skin contact during the first postpartum hour, extended rooming-in 1 day later, both conditions, or the hospital routine of limited contact of nursing at regular intervals. Mother–infant interactions were observed periodically over the first 10 days. The findings suggested that additional early contact may facilitate maternal responsiveness in first-time mothers, although the benefits are very short-lived and occur only in some women. Group differences did not persist beyond the first few days, as mothers in all conditions gained additional experience with their babies. Moreover, the effect was found only for mothers who had desired pregnancy in the first place. Given the structure, support systems and expectations of contemporary Western society which foists homogeneity in child care, biologically significant differences, if they exist, may be obscured.

The long-term effects of additional postpartum contact suggested by Klaus & Kennell (1982) have proved hard to substantiate and replicate (Goldberg, 1983). Nevertheless, according to Fleming *et al.* (1988), mothers who experienced more contact with their infants during the first day also tended to show higher levels of affectionate behavior towards their infants at 3 months postpartum, an effect that may reflect higher levels of maternal motivation.

Hormonal influences

We know that the maternal experience effects are not mediated hormonally in rats, since most females segregated after brief contact have resumed their estrous cycles by day 10. The influence of experience with infants is mediated by many of the same mechanisms known to mediate other types of learning/memory (Davis & Squire, 1984). The consolidation of the experience is apparently dependent on the production of structural proteins (Malenfant *et al.*, 1991) and, in part, on the activation of the noradrenergic system (Moffat, Suh & Fleming, 1993). Which other neurotransmitter systems are important and what brain structures mediate the maternal experience effect are not yet known.

Although hormones do not mediate the effects of maternal experience, it is clear that experiences acquired during the postpartum period are affected by hormones. This can be seen in subsequent pregnancies or maternal reinduction. For example, on reinduction tests 30 days after the initial exposure, animals that had received a 24-hour postpartum maternal experience had shorter latencies than did maternal virgins receiving the same 24 hours of interactive experience. The relatively augmented responsiveness was heightened further when the reinduction test occurred under hormonal priming, during a subsequent pregnancy. These effects can be mimicked in virgins by the administration of a regimen of the maternal hormones, progesterone and estradiol. In this case, exposure to the hormones in the absence of a maternal experience does not facilitate subsequent responsiveness during reinduction tests, even if testing occurs under hormonal priming. However, females that experienced maternal hormones and were behaviorally active during exposure were most responsive at test (Fleming & Sarker, 1990). Thus, experiences acquired under hormones are more easily activated by hormones than by exposure only to the relevant pup stimuli. In this instance the hormones are either further sensitizing the neural systems mediating maternal behavior, or are providing additional concordance with the female's previous state that had been associated with or permitted maternal behavior. There may be an additivity of experiential and hormonal afferents that overcome the dam's fear of pups to reveal the structure of maternal behavior.

General versus specific effects

Is sensitivity to newborns unique in the postpartum female, or is it a manifestation of general sensitivity? Studies from Fleming's laboratory have investigated this issue (Fleming, Kuchera, Lee & Winocur, in press; Fleming, Korsmit & Deller, unpublished results). When compared to virgin animals, postpartum animals show only marginally better performance on both a socially mediated food preference task and on a social recognition task. Yet, after responding maternally to scented pups that still share some maternal odors, dams develop a strong preference for the exposed odor; maternal virgins, similarly exposed and responding similarly, do not. Moreover, in a conditioned place preference task using two distinctive environments, postpartum animals prefer the distinctive environment previously paired with pups over an alternate environment, not so paired. Most nonmaternal virgins either show no preference or actively avoid the pup-paired box. Once maternal, virgins begin to develop a preference for the pup-paired box but this is not as robust as in postpartum dams. Finally, hormones mediate these parity differences: virgins primed with 'maternal' hormones strongly prefer the pup-paired box, whether or not they had prior maternal experience, although experience strengthens that preference. Thus, both maternal experience and the parturitional hormones can heighten the hedonic or reinforcing properties of pups, effects which we believe may be mediated by the same dopaminergic mesolimbic systems that mediate reinforcement within other motivational contexts.

Human mothers are adept at recognizing their own infant odors within hours of birth (Schaal & Porter, 1991). We have recently asked how mothers compare to non-mothers in recognizing a set of infant- and non-infant-related odors to which they have been pre-exposed. In one study (Fleming *et al.*, submitted), groups of day 2 postpartum mothers and female and male non-parent controls were presented with a target stimulus and then asked to identify the target from among three similar stimuli. All groups recognized all the stimuli at better than chance performance (at all the temporal intervals used) and groups did not differ. In the pre-exposure paradigm, mothers did no better at recognizing their own infants' odors than the odors of unfamiliar infants. While few variables were associated with recognition of the unfamiliar infant T-shirt, a number of factors were associated with own infant odor recognition. For instance, in comparison to consistently incorrect or inconsistent responses, mothers who identified their own infants' odors on all three trials experienced earlier and longer contact with their infants after birth, spent more time in close proximal contact with their infants during interactions, and experienced more positive maternal feelings and attitudes. Consistent with the findings of Kaitz & Eidelman (1992), mothers seem no better than non-parents at recognizing either infant odors or other odors to which they have been pre-exposed, indicating that new mothers are not especially primed to acquire information about newborns. Yet early postpartum experience may contribute to mothers' specifically recognizing their own infants' odors. Infant recognition is related to other measures of maternal responsiveness. The direction of causality cannot be specified now.

Conclusions: rat and human maternal behavior

Taken together, these findings show that both rat and human mothers undergo a change in affective state during the early postpartum period which may be hormonally mediated and which can influence the early mother–infant interactions. In the case of a rat dam whose nest environment is homogeneous and who is bred for high yield, postpartum affective state tends to be positive, showing little variability. It should be noted, however, that perturbations to the environment associated with Caesarian section or artificially induced inclement weather conditions often result in cannibalism or litter abandonment. Human mothers, on the other hand, are highly variable on a multitude of environmental and background factors and experience a wide range of emotional states postpartum, although women who exhibit a more positive mood state tend also to exhibit more positive interactions with their infants.

Both rat and human mothers also develop an attraction to odors associated with young. In neither case do we have a very good idea as to what the relevant odors may be, although excrement and urine products seem not to be involved. In the case of humans, newborn infants possess a high concentration of dermal eccrine and sebaceous glands that secrete sweat and sebum. Both types of secretions constitute rich sources for bacteria which rapidly colonize the neonatal skin surface (see Schaal & Porter, 1991). Mothers may find odors attractive that are phenotypically similar to their own, either because of genetic factors or the transfer of odors between mothers and their offspring. With respect to olfactorily based kin recognition, support for each of these mechanisms exists.

Although rat mothers show a preference for infant-related odors in the absence of explicit experience with those odors, it is also clear that with pup experience the preference is stronger and more enduring. Human mothers, in contrast, probably require some experience with the infant for an attraction to infant odors to develop. Heightened attraction was associated with both earlier postpartum contact and more time interacting proximally with the infant. While it is tempting to conclude that mother's attraction to infant odors promotes more rapid maternal attachment, as we suggest may be the case in rats, our data are purely correlational and do not address the issue of causality.

We may speculate at this point that the olfactory bond functions to sustain maternal interest in her fairly unexpressive infant until eye contact and smiling start to appear regularly in the infant at about 4 weeks of age, at which time vision becomes the favored emotional modality. As will be shown below, 4 weeks of age is an important turning point for human infant affect as well.

The infant's perspective

The young must reduce energy expenditure, solicit maternal care and learn that certain signals presented by mother and siblings predict enhanced energy gain, reduced energy loss, or both. We will demonstrate that the various commodities provided by rat and human mothers to their young, through contact and milk delivery, can each serve as unconditioned stimuli that allow infants to learn about their mother and through her about the environment that they will encounter when they leave the nest. Motivational, sensorimotor and regulatory controls over suckling and feeding are available to infants independent of suckling or ingestive experiences (Hall, 1990). Experience, however, allows the infants to select what it is going to ingest and with whom it will interact.

Mothers accurately predict their behavior to their infants. Rodent, carnivore and ungulate mothers invariably establish contact with their young by exciting them through anogenital licking. This excitation has been demonstrated to be highly rewarding to infants. It also elicits nipple search because the mothers, after stimulation, quickly make themselves available to their young for suckling.

In important ways the infant's tasks are complementary to the mother's. Infants must respond appropriately to the mother in order to receive milk, warmth, and a variety of other experiences necessary for growth and adaptive adult behavior. In all mammals that have been studied, nipple rooting reflexes are well developed at birth: infant rats and kittens, for example, vigorously crawl across the mother's ventrum making broad head-scanning movements until nipple contact is established. At that point, the motion changes dramatically. The rear legs no longer propel the infant forward, but are locked in place. Rats change the direction and form of head movement into short, jabbing, semicircular actions to erect and swallow the nipple (MacFarlane, Pedersen, Cornell, & Blass, 1983). The mother settles over the litter and, within 4–10 minutes for rats or instantaneously (generally) for humans and the great apes, milk is let down. This triggers another activity cascade in the infant as it withdraws the available milk from the nipple and swallows it. Infant rats, after milk withdrawal, rapidly abandon the nipple and shift to another one, often one that is milk replete from the letdown (Cramer, Blass & Hall, 1980; Cramer & Blass, 1983). This sequence of milk availability, shifting from nipple to nipple, or in humans from one breast to the next by the mother during a feed, punctuates the otherwise quiescent interactions with a flurry of activity due to postural adjustment, vestibular stimulation, etc.

Eventually, the bout terminates, often in rats because of thermal stress experienced by the mother (Leon, Croskerry & Smith, 1978). Usually it is the mother who breaks ventral contact by either depositing the infants in the nest, absenting herself, or shifting the baby back to the 'papoose' position in humans. The frequency of this sequence varies enormously among species, occurring only once or twice daily in rabbits and some deer, and frequently in rats (Grota & Ader, 1969; Ader & Grota, 1970) and in primates, including humans.

These early interactions have important consequences for the infant in addition to providing nutrition. In rats, handling and anogenital licking cause the pups to urinate and defecate, behaviors otherwise absent until about 10 days of age. Licking also elicits the same elaborate motor patterns as infusions of milk into the mouths of deprived pups (Hall, 1979; Johanson & Hall, 1979), or electrical stimulation of the medial forebrain bundle (Moran, Schwartz & Blass, 1983), which is also the 'reinforcement' pathway. These patterns demonstrate that complex integrated behaviors such as feeding, stretch (as during milk withdrawal), and the female sexual response of lordosis are already encoded in the newborn nervous system (see also Smotherman & Robinson, 1988; Berridge, this volume). For males, licking by the mother further masculinizes them, enhancing sexual differentiation (Moore & Morelli, 1979).

In most mammals, rooting activity improves contact with the mother and, through conduction, infant temperature is brought into equilibrium with maternal temperature. Rooting also provides the infant with tactile and olfactory information and, when its eyes are open, visual information about the particularities of its own mother. Finally, through these movements on the mother's ventrum, the infants may learn something about the consequences of their actions concerning locomotion in space. Maternal contact *per se* is extraordinarily rewarding to rats who, equally, prefer an odor associated with contact alone to one associated with suckling and milk letdown (Alberts & May, 1983). Harlow's (1963) famous studies of rhesus infants preferring a warm furred model that never produced milk to a wire model that did, makes the same point.

Milk letdown – growth and energy conservation

Milk provides obvious nutrients, water, minerals, and immune protection. Through release of endogenous opioids, it also seems to cause energy conservation. As compared to infants receiving water, newborn human infants who received 0.1 ml sucrose once/minute for 5 minutes, showed a cessation of crying and a marked reduction in activity levels and heart rate (from 165 to 130 beats per minute). Both sucrose and milk may cause the release of endogenous opioids by their taste. Sucrose diminishes pain responsivity in humans and rats (Blass & Hoffmeyer, 1991; Blass & Fitzgerald, 1988). All of these taste-induced changes seem to be opioid mediated and all may result in reduced energy expenditure.

Taken together, each episode described here affects proximal infant state in at least four ways through active interactions with mother and siblings: (1) the nervous system is excited; (2) nutrients are gained; (3) energy loss is reduced; and (4) the infant is calmed. These ends are achieved through both general (heat transfer through conduction) and focal stimulation (anogenital licking) by the mother and general and specific patterns of activity by the infants (e.g. locomotion and scanning across the mother's ventrum versus specific probing into the nipple). We maintain that throughout this early developmental period, indeed even during the prenatal period, rat and human fetuses sample their air and aquatic surrounds within the contexts of maternal- and self-induced change, thereby learning about the particularities of their own mother and siblings.

Infant adaptations within the nest

Learning appears to account for the infant rat's first nipple attachment. Early studies by Teicher & Blass (1976) showed that infant rats first suckle a nipple that is coated with amniotic fluid (deposited by the dam during parturition). Washing the nipple eliminates suckling. Returning to the washed nipple, a distillate of the wash, amniotic fluid itself, or the dam's saliva, rich in amniotic fluid, reinstated suckling. More recent studies have analyzed this preference for amniotic fluid and shown that it depends on both the pup's intrauterine experience and its experience with its mother. Pedersen & Blass (1981) injected citral, a tasteless lemon scent, into the amniotic sac on the twentieth day of gestation and delivered the litter on day 21 (normal term) by Cesarean section. They then stimulated the infant's anogenital region with a paintbrush in a fashion that mimicked maternal licking while the infant smelled citral. Pups so treated refused to suckle normal, amniotic fluid-coated nipples. Surrounding normal or washed nipples with citral, however, elicited suckling. Pups that received only prenatal or postnatal exposure to citral, but not both, failed to attach to citral-painted washed nipples. Thus, under normal circumstances, amniotic fluid must gain its efficacy experientially. According to Sullivan, Hofer & Brake (1986), rats exposed to an odor during stimulation subsequently prefer that odor in an open field test. The odor through its pairing with anogenital stimulation takes on general as well as specific properties.

Nest experience contributes to additional changes in olfactory determinants of suckling. Bruno, Teicher & Blass (1980) discovered that, in rats 2–24 days of age, rat pup saliva had to coat the nipple in order for suckling to occur (this held even after eye opening), and Pedersen & Blass (1981) reported that dimethyl disulfide (DMDS) was the critical fraction in rat pup saliva that triggered the suckling response. The mystery was that neither amniotic fluid nor rat mother saliva contained DMDS. Again, the answer had to come from the animal's behavioral experiences in which nipple attachment was embedded in the sequence of suckling events. Pedersen, Williams & Blass (1982) demonstrated in 3-day-old rats that the events that naturally precede nipple attachment, i.e. stroking the pups in the presence of an odor, allows that odor to elicit suckling when placed on washed nipples.

It seems likely that the suckling system is closed by about 5 days of age. First, we were not able to transfer nipple attachment to a new odor in 5 day olds. Second, rats raised by continuous intragastric infusion for 4 days, starting on day 1, would no longer suckle. A 4-day abstinence, starting at day 6, did not disrupt suckling. Third, according to Wood & Leon (1987), changes in central olfactory processing to daily pairing of an odor with stroking are complete by day 8. Thus, anogenital stimulation appears to affect development of olfactory control over nipple attachment only until day 5; for becoming attracted to an odor associated with stroking, day 8 appears to be the limit.

The quieting and energy-conserving effects of milk, fat or sucrose can also gain control over behavior. Intraoral infusions of sucrose or fat paired with an odor cause a preference for that odor that is opioid dependent for both its acquisition and expression (Shide & Blass, 1991). Moreover, the gut hormone cholecystokinin (CCK), which is released by infusions of milk or fatty acid, both quiets isolated rats and is rewarding (Weller & Blass, 1988, 1990). Studies from Johanson's laboratory attribute special qualities to milk odor *per se*. Johanson & Terry (1988), for example, demonstrated that in 24-hour-deprived 3-, 6-, and 9-day-old rats, milk odor was arousing, especially when paired with oral infusions of sucrose. Even water infusions increased activity of rats smelling milk odor. These effects were enduring: milk odor alone increased activity in rats that had received a milk–sucrose pairing. Odor may be associated with the excitement caused by letdown; flavour with milk's calming effects.

Milk's effectiveness reflects the animal's developmental history with milk from the dam. According to Johanson & Terry (1988), rats raised with mothers that ate a diet laced with eucalyptus (which passes through the plasma into the milk) were particularly activated by eucalyptus odor presented alone after privation. Under normal circumstances, milk and its filtrates derived from the mother's diet may cast powerful long-term influences over infant, and later adult behaviors in the domains of food selection and social behaviors. The means used by rats to select the most salient and behaviorally relevant odors are currently under investigation in a number of laboratories and will be discussed below.

Internal changes resulting from milk intake or CCK or opioid release are not the only, or even necessarily the dominant factor supporting acquisition during quiet interactions. According to Alberts and May (1983), 15-day-old rats equally prefer an odor that they experienced while contacting the mother to one experienced while suckling and receiving milk from her. Infant motivational systems, like those of adults, are under multiple controls. In other circumstances, infants also respond to milk consequences. They run faster in an alley when rewarded with milk from a nipple (Amsel, Burdette & Letz, 1976) and, in a Y-maze, choose a lactating over a non-lactating nipple after 15 days of age (Kenny, Stoloff, Bruno & Blass, 1979). Milk qualities and those of the mother's diet affect diet selection by weanling rats (Galef & Sherry, 1973; Galef & Beck, 1990, for review).

The contributions of social interactions during the nest period

The olfactory linkages described for early nipple attachment may be confined to orienting to the dam and to suckling release. Social factors determine olfactory preference starting at about 14 days of age. Alberts & Brunjes (1978) demonstrated that rats spend more time over an odor that had been previously experienced while huddling with conspecifics. The odors did not have to be of a biological origin. Synthetic perfumes were effective when painted on huddling animals. The preference for odors experienced on conspecifics may direct mechanisms that underlie recognition of kin and of individual rats and thereby select with whom later social interactions will occur.

Although the mechanisms that underlie filial recognition in rats have not been adequately explored, considerable information is available on the social preference behavior of a number of species. Cross-fostering studies in thirteen-lined ground squirrels (Holmes, 1988), gray-tailed voles (Boyd & Blaustein, 1985) and spiny mice (Porter, Tepper & White, 1981), all demonstrated the powerful effect of rearing over genetic relatedness in a number of social behaviors. It would seem that for many mammals, including humans, knowledge of kin, in large part, stems from the natal and developmental experiences with them and not necessarily from a commonality provided by a shared genetic marker. This special form of familiarity, referred to as phenotype matching, is a strong motivator. Odors experienced in the presence of conspecifics are considerably more preferred to those experienced for comparable periods of time in isolation, for example.

A powerful demonstration of social recognition has been provided by Leon and his collaborators in weaning rats. At 14–15 days of age, rats start to eat spontaneously from the environment and to make brief forays from the nest. They also start to prefer certain odors emitted by their mother and this preference endures until weaning completion, circa 28 days of age, the time that males disperse from the nest. Coincident with the start of free feeding, the mother increases her own intake markedly in response to the demands of her exponentially growing offspring biomass (recall that feeding by infants is only supplementary at first). In response to increased feeding, the caecum secretes a gelatinous bacteria-rich substance, caeco-trophe, in volumes that exceed the amount needed by the mother for digestion. She excretes the surplus. This odoriferous substance is eaten by the young and may possibly facilitate digestion of the newly found food source by a gastrointestinal tract that has not yet made the transition from a lactose-rich milk diet to the enormously varied diet of the omnivorous rat. Caecotrophe also serves as an attractant to the newly weaned rats. According to Leon, Galef & Behse (1977), weanlings developed the attraction to the odor by experiencing the odor in the mother's presence. Thus, at the time of nest excursions, rats, among other rodents, develop social attractions to mother, siblings, and nest that provide social cohesiveness that endures for the transitional weaning period. For some odors, nest interactions determine adult social structure.

In summary, infant rats, through olfaction, learn about the mother through an early arousal component preceding suckling and a later calming component during each nursing bout. At the start of weaning, they follow the mother and stay close to the nest through an olfactory tether. From the start of weaning, they preferentially interact with siblings. Maternal influences may be short lived and effective only until weaning completion, including the initial feeding bout. Attraction to odors associated with siblings casts a more enduring influence as it extends to affiliational and protective behavior when adults.

Adaptations during and after weaning

Early experiences in the nest also influence later infant adaptations. In a seminal group of experiments, Galef and his colleagues demonstrated two means through which rat mothers could influence infant food selection. The initial bias is provided by the flavour of the mother's milk, although the influence of caecotroph has not been excluded. Regardless of origin, infant omnivores by the eve of weaning have discovered that the foods eaten by the mother are safe. The flavour of her foods is presumably transferred through the milk, which, as a plasma filtrate, allows the passage of relatively large food molecules.

The second strategy is to follow older rats to food sources, observe them eat, and on future occasions go to the same site and eat the same foods (Galef, 1989; Galef & Beck, 1990). Blass has observed this behavior in rat mothers' eating or drinking in an open arena. As soon as they start to eat, they attract all of the infants in the arena who will struggle with them for food, going so far as removing it from the mother's mouth. According to Galef, the olfactory cue utilized by rats to judge the 'reliability' of the informing rat is saliva borne. The infant prepares itself well for this task: Alberts & Cramer (1988), found that up to 80% of the contacts with the mother in 19–20-day-old rats result in oral exploration. As is the case for other forms of olfactory social communication, the mechanisms underlying its acquisition are not known. It is clear, however, that the external excitation that was necessary for the earliest learning is not necessary for learning in older rats. We will return to this issue below.

Mechanisms underlying early olfactory learning

Since the demonstrations in the early 1980s that olfactory learning could readily occur under biologically appropriate conditions, a number of forward-thinking neurophysiologists and intrepid psychobiologists have fused traditional surgical with contemporary autoradiographic and histo-chemical techniques to reveal the olfactory neurology mediating olfactory conditioning.

Selective sectioning of the vomeronasal organ, which detects olfactory qualities of milk aspirated through the soft palate, eliminates the activating effects of milk odor. Anosmia caused by zinc sulphate perfusion also eliminates preference (Johanson & Terry, 1988). This raises the possibility that information received by the vomeronasal organ becomes encoded in the main bulb, perhaps through amygdala projections. Teicher, Stewart, Kauer and Shepherd (1980), using the 2-DG autoradiography technique, were the first to demonstrate a particular focus of activity in the olfactory bulb of rats when suckling natural unmanipulated mothers. The focus was in the juxtaglomerular layer in the bulb. It soon became clear, however, that other locations in the bulb have access to motor and motivational systems subserving suckling. Using a behavioral paradigm, Pedersen *et al.* (1982) found baseline levels of activity in the region of the juxtaglomerular zone but also a substantial amount of activity in the glomeruli. Later studies from Leon's laboratory have revealed different topographic patterns in the bulb for different odors. Thus, different local circuitries in the bulb, delineated by the projections from the receptive fields of the nasal epithelium and therefore selectively activated by different odors, have access to the motivational and motor systems organizing suckling behavior.

Glomerular activity during odor presentation alone was punctate. It occurred in discrete foci tuned to odor characteristics. This is important for three reasons. First, change occurred in the bulb itself. Second, the pattern of glomerular activation by sniffing an odor was experientially selective. It was obtained only in rats that had previously had that odor paired with stroking. Third, the pattern was anatomically selective. In order to investigate the functional characteristics of the system further, Leon and his co-workers studied its neurophysiology. The only units in the bulb that yielded experience-dependent change were those in the areas of high 2-DG activity. Now the system could be further manipulated. Sullivan, Wilson & Leon (1989) did the physiological equivalent of the Pedersen *et al.* (1982) behavioral experiment. They stroked the animal and either gave different doses of the noradrenergic agonist isoproterenol or the antagonist, propranolol. Their electrophysiological findings replicated the behavioral data obtained with stroking and different levels of amphetamine administration. Specifically, at both high or low levels of activation via combination of stroking and norepinephrine (NE), behavioral preferences, 2-DG uptake levels and single unit activity from mitral and tufted cells in the bulb were relatively low. There was a physiologically ideal level of activation. Deviation from that ideal caused behavioral and neurophysiological decrements.

The above experiment establishes important links between two literatures. First, it starts to provide a mechanism for the behavioral changes documented by Pedersen *et al.* (1982) into experiential control over suckling. Second, in implicating NE as a vital transmitter for olfactory-based learning, it extends a substantial literature on developmental changes in other regions of the brain–vision for example (Gordon, Allen & Trombley, 1988).

The Leon studies are noteworthy in another fundamental way. They were informed by behavioral issues and in turn reformulate behavioral orientations. In gaining some insights into the neurophysiology, these investigators advanced our understanding of system characteristics by presenting some animals with modest levels of anogenital stimulation during training and others with more intense levels. All animals received identical odor exposure on the test day. Activity to the constant stimulus on test day reflected the animal's previous history. Activity was relatively exaggerated in the brains of animals with a history of high levels of stimulation and relatively muted in those with lower levels. Thus, the rat's previous behavioral history sets the 'gain' on the system that registers the confluence of olfactory stimulation and arousal. This now poses a challenge to behaviorally oriented scientists to seek naturally occurring developmental phenomena that reflect this central change. It would be of interest to observe graded instrumental behavior to a constant olfactory stimulus that had been associated in different animals with different levels of stimulation.

Another fascinating functional analysis has been undertaken by W. G. Hall and his colleagues, who have investigated neural circuitry caudal to the bulb that underlies behavioral change. For different behavioral tasks during development, rats were trained with one nares occluded and then tested some time later with the original (trained) nares open or with only the untrained nares open, thereby assessing retention and transfer of information within the nervous system. Hall capitalized on the incomplete crossing of the anterior commissure at birth and its slow maturation. There was a complete lack of transfer from one nares to the next for a variety of tasks (e.g. habituation, classical conditioning) until day 10, after which transfer was impressive. Transfer of habituation was disrupted by knife cuts that severed the commissure prior to habituation training, but not several hours after training. Yet commissural cuts did not interfere with classical conditioning. Classical conditioning transfer was blocked by destroying the medial bundle of fibers that exits the olfactory bulb and directly innervates the contralateral pole of the anterior commissure, suggesting ipsilateral storage and contralateral retrieval of information utilized in classical conditioning. For habituation, the information is stored bilaterally after training.

The similarities between the Leon and the Hall results are of interest. In each set of studies, there was anatomical restriction; in Leon's, it was determined by odor characteristics, presumably through the projections from 'odor-sensitive' receptors in the olfactory mucosa. In Hall's case, the information from the different glomerular packets is distributed to different portions of the nervous system in a task-specific way – and this sorting out starts at the level of egress from the bulb, if not sooner. Complexity of processing is further enriched by the odor source. For the Hall and Leon series of studies, olfactory stimulation was airborne and activated afferents in the olfactory bulb. In Johanson's studies, olfactory information from milk arises in the secondary olfactory system (i.e. originates in the vomeronasal organ). The projections of these systems differ and appear to interface originally in the amygdala where the medial and lateral amygdala 'anastomose'. The anastomization may be functional, as suggested by Pedersen *et al.* (1982) for rats exposed to citral odor prenatally and postnatally, during stroking. This provides a mechanism for linking pre-

and postnatal events and also for possibly linking events and associated motivational states between overly determined internal (e.g. catecholaminergic) and external (e.g. odors) sources of stimulation.

Summary

Nested infants, especially omnivores, have a very short period of time to find out about the perils of their environment prior to weaning onset when they start to leave the natal area. The instruction is offered by the mother in all species in which parents do not pair bond. We have presented evidence in neonatal rats for their ability to extract particular maternal olfactory features through the natural events that characterize mother–infant reunions and milk delivery. These events are influential in different ways depending upon infant age and, presumably, state. Investigations of mechanisms underlying changes in behavior based on olfactory experience have started to reveal processing characteristics within the bulb itself and the task-dependent pathways that convey this information.

There already exist a number of lines of evidence that link these behavioral and neural changes with functional consequences. One is in the domain of food selection: the infant, on olfactory bases, preferentially selects foods that familiar-smelling (based on salivary cues) adults eat. Because familiar adults have survived previous experiences with a food, accept it, do not indicate aversiveness, and convey safety accurately, i.e. are not deceitful, the weanling acts as if the food is safe. According to Galef & Beck (1990), this 'learned safety' is a very powerful influence.

A second domain of early influence is in the area of kin recognition and selection. To our knowledge, there have not been specific experiments that have scented kin with a particular odor and traced the effects of that odor into adult behavior in order to demonstrate early causality and identify the determining behavioral events and their underlying mechanisms. An understanding of these events and their neurologies has implications well beyond the specific behaviors and strategies described above. For humans, they hold the promise of gaining access to putative preverbal memories that must involve components of emotions, olfaction and gustation. If proven true, this speculation can provide means of recapturing the past empirically and possibly therapeutically as well as literally (Proust, 1888).

Termination of mother–infant relationships

In rats, starting at about 2 weeks after delivery, the mother distances herself from her rapidly growing and more mobile offspring. She rarely retrieves them, her nest become matted, and she terminates nursing bouts before the infants have had their fill. How does the pup perceive these changes and what might it do to prolong the relationship with its mother?

Michael Stoloff (Stoloff & Blass, 1983), to evaluate weaning from the perspective of food choice, studied rats 17–32 days of age in a Y-maze. Suckling either a nutritive or a non-nutritive nipple of their anesthetized mother could take place on one side, and eating either liquid or solid food in the absence of suckling could take place on the other side of the maze. At first, food was separated from the mother by a gauze shield which presumably allowed maternal olfactory stimuli to reach the infant. Then, because most older infants went to the suckling side of the maze but did not suckle, the test paradigm was changed. The barrier was removed, the dam's nipples involuted, and the animal was able to eat in the mother's presence. Its choices reflected age, milk availability at the nipple, food availability, and maternal availability but not level of deprivation. Even at 17 days of age, maternal context influenced choice. Twenty percent of the rats reached criterion on the food side of the maze where they ate in the mother's presence. By 28 and 32 days of age, rats did not suckle in the maze. Most reached criterion on the food side.

Pfister, Cramer & Blass (1986), by housing individual rats, after weaning, with litters that were 15–21 days of age, were able to sustain sucking in a standardized test in over 40 % of 45-day-old rats and in 15 % of 70-day-old rats. Not only did these older rats attach to a nipple, they 'stretched' when milk was let down. At the least, therefore, sucking termination does not only reflect maturation of serotonergic components of an inhibitory mechanism (Williams, Hall & Rosenblatt, 1980), it also reflects the animal learning something specific about maternal availability. It is not clear why suckling stops in these rats, but we believe that it reflects events that surround the nursing–suckling interaction and not changes in maternal behavior that do not involve nursing. This opinion is based on the following observation. In replicating this experiment with host litters that were 6–12 days of age, Blass observed that the guest rat (male or female) established its own nest in a corner far from the mother's nest and started to transport pups to that site. The mother, predictably, retrieved her pups and smartly punished the intruder. This scenario was carried out repeatedly by the guest rats. It was extinguished by each mother but started anew by the guest upon transfer to a new litter. Yet these severe punishments did not stop suckling by the guest animal, suggesting that suckling naturally stops through specific events that surround the act. These findings are potentially of broad interest because they will inform us about the natural interactions that allow for successful weaning and dispersal. They may also shed light upon abusive parenting and the basis for continued return by the infant.

There is now a substantial primate and human literature on abusive mothering (Eisenberg, 1990; Cicchetti & Carlson, 1989). It is not for us to review here, but simply to point out that the abused infant monkey doggedly returns to the abusing parent (Harlow, 1963). The factors that lead to this persistence, and knowing when the mother is approachable and when not, can be determined in the present paradigm. The present paradigm in conjunction with the one devised by Stoloff hold considerable potential for evaluating infant state, *vis-à-vis* the mother, and the factors that determine that state. The paradigm of the guest rat is potentially important because the guest does not have to suckle or have any contact with the mother at all. Yet it returns.

General summary

We have discussed some of the interfaces between rat and human mothers and their offspring at different phases of the mother–infant cycle. We have seen that complementarity between mothers and their young exists in that sensory stimulation provided by one member of the pair guides the behavior of the other. Thus, rat mothers intensively lick and nuzzle the newborn, thereby stimulating their own chemosensory and perioral somatosensory systems while activating and orienting newborns to the olfactory and tactile features of teats and ventrum. Stimulation of the ventrum by pups in turn activates the neurogenic milk let-down reflex and stimulates somatosensory receptors in her trunk region. Pups, in turn, are labelled by the mother's saliva or body secretions and receive intensive anogenital stimulation, which promotes urination and defecation. The mother, through ingesting the pup urine, recovers some 30 % of the water that is lost in lactation. Both young and mothers gain experience with the somatosensory and chemosensory features of the other during the initiation of nursing, become conditioned to them and presumably, as a result, maintain a specific social and reciprocal attachment.

Against the clear symmetry in our discussions of how behavior in the mother and young is modulated, there was also an asymmetry in our focus; namely, considerable effort was spent on what infants can learn in the nest site but we did not discuss how mothers provide this education. The mother may be completely passive, as in rats. The infant may be extracting information about odors and their relationship to her predicting behaviors. The complexities are undoubtedly greater in other species. Carnivores take the young out to hunt, stun the prey and allow the juveniles to practice. Primate infants assume the social status of their parents, especially their mother. As for biopsychological and biosocial training by human mothers, to our knowledge, formal investigations have not been carried out. They await our attention. Although early delivery in humans may have evolved because full intrauterine growth is not compatible with limited pelvic capacity, infants' remarkable ability to extract information about their environment, especially from situations that naturally occur in the nest, make these early mammalian exchanges a worthy focus of study.

References

Ader, R. & Grota, L. J. (1970). Rhythmicity in the maternal behaviour of *Rattus norvegicus*. *Animal Behaviour*, 18, 144–150.

Alberts, J. R. & Brunjes, P. C. (1978). Ontogeny of thermal and olfactory determinants of huddling in the rat. *Journal of Comparative and Physiological Psychology*, 92, 897–906.

Alberts, J. R. & Cramer, C. P. (1988). Ecology and experience: Sources of means and meaning of developmental change. In *Handbook of Behavioral Neurobiology*, Vol. 9, ed. E. M Blass, pp. 1–39. New York: Plenum Press.

Alberts, J. R. & May, B. (1983). Nonnutritive, thermotactile induction of filial huddling in rat pups. *Developmental Psychobiology*, 17, 161–181.

Amsel, A., Burdette, D. R. & Letz, R. (1976). Appetitive learning, patterned alternation, and extinction in 10 day old rats with non-lactating suckling as a reward. *Nature*, 262, 816–818.

Bauer, J. H. (1983). Effects of maternal state on the responsiveness to nest odors of hooded rats. *Physiology & Behavior*, 30, 229–232.

Blass, E. M. & Fitzgerald, E. (1988). Milk-induced analgesia and comforting in 10-day old rats: Opioid mediation. *Pharmacology, Biochemistry & Behavior*, 29, 9–13.

Blass, E. M., Ganchrow, J. R. & Steiner, J. E. (1984). Classical conditioning in newborn humans 2–48 hours of age. *Infant Behavior and Development*, 7, 223–235.

Blass, E. M & Hoffmeyer, L. B. (1991). Sucrose as an analgesic in newborn humans. *Pediatrics*, 87, 215–218.

Boyd, S. K. & Blaustein, A. R. (1985). Familiarity and inbreeding avoidance in the graytailed vole (*Microtus canicaudus*). *Journal of Mammalogy*, 66, 348–352.

Brewster, J. & Leon, M. (1980). Facilitation of maternal transport by Norway rat pups. *Journal of Comparative and Physiological Psychology*, 94, 115–127.

Bridges, R. S. (1990). Endocrine regulation of parental behavior in rodents. In *Mammalian Parenting: Biochemical, Neurobiological and Behavioral Determinants*, ed. N. A. Krasnegor & R. S. Bridges, pp. 93–117. New York: Oxford University Press.

Bruno, J. P., Teicher, M. H. & Blass, E. M. (1980). Sensory determinants of suckling behavior in weanling rats. *Journal of Comparative and Physiological Psychology*, 94, 115–127.

Cicchetti, D. & Carlson, V. (eds.) (1989). *Child Maltreatment: Theory and Research on the Causes and Consequences of Child Abuse and Neglect.* Cambridge: Cambridge University Press.

Cohen, J. & Bridges, R. S. (1981). Retention of maternal behavior in nulliparous and primiparous rats: Effects of duration of previous maternal experience. *Journal of Comparative and Physiological Psychology*, 95, 450–459.

Corter, C. & Fleming, A. S. (1990). Maternal responsiveness in humans: emotional, cognitive and biological factors. *Advances in the Study of Behavior*, 19, 83–136.

Cramer, C. P. & Blass, E. M. (1983). Mechanisms of control of milk intake in suckling rats. *American Journal of Physiology*, 245, R154–R159.

Cramer, C. P., Blass, E. M. & Hall, W. G. (1980). The ontogeny of nipple-shifting behavior in albino rats: Mechanisms of control and possible significance. *Developmental Psychobiology*, 13, 165–180.

Davis, H. P. & Squire, L. R. (1984). Protein synthesis and memory: A review. *Psychological Bulletin*, 96, 518–559.

Eisenberg, L. (1990). The biosocial context of parenting in human families. In *Mammalian Parenting: Biochemical, Neurobiological and Behavioral Determinants*, ed. N. A. Krasnegor & R. S. Bridges, pp. 9–24. New York: Oxford University Press.

Fleming, A. S. (1987). Psychobiology of rat maternal behavior: How and where hormones act to promote maternal behavior at parturition. In *Reproduction: A Behavioral and Neuroendocrine Perspective*, ed. B. R. Komsaruk, H. I. Siegel, M. F. Cheng & H. H. Feder, pp. 234–251. New York: New York Academy of Sciences.

Fleming, A. S. (1990). Hormonal and experiential correlates of maternal responsiveness in human mothers. In *Mammalian Parenting: Biochemical, Neurobiological and Behavioral Determinants*, ed. N. A. Krasnegor & R. S. Bridges, pp. 184–208. New York: Oxford University Press.

Fleming, A. S. & Corter, C. (1988). Factors influencing maternal responsiveness in humans: Usefulness of an animal model. *Psychoneuroendocrinology*, 13, 189–212.

Fleming, A. S., Corter, C., Franks, P., Surbey, C., Schneider, B. & Steiner, M. (1993). Postpartum factors related to mother's attraction to newborn infant odors. *Developmental Psychobiology*, 26, 115–132.

Fleming, A. S., Kuchera, C., Lee, A. & Winocur, G. (in press). Olfactory-based learning varies as a function of parity in female rats. *Psychobiology*.

Fleming, A. S. & Luebke, C. (1981). Timidity prevents the nulliparous female from being a good mother. *Physiology & Behavior*, 27, 863–868.

Fleming, A. S. & Rosenblatt, J. S. (1974a). Olfactory regulation of maternal behavior in rats: I. Effects of olfactory bulb removal in experienced and inexperienced lactating and cycling females. *Journal of Comparative and Physiological Psychology*. 86, 221–232.

Fleming, A. S. & Rosenblatt, J. S. (1974b). Olfactory regulation of maternal behavior in rats: II. Effects of peripherally-induced anosmia and lesions of the lateral olfactory tract in pup-induced virgins. *Journal of Comparative and Physiological Psychology*, 86, 233–246.

Fleming, A. S., Ruble, D. N., Flett, G. L. & Shaul, D. (1988). Postpartum adjustment in first-time mothers: Relations between mood, maternal attitudes and mother-infant interactions. *Developmental Psychology*, 24, 77–81.

Fleming, A. S. & Sarker, J. (1990). Experience–hormone interactions and maternal behavior in rats. *Physiology & Behavior*, 47, 1165–1173.

Fleming, A. S., Steiner, M. & Anderson, V. (1987). Hormonal and attitudinal correlates of maternal behavior during the early postpartum period. *Journal of Reproduction and Infant Psychology*, 5, 193–205.

Fleming, A. S., Vaccarino, F., Chee, P. & Tambosso, L. (1979). Vomeronasal and main olfactory system modulation of maternal behavior in the rat. *Science*, 203, 372–374.

Formby, D. (1967). Maternal recognition of infant's cry. *Developmental Medicine of Child Neurology*, 9, 292–298.

Galef, B. G., Jr. (1989). Socially-mediated attenuation of taste–aversion learning in Norway rats: Preventing development of food phobias. *Animal Learning and Behavior*, 17, 468–474.

Galef, B. G., Jr. & Beck, M. (1990). Diet selection and poison avoidance by mammals individually and in social groups. In *Handbook of Behavioral Neurobiology: Neurobiology of Food and Fluid Intake*, Vol. 10, ed. E. M. Stricker, pp. 329–352. New York: Plenum Press.

Galef, B. G. & Sherry, D. F. (1973). Mother's milk: A medium for the transmission of cues reflecting the flavour of mother's diet. *Journal of Comparative and Physiological Psychology*, 83, 374–378.

Goldberg, S. (1983). Parent–infant bonding: Another look. *Child Development*, 54, 1355–1382.

Gordon, B., Allen, E. E. & Trombley, R. Q. (1988). The role of norepinephrine in plasticity of visual cortex. *Progress in Neurobiology*, 30, 171–191.

Grossman, K., Thane, K. & Grossman, K. E. (1981). Maternal tactual contact of the newborn after various conditions of mother–infant contact. *Developmental Psychology*, 17, 158–169.

Grota, L. J & Ader, R. (1969). Continuous recording of maternal behavior in *Rattus norvegicus*. *Animal Behaviour*, 17, 722–729.

Hall, W. G. (1979). Feeding and behavioral activation in infant rats. *Science*, 190, 1313–1315.

Hall, W. G. (1990). The ontogeny of ingestive behavior: Changing control of components in the feeding sequence. In *Handbook of Behavioral Neurobiology: Neurobiology of Food and Fluid Intake*, Vol. 10, ed. E. M. Striker, pp. 77–123. New York: Plenum Press.

Hansen, S., Ferreira, A. & Selart, M. F. (1985). Behavioral similarities between mother rats and benzodiazepine-treated non-maternal animals. *Psychopharmacology*, 86, 344–347.

Harlow, H. F. (1963). The maternal affectional system of rhesus monkeys. In *Maternal Behavior in Mammals*, ed. H. L. Rheingold, pp. 254–281. New York: John Wiley & Sons.

Heritage, A. S., Grant, L. D. & Stumpf, W. E. (1977). H3 estradiol catecholamine neurons of the rat brain stem. *Journal of Comparative Neurology*, 176, 607–630.

Holmes, W. G. (1988). Kinship and the development of social preferences. In *Handbook of Behavioral Neurobiology: Developmental Psychobiology and Behavioral Ecology*, Vol. 9, ed. E. M. Blass, pp. 389–414. New York: Plenum Press.

Insel, T. R. (1990). Oxytocin and maternal behavior. In *Mammalian Parenting*:

Biochemical, Neurobiological and Behavioral Determinants, ed. N. A. Krasnegor & R. S. Bridges, pp. 260–280. New York: Oxford University Press.

Jakubowski, M. & Terkel, J. (1986). Establishment and maintenance of maternal responsiveness in postpartum Wistar rats. *Animal Behaviour*, 34, 256–262.

Johanson, I. B. & Hall, W. G. (1979). Appetitive learning in 1-day-old rat pups. *Science*, 205, 419–421.

Johanson, I. B. & Terry, L. M. (1988). Learning in infancy: A mechanism for behavioral change during development. In *Handbook of Behavioral Neurobiology: Developmental Psychobiology and Behavioral Ecology*, Vol. 9, ed. E. M. Blass, pp. 245–281. New York: Plenum Press.

Kaitz, M. & Eidelman, A. I. (1992). Smell-recognition of newborns by women who are not mothers. *Chemical Senses*, 17, 225–229.

Kaitz, M., Lapidot, P., Bronner, R. & Eidelman, A. (1992). Parturient women can recognize their infants by touch. *Developmental Psychology*, 28, 35–39.

Kenny, J. T., Stoloff, M. L., Bruno, J. P. & Blass, E. M. (1979). The ontogeny of preference for nutritive over nonnutritive suckling in albino rats. *Journal of Comparative and Physiological Psychology*, 93, 752–759.

Keverne, E. B. & Kendrick, K. M. (1990). Neurochemical changes accompanying parturition and their significance for maternal behavior. In *Mammalian Parenting: Biochemical, Neurobiological and Behavioral Determinants*, ed. N. A. Krasnegor & R. S. Bridges, pp. 281–304. New York: Oxford University Press.

Klaus, M. H. & Kennell, J. H. (1982). *Parent–Infant Bonding*. New York: Mosby.

Leon, M., Croskerry, P. G. & Smith, G. K. (1978). Thermal control of mother–young contact in rats. *Physiology & Behavior*, 21, 793–811.

Leon, M., Galef, B. G. & Behse, J. H. (1977). Establishment of pheromonal bonds and diet choice in young rats by odor pre-exposure. *Physiology & Behavior*, 18, 387–391.

MacFarlane, B. A., Pedersen, P. E., Cornell, C. E. & Blass, E. M. (1983). Sensory control of suckling-associated behaviours in the domestic Norway rat, *Rattus norvegicus. Animal Behaviour*, 31, 462–471.

Malenfant, S. A., Barry, M. & Fleming, A. S. (1991). The effects of cycloheximide on olfactory learning and maternal experience effects in postpartum rats. *Physiology & Behavior*, 9, 289–294.

Mayer, A. D. (1983). The ontogeny of maternal behavior in rodents. In *Parental Behavior of Rodents*, ed. R. W. Elwood, pp. 1–20. Chichester: Wiley.

Moffat, S. D., Suh, E. J. & Fleming, A. S. (1993). Noradrenergic involvement in the consolidation of maternal experience in postpartum rats. *Physiology & Behavior*, 85, 805–811.

Moore, C. L. (1990). Comparative development of vertebrate sexual behavior: levels, cascades, and webs. In *Contemporary Issues in Comparative Psychology*, ed. D. A. Dewsbury, pp. 278–299. Sunderland, MA: Sinauer.

Moore, C. L. & Morelli, G. A. (1979). Mother rats interact differently with male and female offspring. *Journal of Comparative and Physiological Psychology*, 93, 677–684.

Moran, T. H., Schwartz, G. J. & Blass, E. M. (1983). Organized behavioral responses to lateral hypothalamic stimulation in infant rats. *Journal of Neuroscience*, 3, 10–19.

Morgan, H. D., Fleming, A. S. & Stern, J. M. (1992). Somatosensory control of the onset and retention of maternal responsiveness in primiparous Sprague-Dawley rats. *Physiology & Behavior*, 51, 541–555.

Numan, M. (1988). Maternal behavior. In *Physiology of Reproduction*, Vol. 2., ed. E. Knobil. & J. Neill, pp. 1569–1646. New York: Raven Press.

Orpen, B. G. & Fleming, A. S. (1987). Experience with pups sustains maternal responding in postpartum rats. *Physiology & Behavior*, 40, 47–54.

Pedersen, P. E. & Blass, E. M. (1981). Olfactory control over suckling in albino rats. In *The Development of Perception: Psychobiological Perspectives*, ed. R. N. Aslin, J. R. Alberts & M. R. Petersen, pp. 359–381. New York: Academic Press.

Pedersen, P. E., Williams, C. L. & Blass, E. M. (1982). Activation and odor conditioning of suckling behavior in three day old albino rats. *Journal of Experimental Psychology: Animal Behavior Processes*, 8, 329–341.

Pfister, J., Cramer, C. P. & Blass, E. M. (1986). Suckling in rats extended by continuous living with dams and their preweanling litters. *Animal Behaviour*, 34, 415–420.

Poindron, P. & Levy, F. (1990). Physiological, sensory and experiential determinants of maternal behavior in sheep. In *Mammalian Parenting: Biochemical, Neurobiological and Behavioral Determinants*, ed. N. A. Krasnegor & R. S. Bridges, pp. 133–156. New York: Oxford University Press.

Porter, R. H., Tepper, V. J. & White, L. M. (1981). Experiential influences on the development of huddling preferences and 'sibling' recognition in spiny mice. *Developmental Psychobiology*, 14, 375–382.

Proust, M. (1888). *A la Recherche du Temps Perdu: Du Cote de Chez Swann*.

Rosenblatt, J. S. (1967). Nonhormonal basis of maternal behavior in the rat. *Science*, 156, 1512–1514.

Rosenblatt, J. S. (1990). Landmarks in the physiological study of maternal behavior with special reference to the rat. In *Mammalian Parenting: Biochemical, Neurobiological and Behavioral Determinants*, ed. N. A. Krasnegor & R. S. Bridges, pp. 40–60. New York: Oxford University Press.

Schaal, B. & Porter, R. H. (1991). 'Microsmatic humans' revisited: the generation and perception of chemical signals. *Advances in the Study of Behavior*, 20, 135–199.

Shide, D. J. & Blass, E. M. (1991). Opioid mediation of odor preferences induced by sugar and fat in 6-day-old rats. *Physiology & Behavior*, 50, 961–966.

Smotherman, W. P. & Robinson, S. R. (1988). The uterus as environment: The ecology of fetal behavior. In *Handbook of Behavioral Neurobiology: Developmental Psychobiology and Behavioral Ecology*, Vol. 9, ed. E. M. Blass, pp. 149–196. New York: Plenum Press.

Steiner, M. (1992). Female specific mood disorders. *Clinical Obstetrics and Gynecology*, 35, 594–611.

Stern, J. M. (1983). Maternal behavior priming in virgin and caesarean-delivered Long–Evans rats: effects of brief contact or continuous exteroceptive pup stimulation. *Physiology & Behavior*, 31, 757–763.

Stern, J. M. (1989). Maternal behavior: sensory, hormonal, and neural determinants. In *Psychoendocrinology*, ed. F. R. Brush & S. Levine, pp. 105–226. New York: Academic Press.

Stern, J. M. (1990). Multisensory regulation of maternal behavior and masculine sexual behavior: A revised view. *Neuroscience and Biobehavioral Reviews*, 14, 183–200.

Stoloff, M. L. & Blass, E. M. (1983). Changes in appetitive behavior in weanling age rats: Transitions from suckling to feeding behavior. *Developmental Psychobiology*, 16, 439–454.

Sullivan, R. M., Hofer, M. A. & Brake, S. C. (1986). Olfactory-guided orientation in neonatal rats is enhanced by a conditional change in behavioral state. *Developmental Psychobiology*, 19, 615–623.

Sullivan, R. M., Wilson, D. A. & Leon, M. (1989). Norepinephrine and learning-induced plasticity in infant rat olfactory system. *Journal of Neuroscience*, 9, 3998–4006.

Teicher, M. H. & Blass, E. M. (1976). Suckling in newborn rats: Eliminated by nipple lavage, reinstated by pup saliva. *Science*, 193, 422–425.

Teicher, M. H., Stewart, W. B., Kauer, J. S. & Shepherd, G. M. (1980). Suckling pheromone stimulation of a modified glomerular region in the developing rat olfactory bulb revealed by the 2-deoxyglucose method. *Brain Research*, 194, 530–535.

Weller, A. & Blass, E. M. (1988). Behavioral evidence for cholecystokinin–opiate interactions in neonatal rats. *American Journal of Physiology*, 255, R901–R907.

Weller, A. & Blass, E. M. (1990). Cholecystokinin conditioning in rats: Ontogenetic determinants. *Behavioral Neuroscience*, 104, 199–206.

Williams, C. L., Hall, W. G. & Rosenblatt, J. S. (1980). Changing oral cues in suckling of weaning-age rats: Possible contributions to weaning. *Journal of Comparative and Physiological Psychology*, 94, 472–483.

Woo, C. C. & Leon, M. (1987). Sensitive period for neural and behavioral response development to learned odors. *Developmental Brain Research*, 36, 309–313.

10

Development of behavior systems

JERRY A. HOGAN

I begin this chapter with a brief discussion of what I mean by a behavior system, and then use the dustbathing, hunger, aggression, and sex systems of chickens to illustrate how such systems develop. Using these examples plus comparable information from investigations of mammalian behavior, I next consider the questions of whether there are any general differences between the development of perceptual and motor mechanisms and between social and non-social behavior systems. In the context of social behavior systems, I review some ways in which early experience can have far-reaching effects. Finally, I look at the development of interactions among behavior systems, and ask whether any new principles are necessary to understand this complex process.

Behavior systems

A definition of a behavior system and its components is given in Chapter 1, and a depiction of the concept is presented there in Figure 1.1 (p. 6). A more extensive discussion of this concept and many of the other issues considered in this chapter can be found in Hogan (1988). In brief, a behavior system consists of an organization of its components: perceptual, motor, and central behavior mechanisms. Each of these components is also organized. The study of development comprises: (1) describing the changes in both the organization of the components themselves and the organization of the system as whole (i.e. the connections among the behavior mechanisms), and (2) investigating the causes of those changes. This view of development is a generalization of a proposal of Kruijt (1964), who suggested that the motor components of behavior often function as independent units in young animals, and that only later, after specific experience, do these motor components become integrated into more complex systems, such as hunger, aggression, and sex. This view is also an example of the notion of progressive integration of originally separate elements into merged patterns as discussed by Berridge (this volume).

We have already seen many examples of these concepts and of the causal factors that are responsible for development. We have seen, for example, that songbirds are considered to possess an auditory template, which is a kind of perceptual mechanism. DeVoogd (this volume) has provided evidence showing how this mechanism develops and what factors are required for normal development to occur in a variety of species. Other evidence has shown how the motor mechanisms for song (DeVoogd), locomotion and grooming patterns (Berridge), and social displays (Groothuis) develop. And, the chapters on imprinting (Bischof, Clayton, ten Cate) and maternal behavior (Fleming & Blass) have considered aspects of the development of the sexual system and the maternal system, respectively. I will present some further examples of the development of behavior systems to illustrate the operation of some general principles of development.

In looking for the causal basis of developmental changes, I have found it convenient to make use of the concept of prefunctional (Schiller, 1949; Hogan, 1988). If a behavior mechanism develops prefunctionally, this means that functional experience (or practice) was not necessary for normal development to occur. It should be noted that there is no implication about the role of other kinds of experience. I make this distinction because most people assume that experience means functional experience, and that cause and function go hand in hand. As we have seen in Chapter 1, however, this is often not the case. I shall return to this topic in the section on principles of development.

Development of some behavior systems

Dustbathing

Dustbathing in the adult fowl (and many other species of bird) consists of a sequence of coordinated movements of the wings, feet, head, and body that serve to spread dust through the feathers. It occurs on a regular basis, and bouts of dustbathing last about half an hour (Vestergaard, 1982). When dust is available, dustbathing functions to remove excess lipids from the feathers and to maintain good feather condition (van Liere & Bokma, 1987). Dustbathing also satisfies the definition of a behavior system.

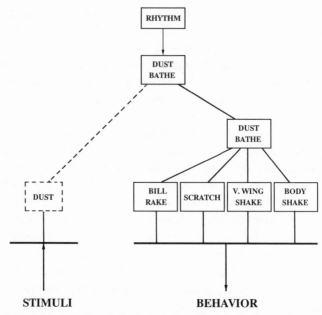

STIMULI   **BEHAVIOR**

Figure 10.1 The dustbathing behavior system of a young chick. Boxes represent putative cognitive (neural) mechanisms: a perceptual mechanism responsible for recognizing dust, a central dustbathing mechanism responsible for integrating input from the perceptual mechanism and other internal influences as well as for coordinating output to the motor mechanisms, and several motor mechanisms responsible for the various motor patterns comprising dustbathing. Solid lines indicate mechanisms and connections among them that develop prefunctionally. Dashed lines indicate mechanisms and connections that develop as the result of specific functional experience. From Vestergaard *et al.* (1990). Reproduced by permission.

Chickens recognize particular substrates (perceptual mechanisms) in which they dustbathe with a coordinated series of motor patterns (motor mechanisms), and the duration and timing of this behavior are controlled by internal factors (central mechanisms). A diagram of the dustbathing system is shown in Figure 10.1.

Dustbathing does not appear fully formed in the young animal. Rather, individual elements of the system appear independently, and only gradually do these elements become fixed in the normal adult form. The motor components of the system include bill raking, ground scratching, vertical wing shaking, head and side rubbing, and body shaking. None of these components is seen the first day after hatching, but by day 12, all the components are present in a chick's repertoire (Kruijt, 1964). According to Kruijt, most of these 'movements are, at their first occurrence after hatching, immediately shown in their characteristic form, even though the chick has not had any opportunity to practice their function' (p. 23).

Vestergaard, Hogan & Kruijt (1990) looked at the influence of the rearing environment on the organization of the motor components of dustbathing. They observed small groups of chicks that were raised either in a normal environment containing sand and grass sod or in a poor environment that had floors covered with wire mesh. A comparison of the dustbathing motor patterns of 2-month-old birds raised in the two environments showed surprisingly few differences. The form of the individual behavior patterns as well as the temporal organization of the elements during extended bouts of dustbathing developed almost identically in both groups. There were some differences in the microstructure of the bouts that could be related to the presence or absence of specific feedback (see also van Liere, Kooijman & Wiepkema, 1990; van Liere, 1992), but the motor mechanisms and their coordination developed essentially normally in chicks raised in a dustless environment. Clearly, the experience of sand in the feathers removing lipids or improving feather quality is not necessary for the integration of the motor components of dustbathing into a normal coordinated sequence.

In adult fowl, the occurrence of dustbathing varies directly with the length of time a bird has been deprived of the opportunity to dustbathe; it also occurs primarily in the middle of the day (Vestergaard, 1982). In young chicks, as soon as dustbathing behavior is seen, at about 1 week of age, it is controlled by the effects of dust deprivation. Hogan, Honrado & Vestergaard (1991) found that deprivation effects could be demonstrated as early as 8 days of age and that they did not change over at least a 4-week period. No specific experience seemed to be necessary for the motivational factors associated with dust deprivation to gain control of dustbathing. Similarly, Hogan & van Boxel (1993) found that a daily rhythm, with most dustbathing occurring in the middle of the day, was seen in chicks at least as young as 14 days of age. However, the occurrence of dustbathing was not as strongly restricted to the middle of the day as in adults, and the length of dustbathing bouts was also shorter in the young birds. Nonetheless, these results suggest that the central mechanism and the connections between it and the motor mechanisms are developed prefunctionally.

Functional experience does play an essential role in the development of the perceptual mechanism for recognizing dust and the connection between it and the central mechanism. Young chicks can be seen engaging in dustbathing movements on almost any surface that is available, ranging from hard ground and stones to sand and dust. In fact, Kruijt (1964) found that making the external situation as favorable as possible for dustbathing was insufficient for releasing the behavior. This result implies that early dustbathing may be controlled exclusively by the internal factors mentioned above. With respect to the behavior system model of dustbathing (see Figure 10.1), it implies that the connection between the dust-recognition perceptual mechanism and the central mechanism is not formed until well after the motor and central mechanisms are functional.

What factors are responsible for the connection between the dust-recognition mechanism and the central mechanism? Vestergaard & Hogan (1992) found that early dustbathing is most likely to occur in whatever substrate is pecked at most. They point out that pecking is a movement that functions as exploratory, feeding, dustbathing, and later aggressive behavior. They suggest that perceptual mechanisms specific to each system develop gradually out of exploratory pecking on the basis of functional experience. Experiments that determine how such experience works have not yet been done, but would provide evidence for progressive differentiation of development (Berridge, this volume).

Some evidence is available for how the dust-recognition mechanism itself develops. The stimulus properties of a substrate are one important factor. For example, Vestergaard & Hogan (1992) found fine black coal dust to be much preferred to white sand, and Petherick & Duncan (1989) and van Liere (1992) found dark peat to be much preferred to sand and wood shavings. In the case of peat, the preference developed gradually, which implies that some aspect of the experience during dustbathing was crucial. It remains to be determined whether removal of lipids, the sensory feedback from the substrate in the feathers, or facilitation of the dustbathing behavior itself is the crucial factor.

Other evidence from the same studies shows that early experience can lead to stable preferences for non-preferred stimuli. As an extreme example, Vestergaard & Hogan (1992) found that some birds developed a stable preference for dustbathing on a skin of junglefowl feathers. This example is important because it shows how a system can develop abnormally and be a cause of the pathological 'feather pecking' seen in many commercial groups of fowl (Vestergaard, Kruijt & Hogan, 1993). Overall, a general conclusion from all these studies is that particular classes of stimuli are more efficacious than others for the development of the perceptual mechanism for the recognition of dust. This conclusion is similar to that reached in studies of the development of perceptual mechanisms for the recognition of conspecific song in some species of songbirds (e.g. Marler,

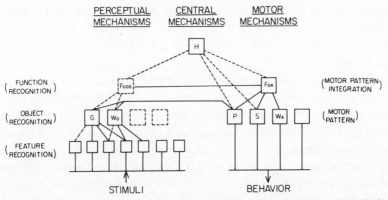

Figure 10.2 The hunger system of a young chick. Perceptual mechanisms include various feature-recognition mechanisms (such as of color, shape, size, and movement), object recognition mechanisms (such as of grain-like objects [G], worm-like objects [Wo], and possibly others), and a function recognition mechanism (Food). Motor mechanisms include those underlying specific behavior patterns (such as pecking [P], ground scratching [S], walking [Wa], and possibly others), and an integrative motor mechanism that could be called foraging (For). There is also a central hunger mechanism (H). Solid lines indicate mechanisms and connections among them that develop prefunctionally. Dashed lines indicate mechanisms and connections that develop as the result of specific functional experience. From Hogan (1988). Reproduced by permission.

1987; DeVoogd, this volume) and perceptual mechanisms for the recognition of conspecifics in imprinting studies in various avian species (Bolhuis, 1991; Bischof, ten Cate, this volume).

Hunger

The hunger system of an adult chicken consists of various perceptual mechanisms that serve a food-recognition function, motor mechanisms that function to locate and ingest food, and a central mechanism that integrates signals from the physiological mechanisms concerned with nutrition and modulates signals from the perceptual mechanisms and to the motor mechanisms. A diagram of the hunger system of a young chick is shown in Figure 10.2.

As with dustbathing, both the individual motor patterns of the system (pecking, ground scratching, walking) and the integration of these patterns into effective foraging behavior appear prefunctionally. Unlike dustbathing, however, the integration of the motor patterns disintegrates in the absence of effective functional experience (Hogan, 1971). Hogan (1988) suggested that new connections were formed between the central hunger mechanism and individual motor mechanisms on the basis of the specific experience of the individual chick, and that these new connections effectively blocked the expression of the original prefunctional connections (see Figure 10.2).

Another difference between the dustbathing and hunger systems is that the central mechanism for hunger does not immediately control the motor mechanisms of the system. A chick begins pecking within a few hours of hatching, but its nutritional state does not influence pecking until about 3 days of age (Hogan, 1971). Early experiments showed that some kind of pecking experience is necessary for this change in control to occur (Hogan, 1973a), and further experiments led to the hypothesis that it is the experience of pecking followed by swallowing that causes the connection between the central hunger mechanism and the pecking mechanism to be formed (Hogan, 1977). In other words, it appears that a chick must learn that pecking is the action that leads to ingestion; once this association has been formed, nutritional factors can directly affect pecking (see Figure 10.2). Subsequent experiments have shown that the association of pecking with ingestion is, indeed, the necessary and sufficient condition for pecking to become integrated into the hunger system (Hogan, 1984).

The development of the perceptual mechanism for recognizing food and of the connection between the food-recognition mechanism and the central hunger mechanism requires functional experience, and is similar to the dustbathing system in this way. The putative food-recognition mechanism in newly hatched chicks must be largely unspecified because of the very wide range of stimuli that are characteristic of items that chicks will come to accept as food. Although certain taste and tactile stimuli are more acceptable than others, these stimuli can be effective only after the chick has the stimulus in its mouth. Chicks learn to associate the visual characteristics of an object with its taste and tactile characteristics as early as 1 day of age, and treat such objects as food or non-food before nutritive factors gain control of pecking on day 3 (see Hogan, 1973b, for review). This means that the food-recognition mechanism is independent of the central mechanism of the developing hunger system. Other evidence shows that the long-term effects of ingestion can also affect the development of the food-recognition mechanism, but only after the chicks are 3 days old (Hogan-Warburg & Hogan, 1981).

The existence of a connection between a perceptual mechanism and a central mechanism can be inferred by demonstrating the existence of 'priming' or 'incentive' effects (Hogan & Roper, 1978, pp. 231–232). For example, presentation of food may make an animal hungrier or presentation of a sexual stimulus may increase its sexual appetite. There is evidence in young chicks that food particles develop incentive value between 3 and 5 days post-hatch, based on the long-term effects of ingestion (Hogan, 1971; Hogan-Warburg & Hogan, 1981). This would then be the time when the connection between the food-recognition and central hunger mechanisms develops (see Figure 10.2). Hogan (1988) has discussed in detail the evidence on which these conclusions are based, and has reviewed similar evidence for the development of a hunger system in rat pups and kittens (see also Baerends-van Roon & Baerends, 1979; Blass, Hall & Teicher, 1979; Hall & Williams, 1983).

Aggression

The aggression system of an adult chicken consists of perceptual mechanisms that serve an 'opponent' recognition function, various motor mechanisms that are used in fighting (including those that control threat display, leaping, wing flapping, kicking, and pecking), and a central mechanism that is sensitive to internal motivational factors (such as testosterone) and that coordinates the activation of the motor mechanisms. Kruijt (1964) showed that fighting develops out of hopping, which is a locomotory pattern that is not initially released by or directed toward other chicks. While hopping, chicks sometimes bump into each other, by accident, and in the course of several days, hopping gradually becomes directed toward other chicks. Frontal threatening starts to occur, and by the age of 3 weeks, pecking and kicking are added to aggressive interactions. Normal, well-coordinated fights are not seen until 2 to 3 months.

The various behavior patterns comprising adult fights can be seen to occur, independently, in the 1- to 2-week-old chick, well before their integration into fighting behavior. This means that functional social experience could be a necessary factor guiding development. This is, however, not the case. In other experiments, Kruijt (1964) raised chicks in social isolation for the first week of life and then placed them together in pairs. Many of these chicks showed aggressive behavior toward each other within seconds. Further, the fights that developed were characteristic of the fights of 1-month-old, socially raised chicks. Such results suggest that the organization of the motor components of the aggression system as well as the connections between the central and motor mechanisms develop prefunctionally, and that the occurrence of aggressive behavior requires only the proper motivational state. Similar results and conclusions apply to the development of aggressive displays in gulls (see Groothuis, this volume). In this way, the aggression system is more like the dustbathing system than it is like the hunger system.

Whether functional social experience ever affects the organization of the motor mechanisms of the aggression system in chickens (as it apparently does in gulls; see Groothuis, this volume) remains an open question. Males raised by Kruijt in social isolation for more than a year still showed reasonably normal aggressive patterns, and the abnormalities that were seen could be accounted for in terms of interference from other systems such as fear. Nonetheless, social experience could be necessary for fighting behavior to develop a high degree of effectiveness.

The opponent-recognition perceptual mechanism must be partially formed prefunctionally because a chick as young as 2.5 days old will respond with frontal threat and aggressive pecks to the stimulus of a 6-cm green wooden triangle moved directly in front of it (Evans, 1968). Likewise, socially isolated chicks showed fully coordinated aggressive behavior when confronted with another chick at the age of 1 week. But isolated chicks of the same age can also direct aggressive behavior to a light bulb hanging in the cage. And older isolated males often come to direct their aggressive behavior to their own tails (Kruijt, 1964). Presumably the complete development of the perceptual mechanism depends on the proper experience at the proper time (just as the templates for song learning in many species), but the experiments necessary to explore this idea have not yet been done.

Sex

The sex system of a normal adult rooster consists of perceptual mechanisms that serve a 'partner' recognition function: motor mechanisms for locomotion, copulation (which includes mounting, sitting, treading, pecking, and tail lowering), and various displays, such as waltzing, wing flapping, tidbitting, and cornering; and a central mechanism that is sensitive to internal motivational factors such as testosterone and that coordinates the activation of the motor mechanisms. In small groups of junglefowl, Kruijt (1964) saw mounting and copulatory trampling (treading) on a model in a sitting position as early as 3–4 days, but such behavior was not common until weeks later. Full copulation with living partners did not occur before the males were 4 months old.

Many of the components of the copulatory sequence, including mounting, sitting, and pecking, are seen independently in young chicks, and there is ample opportunity for social experience to influence the occurrence and integration of these components. As with aggression, however, several lines of evidence suggest that the motor mechanisms are already organized soon after hatching, and that their expression merely requires a sufficiently high level of motivation. For example, Andrew (1966) was able to elicit well-integrated mounting, treading, and pelvic lowering in socially isolated domestic chicks as young as 2 days old by using the stimulus of a human hand moved in a particular manner. Andrew also found that injection of testosterone greatly increased the number of chicks that responded sexually in his tests during the first 2 weeks (see also Groothuis, this volume, for similar evidence on the expression of sexual displays in gulls). Further, junglefowl males that had been raised in social isolation for 6–9 months copulated successfully with females within so few encounters that it was clear that the motor mechanisms had been integrated before testing (Kruijt, 1962).

The occurrence of the courtship displays presents a somewhat different picture. Waltzing, for example, is first seen at 2–3 months of age, when it always appears in the context of fighting. The form of the display develops independently of social experience, whereas the factors controlling the occurrence of waltzing are largely determined by social experience. Waltzing to a female usually has the effect that the female crouches, and a crouching female is the signal for mounting and copulation. Preliminary experiments by Kruijt (1964) showed that the frequency of waltzing increased when mating was contingent on its occurrence and decreased when mating was not allowed. Thus, copulation seems to be the reinforcer that causes the motor mechanism for waltzing to become attached to the central coordinating mechanism for sex. The form of the tidbitting display also develops independently of social experience, and it is likely that the change in control of the display from the hunger system to the sex system is also dependent on social experience, but the necessary experiments have not yet been done (see Moffatt & Hogan, 1992). Similar examples have also been analyzed by Groothuis (this volume) for gulls and other species.

The development of the partner-recognition perceptual mechanism seems to be much more susceptible to the effects of experience than the development of the motor mechanisms of the sex system. For example, junglefowl chicks become sexually dimorphic at about 1 month of age. By about 2 months, young males begin to show incomplete sexual behavior toward conspecifics, but such behavior is directed equally often toward males and females. Only gradually, as a result of specifically sexual experience, does sexual behavior become directed exclusively to females (Kruijt, 1964). The development of the partner-recognition mechanism is, of course, related to the topic of imprinting, and this relationship will be discussed below.

Some principles of behavioral development
Perceptual and motor mechanisms

In our brief review of the development of some behavior systems in the chicken, it may have struck the reader that, without exception, all the motor mechanisms developed prefunctionally, whereas all the perceptual mechanisms required at least some functional experience in order to achieve the normal adult form. In this section I will ask whether there are inherent differences in the way perceptual and motor mechanisms develop. It will be useful to begin by considering some aspects of what is known about neural development.

Brown, Hopkins & Keynes (1991) have divided brain development, at the cellular level, into four major stages: (1) genesis of nerve cells (proliferation, specification, and migration); (2) establishing connections (axon and dendritic growth, and synapse formation); (3) modifying connections (nerve cell death and reorganization of initial inputs); and (4) adult plasticity (learning and nerve growth after injury). Stage 3 is the most relevant to our question.

During fetal development, many more nerve cells are formed than will be found in the adult brain. These nerve cells all send out axons and establish connections with target cells (other neurons and muscle cells), but a large proportion of them die before the synapses become functional. The mechanism underlying this process involves electrical activity in the nerve cells and their targets, but it is still not fully understood (see Oppenheim, 1991). It is thought that neuronal cell death may serve to eliminate errors in the initial pattern of connections.

The axons of the cells that remain are often found to have more extensive branches and to contact more postsynaptic cells than they will in the adult. The mechanisms that bring about the elimination and reorganization of these terminal branches also involve activity in the neurons. These processes are beginning to be understood, and have important implications for understanding the development of behavior mechanisms. In brief, it has been shown that specific spatial and temporal patterns of electrical activity in both the nerve cells and their target cells are necessary for functional connections to form between them: 'cells that fire together wire together' (Shatz, 1992, p. 64).

It was the work of Hubel and Wiesel that established the necessary role of environmental stimulation in the development of the mammalian visual system (see Blakemore, 1973, and Wiesel, 1982, for reviews). Working with kittens and monkeys, they showed that normal development of the connections between the cells of the lateral geniculate nucleus and the visual cortex requires visual stimulation of a certain kind at a specific time during development. Kittens that were raised with total deprivation of visual pattern, with visual patterns presented to only one eye at a time, or with various other kinds of physical and chemical impediments to normal visual stimulation, developed major abnormalities in the organization of the visual cortex. For example, cortical cells in monocularly raised kittens responded only to visual stimuli from one eye, whereas most cortical cells in normally raised kittens responded to visual stimuli from both eyes. These results for ocular dominance were interpreted in terms of the eyes competing for control of cells in the cortex. Further, stimulation, whether normal or abnormal, was effective in determining neural organization only during a specific and relatively short period of time soon after the kitten's eyes opened. These results were important because they showed that the organization of a sensory system was actually driven by stimulation from the environment. They also provided a model for how the perceptual mechanisms underlying bird-song learning and filial and sexual imprinting might develop (see Bolhuis, DeVoogd, Bischof, this volume).

The neural activity responsible for organizing or remodeling connections in the visual cortex was triggered by stimuli originating in the environment after the animal was born and had opened its eyes and could see. Several investigators went on to ask whether neural activity was also necessary for neural connections to form *in utero*, and, if so, how this activity was instigated. Shatz (1992) and her collaborators, for example, looked at axonal remodeling in the lateral geniculate nucleus of the cat, which occurs before birth. They found that the same kind of action-potential activity was necessary for developing normal connections in the lateral geniculate as was later necessary in the cortex. But rather than being instigated by stimulation from the external world, the neural activity was caused by spatial patterns of spontaneous neural firing. How these waves of activity are generated remains to be discovered. What is important in the present context, however, is that the mechanisms for synaptic change are the same before and after birth, and it is irrelevant for the connection being formed whether the neural activity arises from endogenous or exogenous sources (cf. Hogan, 1988, pp. 94–95).

One last piece of information is necessary before we can see whether there are inherent differences in the way perceptual and motor mechanisms develop. Studies on the development of neuromotor connections have shown that a process of axonal remodeling occurs that is very similar to the process in the visual system (see Brown *et al.*, 1991, Chapter 9). In the adult, each muscle fiber is innervated by only one motor neuron, whereas early in development the muscle may be innervated by several motor neurons. A competition for muscle fibers occurs that requires specific temporal patterns of electrical activity. This activity could be instigated either externally or internally, but in many cases it must be generated endogenously. One example is the phenomenon of motor primacy in the chick embryo, which was discovered by Preyer in 1885 (see Oppenheim, 1974). In the chick, the spinal neurons and musculature of the motor system differentiate and begin functioning before the spinal sensory system. Insofar as the mechanisms of neural connection are the same in the chick as in mammals, this means that the organization of neuromotor connections must depend on endogenously generated electrical activity because patterned movements can be seen in the embryo prior to any possibility of sensory input.

These results all suggest that there are no fundamental differences in the causal factors responsible for the development of perceptual and motor mechanisms. In both cases, the organization of neural or neuromotor connections depends on particular spatiotemporal patterns of neural activity that can be generated either endogenously or exogenously. Prior to birth, most of the causal factors would be endogenous, although external stimulation may play a role in some cases (e.g. the auditory system in ducks: Gottlieb, 1978). After birth, both internal and external factors could be important. The fact that most of the motor mechanisms we have considered develop prefunctionally very likely reflects the fact that motor mechanisms generally become organized earlier in development than perceptual mechanisms.

It should be noted that this analysis makes it possible to understand results such as those of Groothuis (this volume) on the development of the oblique posture in the black-headed gull. He found that the display develops normally when a gull is reared in social isolation, but that it sometimes develops abnormally when a gull is reared with only two or three peers. One can suppose that endogenously produced patterns of spontaneous neural firing provide the information necessary to develop the normal connections in the motor mechanism responsible for the form of the display. When conspecifics are present, performance of precursors of the display often leads to reactions by the other gulls. These reactions, in turn, provide additional neural stimulation which could interfere with the endogenously produced patterns and thus lead to different (abnormal) connections being formed in the motor mechanism.

There has been a recent proposal that different neural mechanisms may have evolved for brain systems that serve different functions. In considering the question of how experience can influence the developing brain, Greenough, Black & Wallace (1987) have proposed a categorization scheme based upon the type of information stored. They distinguish experience-expectant from experience-dependent information storage. This distinction is based upon the functional requirements of particular brain systems.

Experience-expectant information storage refers to incorporation of environmental information that is ubiquitous in the environment and common to all species members, such as the basic elements of pattern perception....Experience-dependent information storage refers to incorporation of environmental information that is idiosyncratic, or unique to the individual, such as learning about one's specific physical environment or vocabulary. (p. 539).

They also suggest that experience-expectant processes depend on selection or pruning of overproduced synaptic connections (i.e. axonal remodeling, as discussed above), while experience-dependent processes depend on formation of new synaptic connections.

With respect to the type of environmental information stored, the development of all the perceptual and motor mechanisms we have described would be classified as experience-expectant, and therefore, according to Greenough *et al.*, dependent on synapse pruning. As we have seen above, however, the mechanisms responsible for axonal remodeling are the same regardless of whether the information comes from the environment or from endogenous processes. Thus, the word experience would have to be used broadly so as to include all information originating outside the specific brain structure itself (see Schneirla, 1965; Lehrman, 1970). It does not seem that this use of the word was intended. A further problem with this classification arises with respect to the environmental information stored during imprinting, which is common to all members of the species in the natural environment. As Bolhuis discusses in Chapter 2, the development of perceptual mechanisms during imprinting must also involve experience-dependent processes.

Greenough *et al.* also suggest that their categories offer a new view of phenomena that have previously been labeled critical or sensitive periods. Instead of viewing these phenomena as due 'to the brief opening of a window, with experience influencing development only while the window is open' (cf. Bateson, 1979), their approach 'allows consideration of the evolutionary origins of a process, its adaptive value for the individual, the required timing and character of experience, and the organism's potentially active role in obtaining appropriate experience for itself.' (p. 539). This view proposes a functional explanation for a causal phenomenon, which leads to all the problems discussed in Chapter 1. I would suggest a different way to view critical periods. It may be that neural development that involves the elimination and reorganization of terminal axon branches (axonal remodeling) is essentially irreversible. The critical period then becomes the time that the axonal remodeling occurs; it would depend on all those factors that can affect the timing of the remodeling. This conceptualization includes all experience-expectant processes, but is considerably broader and becomes congruent with the putative neural mechanism underlying it. The production of new synapses continues to occur throughout life, and could modulate the structure of behavior mechanisms after the critical period had passed.

Social and non-social behavior systems

A social behavior system could be defined as one in which the motor patterns belonging to that system are normally directed toward another animal (usually of the same species) and/or in which another animal provides the adequate stimulus for activating the perceptual mechanism(s) belonging to that system. Using this definition, dustbathing and hunger (in chickens) would be non-social systems, whereas aggression and sex would be social systems. The escape/fear system (see Kruijt, 1964; Hogan, 1965) is ambiguous because it is activated in both social and non-social (e.g. with respect to mealworms) situations. However, if the adequate stimulus for the fear system is taken to be any unfamiliar object, it could be classified as non-social. The parental and filial systems, discussed in other chapters, would be social systems. The question to be explored in this section is whether there are any important differences in the development of social and non-social behavior systems.

As we have seen earlier in this chapter in our review of the development of various behavior systems in chickens, the motor mechanisms of all systems tend to develop normally even in the absence of functional experience. This is as true for simple patterns such as pecking and swallowing as it is for the complicated patterns and groups of patterns seen in dustbathing and during aggressive and sexual encounters. Likewise, functional experience, often of an extended and subtle kind, is as necessary for the development of perceptual mechanisms for the recognition of food and dust as it is for the development of perceptual mechanisms for the recognition of the opponent or sex partner. The development of connections between perceptual or motor mechanisms and the central coordinating mechanisms also do not seem to differ between social and non-social systems. For example, in the dustbathing and aggressive systems, most of these connections develop prefunctionally, while in the hunger and sexual systems, functional experience is required. Evidence reviewed in other chapters on the development of the parental and filial systems leads to the same conclusion. Both social and non-social behavior systems develop according to the same rules, and there appear to be no systematic differences between them.

Why then have topics such as imprinting and bonding, which traditionally have been associated with social behavior systems, assumed such an important role in the developmental literature? I think the reason for this interest is related to Lorenz's original conception of imprinting. He defined imprinting as 'the acquisition of the object of instinctive behavior patterns oriented towards conspecifics.' (Lorenz 1935/1970, p. 124). In terms of the concepts used in this chapter, we would say that imprinting refers to the development of a perceptual mechanism (or schema) that is responsible for species recognition, and that is connected to all (or many of) the social behavior systems in the animal. The reason imprinting is so important is that Lorenz's definition implies that a single perceptual mechanism serves a number of different behavior systems.

Current evidence from imprinting studies is usually interpreted to mean that the object-recognition mechanisms for filial and sexual behavior develop separately (Clayton, ten Cate, this volume; see also Bolhuis, 1991, p. 254); and Lorenz himself showed in his studies of jackdaws that the objects of the various functional systems he discussed (infant, sexual, social) might be different, and might develop at different periods in the animal's life. Thus, the implication of Lorenz's definition may generally not be true. Nonetheless, the idea that early experience has far-reaching, general effects on later social behavior has remained influential, and is supported by a wide variety of evidence. The question is, how do these effects come about if the perceptual mechanisms of the various social behavior systems develop independently?

One recent suggestion is likely to be widely applicable. Hofer (1987) has studied the processes of early social attachment in young rats and their responses to separation from their mothers. His results show that separation has extensive effects on the young rats' behavior, similar to (though not as dramatic as) the effects of maternal separation on the behavior of young rhesus monkeys (Harlow & Harlow, 1962; see also Hinde & Spencer-Booth, 1971). Hofer analyzes these effects into two components. The first involves the formation of an attachment system, which has similarities to the one proposed by Bowlby (1969) for human infants and to the filial system implicated in imprinting studies in birds. This system develops as the young rat learns the characteristics of its mother; when the infant is separated from her it shows distress reactions, and shows relief when it is later returned to her. If one substitutes an alternative 'caregiver' for the mother, such as an inanimate object or another rat pup, Hofer's results show that the attachment system still seems to function normally.

The novel aspect of Hofer's analysis is the second component: the behavioral and physiological effects that occur during long-term separation from the mother are shown to depend on specific aspects of the mother–infant interaction. Hofer has isolated a number of regulators including body warmth, tactile and olfactory stimulation, stimulation peculiar to the suckling situation, etc. Many of the specific effects of these factors have been described in some detail by Fleming & Blass (this volume). A real mother provides all the necessary regulators, but alternative caregivers do not. Under such circumstances, various behavioral and physiological abnormalities will develop.

Very recently, Kraemer (1992) has interpreted the development of primate social attachment in similar terms to those of Hofer. He points out that a young rhesus monkey may become 'attached' to an abusive mother or to a peer, and that such young monkeys can be seen in many ways to have a normally developed attachment system. But such monkeys also develop abnormally in many other ways. Kraemer provides evidence that absence of an adequate caregiver (i.e. a normal mother monkey) leads to aberrant development of brain biogenic amine systems which are implicated in the control of sensorimotor integration and emotion: 'If the attachment process fails, or if the caregiver is incompetent as a member of the species, the developing infant will also fail to regulate its social behavior and may be dysfunctional in the social environment.' (p. 493).

Lorenz's and Hofer's theories are the same in that both postulate that a representation of the imprinting object or caregiver (perceptual mechanism) is formed early in ontogeny. In Lorenz's theory, that representation controls a number of social behavior systems; and long-term effects are seen because each system matures at its own time in the life of the animal. In Hofer's theory, the representation controls only the attachment system; long-term effects are seen because the object to which the animal is attached provides the necessary conditions for various biochemical and neural changes that are indispensable for normal development of other systems.

Interactions among behavior systems

A basic tenet of ethological theory is that various behaviors of an animal are the expression of the activation of not just a single behavior system, but of the interaction of two or more behavior systems that are activated simultaneously. This conflict hypothesis was proposed by Tinbergen (1952), and has been discussed by Kruijt (1964), Baerends (1975), and Groothuis (this volume). One can ask whether any new principles are necessary to understand the development of such interactions. I will restrict my discussion to the behavior of chickens, but other work on some of these questions has been carried out by Groothuis with respect to the development of displays in black-headed gulls. This work is described and discussed in Chapter 8.

Kruijt's (1964) results show that the major behavior systems of escape, aggression, and sex develop in chickens in that order. Further, activation of a system already developed inhibits the expression of systems that are just beginning to develop. For example, a young chick that shows frontal threatening and jumping to another chick may immediately stop this early aggressive behavior if it bumps into the other too hard. However, as the chicks grow older, the causal factors for aggression become stronger and such escape stimuli no longer stop aggressive behavior. Instead, attack and escape begin to occur in rapid alternation, and various irrelevant movements start to appear during fighting. Similarly, early sexual behavior is immediately interrupted if either the attack or the escape system is activated, but later, behavior containing components of attack, escape, and sex can be seen simultaneously. As we have seen earlier, there is evidence that the basic organization of these major behavior systems is formed prefunctionally, and that their expression merely requires a sufficiently high level of causal factors. The gradual appearance of more complex interactions can be interpreted as reflecting changes in the strength of causal factors rather than changes in the connections among central mechanisms.

The fact that another member of the species is the adequate stimulus for activating the escape, aggression, and sex systems means that all these systems must normally be activated when a conspecific is present. Kruijt (1964) points out that the precise state of activation of these systems at any moment depends on the previous history of the male and on the appearance, distance, and behavior of the other bird. He suggests that the appearance of smooth, typical, adult courtship behavior depends on an increasing activation and mutual inhibition of the attack and escape systems, and the relationship between attack and escape is stabilized by the activation of the sexual system. He posits a stabilizing factor in order to explain why the adult animals do not constantly switch quickly from performing one type of behavior to performing another.

The stabilizing influence of sex on the agonistic systems of escape and aggression is not merely a consequence of increasing hormone levels as the animals grow older. Experience also plays a major role. Kruijt (1964) found that junglefowl males reared from hatching in social isolation for more than 9 months showed serious and apparently irreversible abnormalities in their courtship and sexual behavior. To a large extent, these abnormalities could be characterized as switching too quickly among escape, aggressive and sexual behavior. In other words, the stabilizing influence of sex was present only after experience of the sort that would occur during normal early development. Kruijt also found that as little as 2.5 months of normal social experience immediately after hatching was sufficient to obviate the effects of subsequent social isolation for periods of at least 16 months. These results are difficult to interpret because during the first 2.5 months of life chicks show essentially no sexual behavior. Thus the chicks could not be learning anything specific about sexual behavior. Instead, it would seem that the experience a chick gains during normal encounters early in life provides the information necessary for it to stabilize its agonistic systems, and that normal sexual behavior can only occur if the agonistic systems are already stabilized.

How a stabilizing influence develops has not been studied. We have seen that some of the experiences of a normally raised young chick, such as bumping into other chicks or being pecked at as a result of pecking another chick, are not necessary for normal aggressive behavior to develop. Such experiences might, however, be crucial for developing a normal attack–escape relationship. The fact that this experience must occur early in life implies that axonal remodeling-type processes are involved. In effect, it could be that various perceptual neurons are competing for connections with the attack and escape systems, and that a stable attack–escape balance depends on the pattern of connections that finally develops. This is a very speculative suggestion, but it does fit in well with what is known about development of neural connections at earlier stages. Such a suggestion also implies that no new principles of development are required to understand the development of behavior system interaction.

Conclusions

The development of behavior systems is a very complex process, involving intricate interactions of external and internal causal factors with the genes and their products at every stage. Yet the principles involved in this process seem relatively simple. Specific patterns of neural activity are responsible for the formation of the basic behavioral mechanisms and many of the connections between them, probably through the mechanism of synapse pruning or axonal remodeling. Later stimulation causes the formation of new synapses which probably underlie the modification of behavioral mechanisms and the formation of new connections between them; new synapses are probably also important in the development of new representations (cognitive structures). These two neural mechanisms are, in fact, sufficient for understanding a wide range of developmental phenomena such as canalization, critical periods and irreversibility.

Yet understanding the neural mechanisms that determine development tells us nothing about how a particular system will develop in a particular animal. The development of any specific system and of its interactions with other systems will need to be studied in each specific case. Still, the examples considered in this chapter show that the general principles are the same for all kinds of behavioral mechanisms (perceptual and motor) and for all kinds of behavior systems (social and non-social). This means that any insights gained while studying one system are very likely to generate useful hypotheses when studying other systems.

Acknowledgements

This chapter is dedicated to Jaap Kruijt on the occasion of his retirement. It was he who introduced me to the study of development and, as should be clear from the contents, he has continued to have a very important influence on my thinking. The research from my laboratory reported in this chapter was supported by a grant from the Natural Sciences and Engineering Research Council of Canada. I thank Johan Bolhuis and Lidy Hogan-Warburg for their many suggestions for improvements to the manuscript.

References

Andrew, R. J. (1966). Precocious adult behaviour in the young chick. *Animal Behaviour*, 14, 485–500.
Baerends, G. P. (1975). An evaluation of the conflict hypothesis as an

explanatory principle for the evolution of displays. In *Function and Evolution in Behaviour*, ed. G. P. Baerends, C. Beer & A. Manning, pp. 187–227. London: Oxford University Press.

Baerends-van Roon, J. M. & Baerends, G. P. (1979). The morphogenesis of the behaviour of the domestic cat, with a special emphasis on the development of prey-catching. *Verhandelingen der Koninklijke Nederlandse Akademie van Wetenschappen, Adf. Natuurkunde*, Tweede Reeks (*Proceedings of the Royal Netherlands Academy of Sciences*, Section Physics, Second Series), Part 72.

Bateson, P. P. G. (1979). How do sensitive periods arise and what are they for? *Animal Behaviour*, 27, 470–486.

Blakemore, C. (1973). Environmental constraints on development in the visual system. In *Constraints on Learning*, ed. R. A. Hinde & J. Stevenson-Hinde, pp. 51–73. London: Academic Press.

Blass, E. M., Hall, W. G. & Teicher, M. H. (1979). The ontogeny of suckling and ingestive behaviors. *Progress in Psychobiology and Physiological Psychology*, 8, 243–299.

Bolhuis, J. J. (1991). Mechanisms of avian imprinting: a review. *Biological Reviews*, 66, 303–345.

Bowlby, J. (1969). *Attachment and Loss: Attachment*. New York: Basic Books.

Brown, M. C., Hopkins, W. G. & Keynes, R. J. (1991). *Essentials of Neural Development*. Cambridge: Cambridge University Press.

Evans, R. M. (1968). Early aggressive responses in domestic chicks. *Animal Behaviour*, 16, 24–28.

Gottlieb, G. (1978). Development of species identification in ducklings: IV. Change in species-specific perception caused by auditory deprivation. *Journal of Comparative and Physiological Psychology*, 92, 375–387.

Greenough, W. T., Black, J. E. & Wallace, C. S. (1987). Experience and brain development. *Child Development*, 58, 539–559.

Hall, W. G. & Williams, C. L. (1983). Suckling isn't feeding, or is it? A search for developmental continuities. *Advances in the Study of Behavior*, 13, 219–254.

Harlow, H. F. & Harlow, M. K. (1962). Social deprivation in monkeys. *Scientific American*, 207(5), 136–146.

Hinde, R. A. & Spencer-Booth, Y. (1971). Towards understanding individual differences in rhesus mother–infant interaction. *Animal Behaviour*, 19, 165–173.

Hofer, M. A. (1987). Early social relationships: A psychobiologist's view. *Child Development*, 58, 633–647.

Hogan, J. A. (1965). An experimental study of conflict and fear: An analysis of behavior of young chicks to a mealworm. Part I. The behavior of chicks which do not eat the mealworm. *Behaviour*, 25, 45–97.

Hogan, J. A. (1971). The development of a hunger system in young chicks. *Behaviour*, 39, 128–201.

Hogan, J. A. (1973a). Development of food recognition in young chicks. I. Maturation and nutrition. *Journal of Comparative and Physiological Psychology*, 83, 355–366.

Hogan, J. A. (1973b). How young chicks learn to recognize food. In *Constraints on Learning*, ed. R. A. Hinde & J. G. Stevenson-Hinde, pp. 119–139. London: Academic Press.

Hogan, J. A. (1977). The ontogeny of food preferences in chicks and other animals. In *Learning Mechanisms in Food Selection*, ed. L. M. Barker, M. Best & M. Domjan, pp. 71–97. Waco, Texas: Baylor University Press.

Hogan, J. A. (1984). Pecking and feeding in chicks. *Learning and Motivation*, 15, 360–376.

Hogan, J. A. (1988). Cause and function in the development of behavior systems. In *Handbook of Behavioral Neurobiology*, Vol. 9, ed. E. M. Blass, pp. 63–106. New York: Plenum Press.

Hogan, J. A., Honrado, G. I. & Vestergaard, K. (1991). Development of a behavior system: Dustbathing in the Burmese red junglefowl (*Gallus gallus spadiceus*): II. Internal factors. *Journal of Comparative Psychology*, 105, 269–273.

Hogan, J. A. & Roper, T. J. (1978). A comparison of the properties of different reinforcers. *Advances in the Study of Behavior*, 8, 155–255.

Hogan, J. A. & van Boxel, F. (1993). Causal factors controlling dustbathing in Burmese red junglefowl: Some results and a model. *Animal Behaviour*, 46, 627–635.

Hogan-Warburg, A. J. & Hogan, J. A. (1981). Feeding strategies in the development of food recognition in young chicks. *Animal Behaviour*, 29, 143–154.

Kraemer, G. W. (1992). A psychobiological theory of attachment. *Behavioral and Brain Sciences*, 15, 493–511.

Kruijt, J. P. (1962). Imprinting in relation to drive interactions in Burmese red junglefowl. *Symposia of the Zoological Society, London*, 8, 219–226.

Kruijt, J. P. (1964). Ontogeny of social behaviour in Burmese red junglefowl (*Gallus gallus spadiceus*). *Behaviour*, Supplement 9, 1–201.

Lehrman, D. S. (1970). Semantic and conceptual issues in the nature-nurture problem. In *Development and Evolution of Behavior*, ed. L. R. Aronson, E. Tobach, D. S. Lehrman & J. S. Rosenblatt, pp. 17–52. San Francisco: Freeman.

Lorenz, K. (1970). Companions as factors in the bird's environment (1935). In *Studies in Animal and Human Behaviour*, tr. R. Martin, Vol. 1, pp. 101–258. London: Methuen.

Marler, P. (1987). Sensitive periods and the roles of specific and general sensory stimulation in birdsong learning. In *Imprinting and Cortical Plasticity*, ed. J. P. Rauschecker & P. Marler, pp. 99–135. New York: John Wiley.

Moffatt, C. & Hogan, J. A. (1992). Ontogeny of chick responses to maternal food calls in the Burmese red junglefowl. *Journal of Comparative Psychology*, 106, 92–96.

Oppenheim, R. W. (1974). The ontogeny of behavior in the chick embryo. *Advances in the Study of Behavior*, 5, 133–172.

Oppenheim, R. W. (1991). Cell death during development of the nervous system. *Annual Review of Neuroscience*, 14, 453–501.

Petherick, J. C. & Duncan, I. J. H. (1989). Behaviour of young domestic fowl directed towards different substrates. *British Poultry Science*, 30, 229–238.

Schiller, P. H. (1949). Manipulative patterns in the chimpanzee. In *Instinctive Behavior*, ed. C. H. Schiller, pp. 264–287. New York: International Universities Press.

Schneirla, T. C. (1965). Aspects of stimulation and organization in approach/withdrawal processes underlying vertebrate behavioral development. *Advances in the Study of Behavior*, 1, 1–74.

Shatz, C. J. (1992). The developing brain. *Scientific American*, 267(3), 60–67.

Tinbergen, N. (1952). Derived activities: Their causation, biological significance, origin and emancipation during evolution. *Quarterly Review of Biology*, 27, 1–32.

van Liere, D. W. (1992). The significance of fowls' bathing in dust. *Animal Welfare*, 1, 187–202.

van Liere, D. W. & Bokma, S. (1987). Short-term feather maintenance as a function of dust-bathing in laying hens. *Applied Animal Behaviour Science*, 18, 197–204.

van Liere, D. W., Kooijman, J. & Wiepkema, P. R. (1990). Dustbathing behaviour of laying hens as related to quality of dustbathing material. *Applied Animal Behaviour Science*, 26, 127–141.

Vestergaard, K. (1982). Dust-bathing in the domestic fowl – Diurnal rhythm and dust deprivation. *Applied Animal Ethology*, 8, 487–495.

Vestergaard, K. & Hogan J. A. (1992). The development of a behavior system: Dustbathing in the Burmese red junglefowl. III. Effects of experience on stimulus preference. *Behaviour*, 121, 215–230.

Vestergaard, K., Hogan, J. A. & Kruijt, J. P. (1990). The development of a behavior system: Dustbathing in the Burmese red junglefowl: I. The influence of the rearing environment on the organization of dustbathing. *Behaviour*, 112, 99–116.

Vestergaard, K., Kruijt, J. P. & Hogan, J. A. (1993). Feather pecking and chronic fear in groups of red junglefowl: Its relations to dustbathing, rearing environment and social status. *Animal Behaviour*, 45, 1117–1126.

Wiesel, T. N. (1982). Postnatal development of the visual cortex and the influence of the environment. *Nature*, 299, 583–591.

Part four
Development of cognition

11

Cortical mechanisms of cognitive development

MARK H. JOHNSON

Introduction

The region of the primate brain which shows the greatest extent of postnatal development is the cerebral cortex, with detectable changes occurring within the region in humans until the teenage years. Not unrelatedly, the cerebral cortex is also the region of the mammalian brain most susceptible to the effects of postnatal experience. While the exact role of cerebral cortex in psychological processes is still unclear, several authors have argued that extent of the cerebral cortex may be correlated with 'intelligence' across species (e.g. MacPhail, 1982). Thus, the main evolutionary development within the brain across mammals is the relative expansion of the area of cerebral cortex. For example, the area of the cortex in the cat is about 100 cm, whereas that of the human is about 2400 cm (24 times the size). This suggests that the extra cortex possessed by primates, and especially humans, is related to the higher cognitive functions they possess. The aim of this chapter is to describe some of the progressive and regressive events which occur during the postnatal development of the primate cortex, and to discuss how these neural developments may relate to advances in perceptual, attentional, and memory abilities in human infants.

Until recently, the study of the development of cognitive abilities such as attention, language, and object recognition had proceeded largely independently of any considerations of their neural concomitants. This lack of interest in the brain by cognitive developmentalists may be due to an implicit assumption that identifiable neural developments which correspond in time or age to a cognitive change may allow the inference that the cognitive change was caused by the maturation of a neural structure (see discussion in Johnson & Morton, 1991). As will become clear later in this chapter, this assumption is not always correct, since many aspects of cortical development appear to be extremely sensitive to the effects of experience. Furthermore, the particular experiences that the animal has early in life are partly determined by its existing cognitive abilities.

Both progressive and regressive events are characteristic of the postnatal development of the cortex, and both classes of event may have their effects on psychological abilities. In this chapter, I shall suggest that while most of the progressive events in postnatal cortical development are relatively impervious to experiential factors, the majority of regressive events are shaped by sensory input. Further, I shall argue that the relation between neural and cognitive development is two-way. Not only do neural developments bring about cognitive changes, but aspects of neural development may be shaped by the interaction between neurocognitive systems within the organism and its external environment.

The postnatal development of the cerebral cortex

The cortex is basically a thin flat sheet (about 3–4 mm thick) which becomes increasingly convoluted with both phylogenetic and ontogenetic development. It has a fairly consistent internal structure throughout its extent. This does not mean, however, that this structure is simple. It is not. In fact, we still have no widely accepted computational theories of how the cortex works. Figure 11.1 shows a schematic section through an area of primate cortex, the primary visual cortex of the adult macaque monkey. This section is cut at right-angles to the surface of the cortex, and reveals its laminar structure. Each of the laminae has particular cell types within it, and has particular patterns of inputs and outputs. For example, with regard to inputs and outputs, layer IV is where inputs to the cortex from thalamic regions terminate. This is the layer where most of the inputs from regions other than the cortex arrive. The deeper layers, V and VI, often project back to subcortical regions that project to the corresponding layer IV, as well as other subcortical areas. More superficial layers, II and III, commonly project forward to neighbouring regions of cortex. Layer I is very diffuse and may be involved in long-range cortico–cortico and cortico–limbic connections. We shall see later that the differential structure of layers in the cortex may be used to analyse developmental changes at the cognitive level.

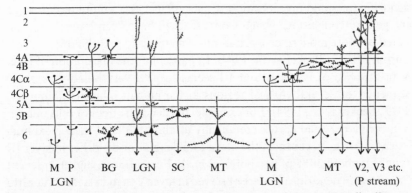

Figure 11.1 A schematic representation of primary visual cortex. Not all cell types or connections are shown. Abbreviations: LGN = lateral geniculate nucleus; SC = superior colliculus; BG = basal ganglia; MT = middle temporal area; M = broad band (magnocellular) stream; P = colour opponent (parvocellular) stream. (Reprinted from *Journal of Cognitive Neuroscience* (Johnson, 1990) by permission of the MIT Press.)

Progressive events in cortical development

Rakic (1988) has proposed a 'radial unit model' of neocortical development which gives an account of how the structure of the mammalian cerebral cortex arises during ontogeny. According to this model, the vertical columnar organisation of the cerebral cortex is determined by the fact that each proliferative unit (a cell that gives birth to many others in the subventricular zone, just below the region that subsequently becomes the cortex) gives rise to approximately 100 neurones that all migrate up the same radial glial fibre, with the latest to be born travelling past their older relatives. This results in an inside-out pattern of growth. Thus, radial glial fibres act like a climbing rope to ensure that cells produced by one proliferative unit all contribute to one ontogenetic column. Rakic has speculated that regulatory genes, known as homeobox genes, may be involved in parcellating the proliferative zone into a basic protomap of cortical cytoarchitectonic areas. The scaffolding provided by the radial glial fibres simply enables the translation of this two-dimensional map from one location (the thalamus) to another (the cortex).

Although most cortical neurones are in their appropriate locations by the time of birth in the primate, the 'inside-out' pattern of growth extends into postnatal development. Extensive descriptive neuroanatomical studies of cortical development in the human infant by Conel over a 30-year period led him to the conclusion that the postnatal growth of cortex proceeds in an 'inside-out' pattern with regard to the extent of dendrites, dendritic trees and myelinisation (Conel, 1936–67). In more recent years, the general conclusions which Conel reached have been largely substantiated by more modern neuroanatomical methods (e.g. Rabinowicz, 1979; Purpura, 1975; Becker *et al.*, 1984). In particular, the maturation of layer V in advance of layers II and III seems to be a very reliably observed sequence for several cortical regions in the human infant (Rabinowicz, 1979; Becker *et al.*, 1984). For example, the dendritic trees of cells in layer V of primary visual cortex are already at about 60% of their maximum extent at birth. In contrast, the mean total length for dendrites in layer III is only at about 30% of maximum at birth. Furthermore, higher orders of branching in dendritic trees are observed in layer V than in layer III at birth (Becker *et al.*, 1984; Huttenlocher, 1990). Interestingly, this inside-out pattern of growth is not evident in the later occurring rise and fall in synaptic density. For this measure, there are no clear differences between cerebral cortical layers. We will return to this issue later.

The differential development of cortical areas

In many of the neuroanatomical variables mentioned so far, there appears to be differential development not only between different brain structures, but also between regions within the same structure. For example, Huttenlocher (1990) reports clear evidence of a difference in the timing of postnatal neuroanatomical events between the primary visual cortex and the frontal cortex in human infants, with the latter reaching the same developmental landmarks considerably later in postnatal life than the former. It is worth noting that this differential development within the cerebral cortex has not always been reported in other primate species (Rakic *et al.*, 1986).

Consistent with the reports from postmortem tissue, a study in which the functional development of the human brain was investigated by positron emission tomography (PET) has also found differential development between regions of cortex (Chugani & Phelps, 1986; Chugani, Phelps & Mazziotta, 1987). In infants under 5 weeks of age, glucose uptake was highest in sensorimotor cortex, thalamus, brainstem and the cerebellar vermis. By 3 months of age there were considerable rises in the parietal, temporal and occipital cortices, basal ganglia, and cerebellar cortex. Maturational rises were not found in the frontal and dorsolateral occipital cortex until approximately 6 to 8 months.

Regressive events in cortical development

Regressive events are commonly observed by those studying the development of nerve cells and their connections in the brain. Within the cortex, these events commonly occur some time after the layer-wise dendritic development discussed earlier. For example, in primary visual cortex the mean density of synapses per neurone only starts to decrease at the end of the first year of life (e.g. Huttenlocher, 1990). In humans, most cortical regions and pathways undergo this postnatal loss, which can sometimes exceed 50 % (for reviews see: Clarke, 1985; Cowan *et al.*, 1984; Hopkins & Brown, 1984; Janowsky & Findlay, 1986; Purves & Lichtman, 1985).

That selective loss is a significant influence on postnatal primate brain development is evident in a number of different quantitative measures. For example, in the PET study alluded to above, the authors found a somewhat surprising result. This was that the absolute rates of glucose metabolism rise postnatally until they *exceed* adult levels, before returning to adult levels after about 9 years of age. For most cerebral cortical regions the levels reached are about double those found in the adult. One possibility is that this developmental pattern of glucose uptake reflects the over-production of synapses known to occur in many primate brain regions (Chugani *et al.*, 1987).

Consistent with the PET findings of Chugani and colleagues, Huttenlocher (1990) reports quantitative neuroanatomical evidence from two comparatively well-studied regions of the human brain – the primary visual cortex and a part of the prefrontal cortex. In both regions the density of synapses shows a characteristic increase to levels about twice that found in the adult. This is then followed by a period of synaptic loss. Huttenlocher suggests that this initial overproduction of synapses may have an important role in the apparent plasticity of the young brain, a matter which will be discussed in more detail later. There is no strong evidence for this pattern of rise and fall for either density of dendrites or for the number of neurones themselves in humans or other primates. However, in rodents and other vertebrates, postnatal cell loss may be more significant.

Support for the proposal that the peak in density of synapses coincides with the peak in glucose uptake as identified by PET has been obtained in a recent developmental study conducted with cats (Chugani *et al.*, 1991). In this study, the peak of glucose uptake in cat visual cortex was found to coincide with the peak in overproduction of synapses in this region. However, when similar data from human visual cortex are plotted together

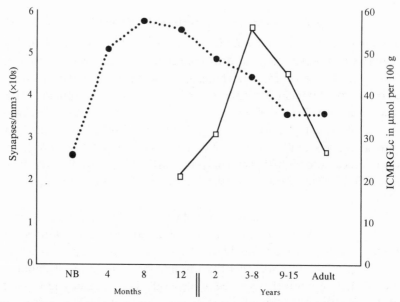

Figure 11.2 Graph showing the development of density of synapses in human primary visual cortex (dotted line - data taken from Huttenlocher, 1990), and resting glucose uptake in the occipital cortex as measured by positron emission tomography (solid line - data taken from Chugani *et al.*, 1987). (ICMRGlc refers to the local cerebral metabolic rates for glucose.)

(Figure 11.2), it is apparent that the peak of glucose uptake lags behind synaptic density. Clearly, further work on the relation between these two measures of cortical development is required.

Cortical growth and the development of perceptual and cognitive abilities

Progressive events and perceptual development

The subcortical to cortical shift

Since the classic studies on motor development by McGraw (1943) and others, it has been argued that development in early infancy can be characterised as a shift from subcortical to cortical control over behaviour. Bronson (1974, 1982) proposed that the primary visual pathway does not gain control over visuomotor behavior in the human infant until between 2 and 4 months after birth. Prior to this age, the infant was thought to respond to visual stimuli primarily by means of the subcortical visual pathway. The detailed evidence in support of this contention has been discussed and extensively reviewed elsewhere (Atkinson, 1984; Bronson, 1974, 1982; Johnson, 1990). Briefly, a variety of neuroanatomical, electrophysiological and lesion evidence indicates that structures on the subcortical visual pathway are developmentally in advance of those on the cortical visual pathway in early infancy. Bronson used this evidence to account for the differences between visually guided behaviour in the newborn infant, as compared to the infant of 3 or 4 months old. The general claim of a shift from subcortical to cortical processing has been reinforced by evidence from the PET studies discussed earlier (Chugani *et al.*, 1987). Assuming that there is some validity to the general claim that there is a shift from subcortical to cortical processing in early infancy, then this should have implications not only for simple aspects of visually guided behaviour, but also for more 'cognitive' aspects of perception, such as the ability to recognise faces. With this in mind, my colleagues and I investigated changes in face-recognition abilities that coincide with the period of rapid cortical visual pathway maturation around 2 months of age. When we began our work on this topic, the existing literature appeared somewhat contradictory. The prevailing view, and most of the evidence, supported the contention that it takes the infant about 2 or 3 months to learn about the arrangement of features that compose a face. Clearly, this is consistent with the idea that this ability is subserved by cortical mechanisms. However, one study provided evidence that newborn infants about 10 minutes old would track (by means of head and eye movements) a face-like pattern further than various 'scrambled' face patterns (Goren, Sarty & Wu, 1975). My colleagues and I attempted to replicate the claims made by Goren *et al.* As in the original study, newborn infants (around 30 minutes old) were required to track different visual stimuli. This testing procedure differs markedly from that employed by the majority of other investigators. Rather than the infant viewing one or more stimuli in static locations and measuring the length of time spent looking at the stimuli, in the Goren *et al.* tracking procedure the dependent measure is how far the infant will turn its head and eyes in order to keep a moving stimulus in view. The stimulus is moved slowly away from the midline and the angle at which the infant disengages its eyes from the stimulus is recorded. We used similar stimuli to those used in the original study, and replicated the effect reported (Johnson, Dziurawiec, Ellis & Morton, 1991: Experiment 1).

The replication of the Goren *et al.* effect raised two questions. First, how specifically face-like does the test stimulus require to be in order to elicit the preferential tracking? Second, what is the developmental time course of this preferential tracking response? In an attempt to begin to address the first of these two questions, we performed a series of experiments using similar procedures to the one just mentioned except that we expanded the set of stimuli to include various forms of 'degraded' face stimuli and optimal spatial frequency patterns (Johnson *et al.*, 1991: Experiment 2; see Chapter 5 of Johnson & Morton, 1991). Although there is much work still to be done in defining the exact characteristics of the face-like patterns that are important, at present it appears that high-contrast 'blobs' in the correct relative spatial locations for eyes and a mouth are sufficient. Spatial frequency components of the stimuli alone cannot account for the preference (see Morton, Johnson & Maurer, 1990).

The second question raised by the newborn findings related to the time course of the preferential tracking response. Although it is not feasible to test older infants with exactly the same procedure we used with newborns, we devised an equivalent situation in which the infant was still required to track similar stimuli by means of head and eye turning (Johnson *et al.*, 1991: Experiment 3). Using this procedure, the results indicated that the preferential tracking of faces declines sharply between 4 and 6 weeks after birth.

Why should this decline occur at this age? We have proposed that the preferential tracking response in newborns is primarily mediated by the subcortical visual pathway (Johnson, 1988; Johnson & Morton, 1991; Morton & Johnson, 1991). Apart from considerations of developmental neuroanatomy and neurophysiology alluded to earlier, the time course of the preferential tracking is similar to that of other responses of newborns thought to be mediated by subcortical circuits, e.g. the imitation of facial gestures (Vinter, 1986) or prereaching (Von Hofsten, 1984). The disappearance of these early reflex-like behaviours in the second month of life has been proposed to be due to inhibition of subcortical circuits by developing cortical circuits (see also Muir, Clifton & Clarkson, 1989).

Other sources of evidence support the conclusion that the newborns' face tracking is primarily mediated by subcortical circuits. The temporal visual field feeds more directly into the subcortical pathway, whereas the nasal feeds mainly into the cortical pathway. In a tracking task such as that described, the stimulus is continually moving out of the central visual field and toward the periphery. It may be that this movement into the temporal field initiates a saccade to re-foveate the stimulus in newborns (for details see Bronson, 1974; Johnson, 1990). This consistent movement toward the periphery would not necessarily arise with static presentations, and therefore preferences will rarely be elicited. Thus, the tracking task may effectively tap into the capacities of subcortical structures such as the superior colliculus.

It appears, then, that the development of face recognition in human infants may be accounted for in terms of two mechanisms with independent time courses of neural development. First, there is a system accessed via the subcortical visual pathway that underlies the preferential tracking in newborns, but whose influence wanes by 6 weeks. Second, there is a system dependent upon cortical maturity and requiring exposure to faces that appears around 8 weeks of age. The notion of a shift from subcortical to cortical processing over the first few months of life, therefore, may be useful when considering even apparently complex perceptual abilities.

In recent years, the notion of a shift from subcortical to cortical visual processing in early infancy has come under criticism for several reasons. First, the increasing evidence for apparently sophisticated perceptual abilities in early infancy (e.g. Slater, Morison & Somers, 1988; Bushnell, Sai & Mullin, 1989). Second, recent knowledge about cortical visual processing has brought into question the notion of one cortical visual pathway: recent evidence suggests that multiple distinct cortical pathways should be considered (e.g. Van Essen, 1985; de Yoe & Van Essen, 1988). These considerations have led several authors to propose some form of partial cortical functioning in the newborn (e.g. Maurer & Lewis, 1979; Posner & Rothbart, 1980; Atkinson *et al.*, 1988). As yet, however, this partial cortical functioning remains poorly specified. This consideration motivated some recent attempts to relate differential maturation intrinsic to the cortex to aspects of perceptual and cognitive development. One of my own attempts in this direction is discussed in the following section.

The 'inside-out' pattern of growth and the development of visual attention

The inside-out pattern of postnatal growth of the cortex is crucial for the proposals reviewed in this section. This is because, as mentioned earlier, particular afferents and efferents to any cortical area terminate and depart from particular layers. Thus, if only some layers are functionally mature, then there is a restriction on the afferent and efferent connectivity of the area concerned. Figure 11.1 is a highly schematised diagram of the inputs and outputs of primary visual cortex of the macaque. Several things are worth noting about this section through the visual cortex. As development proceeds upwards through the layers, more output pathways will come 'on line'. In fact, in primary visual cortex this effect may be exacerbated since the afferents from the lateral geniculate nucleus (LGN) slowly grow through the deeper layers (possibly forming temporary synapses on the way) until they reach their adult termination sites in layer 4 at about 2 months of age in the human infant (Conel, 1939–1967). Therefore, as the innervation passes up through the layers, different output pathways may start to feed-forward information to other cortical areas. Since the primary visual cortex is the 'gateway' to nearly all cortical visual processing (Schiller, 1985; but see Johnson, in press a, for caveats to this), this differential maturation has some profound consequences for the control of visually guided behaviour.

Schiller (1985) proposed, on the basis of electrophysiological and lesion data, that there are four pathways underlying oculomotor control in adult primates These four pathways are: (i) a pathway from the retina to the superior colliculus, the *SC pathway*, thought to be involved in the generation of eye movements toward simple, easily discriminable stimuli, and fed mainly by the peripheral visual field; (ii) a cortical pathway that goes both directly to the superior colliculus from the primary visual cortex and also via the middle temporal area (MT), the *MT pathway*; (iii) a cortical pathway where both magnocellular and parvocellular streams of processing converge in the frontal eye fields, the *FEF pathway*, and which is involved in the detailed and complex analysis of visual stimuli such as the temporal sequencing of eye movements within complex arrays. (Neurophysiological and psychophysical evidence suggests that there are two relatively independent pathways for visual signals from the retina, through the LGN, and within some cortical structures. One of these pathways, the magnocellular pathway, is thought to be concerned with motion processing, while the other, the parvocellular pathway, may be more concerned with colour and form processing.) (iv) The fourth pathway for the control of eye movements is an inhibitory input to the colliculus from several cortical areas via the substantia nigra, the *inhibitory pathway*. Schiller proposes that this final pathway ensures that the activity of the colliculus can be regulated.

The specific proposal made by Johnson (1990) is that, first, the characteristics of visually guided behaviour in the human infant at particular ages is determined by which of these pathways is functional, and second, which of these pathways is functional is determined in turn by the maturational state of the visual cortex. In this chapter only one of the transitions will be illustrated, that which is attributed to the onset of the FEF pathway between 2 and 4 months of age.

There are two types of transitions observed between 2 and 4 months of age which are indicative of FEF pathway functioning. The first of these is that the control over eye movements becomes less input-driven, and more anticipatory in nature, while the second concerns the onset of the ability to perform so-called anti-saccades.

(i) *The onset of anticipatory eye movements.* This onset of anticipatory eye movements can be demonstrated both in tasks which require the tracking of a moving object, and also in tasks where the infant shows savings to react to predictable sequences of stimuli. By around 2 months, infants begin to be able to track moving objects smoothly and respond readily to stimuli placed in the nasal visual field (Aslin, 1981). Only after 2 months, however, do infants start to show anticipatory tracking in that their eye movements 'predict' the trajectory and speed of a stimulus which moves in a regular repeating fashion (see Johnson, 1990). Other studies, which involve presenting stimuli on one of two different screens in regular or irregular sequences, have shown that only after about 2 months of age do infants show faster reaction times to orient toward stimuli when they are part of regular repeating sequences of presentation (Haith, Hazan & Goodman, 1988). Using a different procedure, Johnson, Posner & Rothbart (1991) demonstrated that only by 4 months of age do infants show clear evidence of the ability to make predictive eye movements.

(ii) *The onset of anti-saccades.* The frontal eye fields contain neurones which often discharge in relation to saccadic eye movements, and increase their response to targets within their visual field when that stimulus serves as a target for a subsequent saccade (Goldberg & Bushnell, 1981; Bruce, 1988). A proportion of cells respond prior to a saccade being made. These observations have led to the proposal that the region is involved in the planning of saccadic eye movements. Consistent with this suggestion is the finding that frontal cortex damage in humans results in an inability to suppress involuntary saccades toward targets, and an apparent inability to control volitional saccades (Guitton, Buchtel & Douglas, 1985; Fischer & Breitmeyer, 1987). Guitton *et al.* (1985) studied patients with frontal lobe lesions, temporal lobe lesions, and normal subjects in an 'anti-saccade' task. In such a task, subjects are instructed not to look at a briefly flashed cue, but rather to look in the opposite direction. Guitton *et al.* (1985) reported that while normal subjects, and patients with temporal lobe damage, could do this easily, patients with frontal damage, and especially those with damage around the FEF, were severely impaired. The frontal

patients had difficulty in suppressing the unwanted saccades
toward the cue stimulus.

Clearly, one cannot give verbal instruction to a young infant to look to
the side opposite from where the cue stimulus appears. Fortunately,
however, as discussed in the previous section, infants of around 4 months
or older will rapidly acquire expectations for sequences of stimuli. In order
to test for anti-saccades in young infants, I trained them on a series of trials
with a predictable sequence of presentation. Figure 11.3 illustrates the
sequence of stimuli used by Johnson (in preparation). The infant faces
three video monitors. At the beginning of each trial, infants are presented
with an attractive fixation stimulus on the centre of the three screens. Once
the infant is gazing at this stimulus, a second stimulus is briefly flashed on
one of the two side screens. If this stimulus is presented for long enough,
infants will make a saccade toward it. After a brief gap in which no stimuli
are presented, an attractive, dynamic, target stimulus is presented on the
opposite screen to that on which the brief stimulus had appeared. While
infants will commonly make a saccade toward the briefly presented
stimulus over the first few trials to which they are exposed, they rapidly
learn to ignore this less interesting stimulus and make saccades directly
from the central fixation stimulus to the interesting target stimulus. That is,
despite the presence of a stimulus that we know normally elicits a saccade,
4-month-old infants are able to suppress an eye movement toward it in
order to look toward a more attractive target stimulus.

The differential development of cortical areas

As discussed earlier in the chapter, the frontal area is, in general, the latest
region of the cortex to develop postnatally, with detectable changes
occurring until the teenage years in the human. This suggests that many
aspects of cognitive development may have their neural concomitants in
this region. One aspect of cognitive development which has been related to
the development of the prefrontal cortex is the ability to retrieve an object
in certain situations.

Piaget (1954) discovered a phenomenon which has become known as the
'A not B error'. In this object-retrieval task, infants watch a toy being
hidden in one of two wells, or under one of two cups. After a delay of
several seconds the infant is allowed to reach toward where he or she thinks
that the toy is. From an early age infants will successfully reach to the
appropriate location and retrieve the toy. If, however, after a number of
such trials, the experimenter switches the side on which the toy is hidden in

Training trials

Fixation stimulus

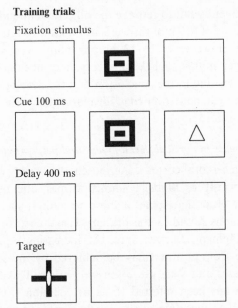

Figure 11.3 The sequence of stimulus presentations for each trial in the 'anti-saccade' experiment. (See text for details.)

full view of the infant, infants will continue to reach to the location in which they originally saw the toy being hidden. That is, they make a perseverative error. Once infants are about 8 to 10 months old, they begin to be able to perform this task quite successfully.

Diamond & Goldman-Rakic (1983, 1985; Diamond, 1988) have gathered evidence from a variety of sources in support of the contention that this advance in infant behaviour is mediated by developments in the dorsolateral prefrontal cortex. These authors noticed the similarity between some of the delayed response tests sensitive to prefrontal damage in human adults and monkeys on the one hand, and the object permanence test devised by Piaget on the other. When they administered a monkey version of Piaget's task, they found that infant monkeys appeared to show the same developmental progression between 1 and 4 months as human infants show between 7 and 12 months.

The next question was how human infants would behave in a task known to be sensitive to frontal cortex damage in monkeys. In this simple task the subject is required to reach for an object (food item or toy). The object is in a box with transparent plastic walls, except that one of the side walls is open. An earlier study had shown that monkeys with frontal damage attempt to reach straight for the food through the solid perspex walls, and thus will consequently fail to retrieve the object. Diamond (1988) reports that human infants of 6–7 months were also unable to retrieve the reward in this situation. However, by 11–12 months infants were successful at the task. The younger infants, then, behave very much like the brain-damaged adult monkeys, leading to the conclusion that the developmental changes seen in the infants can be attributed to the maturation of the frontal cortex.

Regressive events in cortical development and perceptual development

The mammalian cerebral cortex is composed of a variety of vertical units referred to variously as blobs, columns, stripes, and modules. It has recently become clear that some of these structures, such as the ocular dominance columns found in the primary visual cortex, emerge during postnatal development. Furthermore, the increasing separation of one cortical column from another may be achieved by the selective loss of dendritic processes and synapses discussed earlier. This functional segregation, which has been referred to as parcellation (Ebbesson, 1984), often results in the separation of previously combined projections. What effects does this process in cortical development have on infant perception?

Ocular dominance columns and binocular vision

Held (1985) has reviewed converging evidence that binocular vision comes in toward the end of the fourth month of life in human infants (although it is worth noting that this can vary from 2 months to 5 months in different infants). One of the abilities associated with binocular vision, stereoacuity, increases very rapidly from the onset of stereopsis, such that it reaches adult levels within a few weeks. This is in contrast to other measures of acuity, such as grating acuity, which increase much more gradually. Held suggests that this very rapid, sudden spurt in stereoacuity requires some radical change in the neural substrate supporting it. On the basis of evidence from animal studies, he proposes that this substrate is the development of ocular dominance columns found in layer IV of the primary visual cortex.

Neurophysiological studies in monkeys and cats have demonstrated that the geniculocortical afferents from the two eyes are initially mixed so that they synapse on common cortical neurones in layer IV. These layer IV cells project to disparity selective cells (possibly in cortical layers II and III). During ontogeny, geniculate axons originating from one eye withdraw from the region leaving behind axons from the other eye. Thus, clusters of layer IV cells become 'captured' by one of the two eyes. Held posits that it is these events at the neural level that give rise to the sudden increase in stereoacuity observed by behavioural measures at around 4 months of age in the human infant.

More recently, Held (1993) explores implications of the fact that prior to segregation of neuronal input, both eyes project to the same cells in layer IV of the primary visual cortex. Thus, there will be a certain degree of integration between the eyes that will decline once each neurone only receives innervation from one eye. This is elegantly shown in an experiment in which Held and colleagues demonstrate that younger infants (under 4 months) can perform certain types of integration between the two eyes that older infants cannot (Shimojo, Bauer, O'Connell & Held, 1986).

Held and colleagues first of all demonstrated that infants during the first few months of life prefer to look at grid patterns over grating patterns of comparable spatial frequencies. The logic of the experiment depended on the fact that when two orthogonal grating patterns are combined together they form a grid pattern. If inputs from the two eyes are combined, then a grating in one eye orthogonal to that in the other will form a gridlike representation. In the experiment, interocularly orthogonal gratings were paired with parallel ones in a two-alternative preferential looking test. Infants might be expected to prefer the orthogonal gratings before segregation of ocular dominance columns since they summate to some approximation of the preferred grid pattern. However, upon the formation of ocular dominance columns, the infants should prefer binocularly fusible stimuli over competing ones. The results of the experiment bore out the prediction. The infants shifted suddenly, at about the same age as they acquire stereopsis, from a preference for the orthogonal gratings to one for the parallel gratings. In fact, this shift in preference often occurred from one testing session to another only 1 or 2 weeks apart.

Regression of sensory afferents and visuo-spatial processing

In an attempt to account for the loss of processes and synapses that occur during cortical development, Changeux has proposed a mechanism hypothesised to underlie the ontogeny of neural circuits (Changeux, Courrege & Danchin, 1973; Changeux & Dehaene, 1989). In brief, he argues that connections between classes of cells are specified genetically, but are initially labile. Synapses in their labile state may either become stabilised or regress, depending on the total activity of the postsynaptic cell. The activity of the postsynaptic cell is, in turn, dependent upon its input. Initially, this input may be the result of spontaneous activity in the network, but rapidly it is evoked by environmental input. The critical concept here is that of *selective stabilisation*. In other words, Changeux proposes that: 'to learn is to eliminate', as opposed to learning taking place by instruction. What might the consequence of such a neural process be?

Neville (1991) has attempted to provide evidence concerning the mechanisms proposed by Changeux, the elimination of exuberance by sensory input, by using event-related potentials (ERPs) to study inter-sensory competition. In her experiments, she provides evidence from visual and spatial tasks with deaf subjects that cortically mediated visual processing is different among subjects reared in the absence of auditory input. Specifically, the congenitally deaf seem more sensitive to events in the peripheral visual field than are hearing subjects. ERPs recorded over classical auditory regions, such as the temporal lobe, are two or three times larger for deaf than for hearing subjects following peripheral visual field stimulation.

Neville suggests that her data support the idea that, in the absence of auditory input to auditory and polysensory areas of cortex, visual afferents that are normally lost, stabilise or increase. This is clearly consistent with the general position put forward by Changeux and colleagues. In particular, Changeux & Dehaene (1989) proposed that the stabilisation of afferents requires neural activity. Consistent with this, Neville argues that the acquisition of competence in grammar is essential for the left hemisphere specialisation in language. This specialisation may not other-wise emerge. The evidence for this comes from ERP studies of language and hemispheric specialisation. ERP studies reveal greater activation of areas of the left hemisphere in normal hearing subjects during a word-reading task. In contrast to this, deaf subjects did not show this lateralisation. A group of subjects that had learned sign language as their first language, but who were not themselves deaf, showed the lateralisation found in the normal hearing subjects, suggesting that it is not the acquisition of sign language *per se* that gives rise to this difference. Rather, the difference may index the activity of processes involved in either phonological or grammatical decoding. However, the lateralisation effect was found in deaf signers when they were presented with sign language, indicating that the left hemisphere becomes specialised for language regardless of the modality involved, and may develop in association with the acquisition of competence in the grammar.

Constructive processes in neurocognitive development

While much of the discussion so far has centered on cases in which a development in cortical structure may be said to give rise to, or at least provide the necessary substrate for, a cognitive or perceptual advance, I have also alluded to the fact that many aspects of postnatal cortical development are extremely sensitive to the effects of experience. In this concluding section I will review evidence in support of two contentions about the relation between cortical growth and cognitive development:

(i) regressive events in neural development, such as synaptic loss, are those most sensitive to postnatal experience;
(ii) while in some cases neural developments in the cortex may be said to cause or allow cognitive changes, in other cases neurocognitive systems may shape the pattern of neural developments within the cortex. That is, there is a two-way interaction between neural and cognitive development.

Regressive events in the cortex and plasticity

As mentioned earlier, Changeux and others (Changeux *et al.*, 1973; Changeux & Dehaene, 1989; Held, 1985, 1993; Johnson & Karmiloff-Smith, 1992) have argued that regressive events in neural development, such as the 'pruning' of synapses, are a crucial substrate for the plasticity of the cortex. For example, several authors have proposed that a distinction should be made between effects of experience that are common to all members of a species, which the genes of the species have evolved to expect, and changes which are caused by experience likely to be unique to an individual, 'learning' (Greenough, Black & Wallace, 1987; Johnson & Morton, 1991). Greenough and colleagues have referred to effects of experience common to most or all members of a species as 'experience-expectant', and have provided evidence from rats reared in impoverished and enriched environments that this is associated with the blooming and pruning of synapses. (Johnson & Morton's (1991) term 'species-typical environment' corresponds to Greenough's 'experience-expectant', with the difference that the former refers to an aspect of environmental input, while the latter also refers to the neural consequence of that experience.)

While Greenough and colleagues (Greenough *et al.*, 1987; Greenough, 1993) have argued that progressive events in cortical development such as dendritic growth may be influenced by individual learning, or 'experience-dependent' information storage, most of the examples of progressive developments discussed in this chapter have been associated with cognitive and perceptual changes which may be comparatively impervious to the effects of postnatal experience. For example, the development of components of visual attention which were accounted for by the inside-out growth of the primary visual cortex over the first few months of life (Johnson, 1990) may be only accelerated or decelerated to a small extent by environmental factors (Johnson, 1992). Diamond (1991) similarly allows for some limited role for experience in the behavioural transitions which she attributes to progressive developments in the prefrontal cortex. She reports that repeated testing on any of the tasks can accelerate the performance of human infants by 2 or 3 weeks. However, she argues that these accelerations are limited by the maturational state of the brain, since transitions attributed to frontal development occur at around the same age in three tasks which appear to require very different forms of experience. Diamond suggests that the underlying maturation of neural circuitry allows certain types of experience to have effects at certain points only. One piece of evidence for this is that in an object-retrieval task, the infants always go through the same sequence of stages before finally succeeding. Experience cannot push them straight to the final behaviour.

While distinctions between forms of plasticity in the cortex are unlikely to be absolute (Johnson & Morton, 1991), the examples discussed in this chapter lead to the view that while progressive growth of the cortex can only be accelerated or retarded by effects of experience, regressive events such as the 'pruning' of synapses may be more sensitive to postnatal experience (see also Johnson & Karmiloff-Smith, 1992).

The role of neurocognitive systems in cortical development

While the traditional approach to understanding the relation between brain development and behaviour has been to assume that the former causes, or allows, the latter, it has recently become evident that behaviour can affect brain development also (Johnson & Morton, 1991; Johnson, 1992). Let us take the example of the development of face recognition which was discussed earlier.

Evidence was presented consistent with the notion that there are two independent systems involved in face recognition in early infancy; first, a predisposition for newborns to orient toward face-like stimuli (probably subserved by subcortical structures in the brain), and second, a cortically mediated attention toward faces that emerges around the second month of life. Thus, the first system may be said to 'tutor', or select the appropriate input for, the second. Johnson & Morton (1991) suggest that this 'bootstrapping' phenomenon may also apply to other domains of perceptual and cognitive development. By this view it is clear that orienting and attention systems have a vital role in development (Johnson, 1992). By orienting the sensory organs toward particular classes of stimuli while the cortical circuits are still developing, these neurocognitive systems ensure that the cortex develops specialisations for processing information about biologically relevant classes of stimuli.

While the view that newborn predispositions 'tutor' the developing cortex has also been put forward by others (e.g. Horn, 1985; Gibson, 1991), one challenge for the future lies in understanding the influence of cortically mediated components of visual and auditory attention on other aspects of cortical development.

Acknowledgements

Sections of this chapter were begun while I was a Research Fellow at the Center for Cognitive Neuroscience, University of Oregon. I wish to thank Johan Bolhuis, Annette Karmiloff-Smith, Mike Posner, and Mary Rothbart for their comments on earlier versions of this chapter, and Leslie Tucker and Kathy Sutton for their assistance in its preparation. Financial assistance from the Human Frontiers Scientific Foundation, Carnegie Mellon University, and NSF (grant DBS-9120433) is gratefully acknowledged.

References

Aslin, R. N. (1981). Development of smooth pursuit in human infants. In *Eye Movements: Cognition and Visual Perception*, ed. D. F. Fisher, R. A. Monty, & J. W. Senders, pp. 31–51. Hillsdale, NJ: Lawrence Erlbaum.

Atkinson, J. (1984). Human visual development over the first six months of life: a review and a hypothesis. *Human Neurobiology*, 3, 61–74.

Atkinson, J., Hood, B., Wattam-Bell, J., Anker, S. & Tricklebank, J. (1988). Development of orientation discrimination in infants. *Perception*, 17, 587–595.

Becker, L. E., Armstrong, D. L., Chan, F. & Wood, M. M. (1984). Dendritic development on human occipital cortex neurones. *Brain Research*, 315, 117–124.

Bronson, G. W. (1974). The postnatal growth of visual capacity. *Child Development*, 45, 873–890.

Bronson, G. W. (1982). Structure, status and characteristics of the nervous system at birth. In *Psychobiology of the Human Newborn*, ed. P. Stratton, pp. 99–104. Chichester: John Wiley & Sons.

Bruce, C. J. (1988). Single neuron activity in the monkey's prefrontal cortex. In

Neurobiology of Neocortex, ed. P. Rakic & W. Singer, pp. 297–331. Chichester: John Wiley & Sons.

Bushnell, I. W. R., Sai, F. & Mullin, J. T. (1989). Neonatal recognition of the mother's face. *British Journal of Developmental Psychology*, 7, 3–15.

Changeux, J.-P., Courrege, P. & Danchin, A. (1973). A theory of the epigenesis of neuronal networks by selective stabilization of synapses. *Proceedings of the National Academy of Sciences of the USA*, 70, 2974–2978.

Changeux, J.-P. & Dehaene, S. (1989). Neuronal models of cognitive functions. *Cognition*, 33, 63–109.

Chugani, H. T., Hovda, D. A., Villablanca, J. R., Phelps, M. E. & Xu, W-F. (1991). Metabolic maturation of the brain: A study of local cerebral glucose utilization in the developing cat. *Journal of Cerebral Blood Flow and Metabolism*, 11, 35–47.

Chugani, H. T. & Phelps, M. E. (1986). Maturational changes in cerebral function infants determined by (18)FDG positron emission tomography. *Science*, 231, 840–843.

Chugani, H. T., Phelps, M. E. & Mazziotta, J. C. (1987). Positron emission tomography study of human brain functional development. *Annals of Neurology*, 22, 487–497.

Clarke, P. G. H. (1985). Neuronal death in the development of the vertebrate nervous system. *Trends in Neurosciences*, 8, 345–349.

Conel, J. L. (1939–1967). *The Postnatal Development of the Human Cerebral Cortex*, Vols. I–VIII. Cambridge, Mass.: Harvard University Press.

Cowan, M. W., Fawcett, J. W., O'Leary, D. M., & Stanfield, B. B. (1984). Regressive events in neurogenesis. *Science*, 225, 1258–1265.

de Yoe, E. A. & Van Essen, D. C. (1988). Concurrent processing streams in monkey visual cortex. *Trends in Neurosciences*, 11, 219–226.

Diamond, A. (1988). Differences between adult and infant cognition: is the crucial variable presence or absence of language? In *Thought without Language*, ed. L. Weiskrantz, pp. 335–366. Oxford: Clarendon Press.

Diamond, A. (1991). Neuropsychological insights into the meaning of object concept development. In *The Epigenesis of Mind: Essays on Biology and Cognition*, ed. S. Carey & R. Gelman, pp. 67–107. Hillsdale, NJ: Lawrence Erlbaum.

Diamond, A. & Goldman-Rakic, P. S. (1983). Comparison of performance on a Piagetian object permanence task in human infants and rhesus monkeys: evidence for involvement of prefrontal cortex. *Society for Neuroscience Abstracts*, 9, 641.

Diamond, A. & Goldman-Rakic, P. S. (1985). Evidence for the involvement of the prefrontal cortex in cognitive changes during the first year of life. *Society for Neuroscience Abstracts*, 11, 832.

Ebbesson, S. O. (1984). The evolution and ontogeny of neural circuits. *Behavioral and Brain Sciences*, 7, 321–326.

Fischer, B. & Breitmeyer, B. (1987). Mechanisms of visual attention revealed by saccadic eye movements. *Neuropsychologia*, 25, 73–83.

Gibson, K. R. (1991). New perspectives on instincts and intelligence: Brain size and the emergence of hierarchical mental constructional skills. In *'Language' and Intelligence in Monkeys and Apes*, ed. S. T. Parker & K. R. Gibson, pp. 97–128. Cambridge: Cambridge University Press.

Goldberg, M. E. & Bushnell, M. C. (1981). Behavioral enhancement of visual responses in monkey cerebral cortex II. Modulation in frontal eye fields specifically related to saccades. *Journal of Neurophysiology*, 46, 773–787.

Goren, C. C., Sarty, M. & Wu, P. Y. K. (1975). Visual following and pattern discrimination of face-like stimuli by newborn infants. *Pediatrics*, 56, 544–549.

Greenough, W. T. (1993). Brain adaptation to experience: An update. In *Brain Development and Cognition: A Reader*, ed. M. H. Johnson, pp. 319–322. Oxford: Blackwell.

Greenough, W. T., Black, J. E. & Wallace, C. S. (1987). Experience and brain development. *Child Development*, 58, 539–559.

Guitton, H. A., Buchtel, H. A. & Douglas, R. M. (1985). Frontal lobe lesions in man cause difficulties in suppressing reflexive glances and in generating goal-directed saccades. *Experimental Brain Research*, 58, 455–472.

Haith, M. M., Hazan, C. & Goodman, G. S. (1988). Expectation and anticipation of dynamic visual events by 3.5-month old babies. *Child Development*, 59, 467–479.

Held, R. (1985). Binocular vision – Behavioral and neural development. In *Neonate Cognition: Beyond the Blooming, Buzzing Confusion*, ed. J.Mehler & R.Fox, pp. 37–44. Hillsdale, NJ: Lawrence Erlbaum

Held, R. (1993). Development of binocular vision revisited. In *Brain Development and Cognition: A Reader*, ed. M. H. Johnson, pp. 159–165. Oxford: Blackwell.

Hopkins, W. G. & Brown, M. C. (1984). *Development of Nerve Cells and their Connections*. Cambridge: Cambridge University Press.

Horn, G. (1985). *Memory, Imprinting and the Brain*. Oxford: Clarendon Press.

Huttenlocher, P. R. (1990). Morphometric study of human cerebral cortex development. *Neuropsychologia*, 28, 517–527.

Janowsky, J. S. & Findlay, B. L. (1986). The outcome of perinatal brain damage: The role of normal neuron loss and axon retraction. *Developmental Medicine and Child Neurology*, 28, 375–389.

Johnson, M. H. (1988). Memories of mother. *New Scientist*, 1600, 60–62.

Johnson, M. H. (1990). Cortical maturation and the development of visual attention in early infancy. *Journal of Cognitive Neuroscience*, 2, 81–95.

Johnson, M. H. (1992). Cognition and development: Four contentions about the role of visual attention. In *Cognitive Science and Clinical Disorders*. ed. D. J. Stein & J. E. Young, pp. 45–62. New York: Academic Press.

Johnson, M. H. (In press a). Dissociating components of visual attention: A neurodevelopmental approach. In *The Neural Basis of High-level Vision*, ed. M. Farah & G. Radcliffe. Hillsdale, NJ: Lawrence Erlbaum.

Johnson, M. H. (In press b). The development of attention. In *The Cognitive Neurosciences*, ed. M. S. Gazzaniga. Cambridge, Mass: MIT Press.

Johnson, M. H., Dziurawiec, S., Ellis, H. D. & Morton, J. (1991). Newborns' preferential tracking of face-like stimuli and its subsequent decline. *Cognition*, 40, 1–19.

Johnson, M. H. & Karmiloff-Smith, A. (1992). Can neural selectionism be applied to cognitive development and its disorders? *New Ideas in Psychology*, 10, 35–47.

Johnson, M. H. & Morton, J. (1991). *Biology and Cognitive Development: The Case of Face Recognition*. Oxford: Blackwell.

Johnson, M. H., Posner, M. I. & Rothbart, M. (1991). The development of visual attention in infancy: Contingency learning, anticipations and disengaging. *Journal of Cognitive Neuroscience*, 3, 335–344.

MacPhail, E. (1982). *Brain and Intelligence in Vertebrates*. Oxford: Oxford University Press.

Maurer, D. & Lewis, T. L. (1979). A physiological explanation of infants' early visual development. *Canadian Journal of Psychology*, 33, 232–252.

McGraw, M. B. (1943). *The Neuromuscular Maturation of the Human Infant.* New York: Columbia University Press.

Morton, J. & Johnson, M. H. (1991). Conspec and Conlern: A two-process theory of infant face recognition. *Psychological Review*, 98, 164–181.

Morton, J., Johnson, M. H. & Maurer, D. (1990). On the reasons for newborns responses to faces. *Infant Behavior & Development*, 13, 99–103.

Muir, D. W., Clifton, R. K. & Clarkson, M. G. (1989). The development of a human auditory localization response: a U-shaped function. *Canadian Journal of Psychology*, 43, 199–216.

Neville, H. (1991). Neurobiology of cognitive and language processing: Effects of early experience. In *Brain Maturation and Cognitive Development*, ed. K. R. Gibson & A. C. Peterson, pp. 335–380. Hawthorne, NY: Aldine de Gruyter.

Piaget, J. (1954). *The Construction of Reality in the Child.* New York: Basic Books.

Posner, M. I. & Rothbart, M. K. (1980). The development of attentional mechanisms. In *Nebraska Symposium on Motivation*, ed. J. H. Flower, pp. 1–52. Lincoln, Nebraska: University of Nebraska Press.

Purpura, D. P. (1975). Normal and aberrant neuronal development in the cerebral cortex of human fetus and young infant. In *Brain Mechanisms of Mental Retardation*, ed. N. A. Buchwald & M. A. B. Brazier, pp. 141–169. NewYork: Academic Press.

Purves, D. & Lichtman, J. W. (1985). *Principles of Neural Development.* Sunderland, Mass.: Sinauer.

Rabinowicz, T. (1979). The differential maturation of the human cerebral cortex. In *Human Growth* Vol. 3. *Neurobiology and Nutrition*, ed. F. Falkner & J. M. Tanner, pp. 141–169. New York: Plenum Press.

Rakic, P. (1988). Intrinsic and extrinsic determinants of neocortical parcellation: A radial unit model. In *Neurobiology of Neocortex*, ed. P. Rakic & W. Singer, pp. 5–27. New York: John Wiley & Sons.

Rakic, P., Bourgeois, J.-P., Eckenhoff, M. F., Zecevic, N. & Goldman-Rakic, P. S. (1986). Concurrent overproduction of synapses in diverse regions of the primate cerebral cortex. *Science*, 232, 232–234.

Schiller, P. H. (1985). A model for the generation of visually guided saccadic eye movements. In *Models of the Visual Cortex*, ed. D. Rose & V. G. Dobson, pp. 62–71. Chichester: John Wiley & Sons.

Shimojo, S., Bauer, J. A., O'Connell, K. M. & Held, R. (1986). Pre-stereoptic binocular vision in infants. *Vision Research*, 26, 501–510.

Slater, A., Morison, V. & Somers, M. (1988). Orientation discrimination and cortical function in the human newborn. *Perception*, 17, 597–602.

Van Essen, D. C. (1985). Functional organisation of primate visual cortex. In *Cerebral Cortex*, Vol. 3, ed. A. Peters & E. G. Jones, pp. 259–329. New York: Plenum Press.

Vinter, A. (1986). The role of movement in eliciting early imitations. *Child Development*, 57, 66–71.

Von Hofsten, C. (1984). Developmental changes in the organisation of prereaching movements. *Developmental Psychology*, 20, 378–388.

12

Cognitive development in animals
DAVID F. SHERRY

Young organisms cannot do many of the things that adults can, and some of these things are cognitive in nature. Very young humans cannot understand or produce speech very well. Young animals cannot recognize predators or collect food at the adult level of proficiency. Young humans, however, develop the ability to understand and produce speech at a rate that is scarcely believable, and young animals can learn to forage and avoid predators, as well as form social attachments, learn to communicate, and learn to orient themselves in the world, sometimes by means that are not available to adults. This chapter offers a highly selective review of cognitive development in animals, concentrating on a few questions. The two fundamental questions are, do animals have 'cognitions', and do they change ontogenetically? The next question concerns the nature of cognitive development. Cognitive development clearly involves a multitude of changes in behaviour and the nervous system. This discussion will emphasize that there is an important difference between the acquisition of information by naive animals and developmental change in cognitive mechanisms. Several examples, drawn from foraging, aversion learning, and spatial behaviour illustrate this distinction, and allow for some discussion in passing of developmental changes in the brain that are correlated with cognitive development.

Do animals have cognitions?

The social psychologist Leon Festinger had no doubts on the matter (Festinger, 1961):

All that is meant by cognition is knowledge or information. It seems to me that one can assume that an organism has cognitions or information if one can observe some behavioral difference under different stimulus conditions. If the organism changes his behaviour when the environment changes, then obviously he uses information about the environment and, equally obviously, can be said to have cognitions (p.4).

The 'cognitive revolution' in animal learning and behaviour has been under way longer than is generally appreciated! (cf. Terrace, 1984; Roitblat & Von Fersen, 1992). The issue is, however, more complex than Festinger's analysis suggests. Changes in stimulus conditions can produce changes in behaviour for many reasons, and the prevailing view during the heyday of S–R theorizing was that stimulus conditions changed the frequency of behaviour and these stimulus conditions controlled behaviour. The cognitive view, and here Festinger identified the crucial element, is that animals possess knowledge or information, and are not simply controlled by external stimulus conditions and their past experience with them. Recent experimental work shows that many animals possess information that is structured and organized to a degree that strict stimulus–response accounts become unwieldy and, paradoxically, unparsimonious. Baker & Mercier (1989) have described the essential difference between S–R theories, like those of Thorndike, Guthrie, and Hull, and cognitive theories, like those of Tolman and his successors. Both are associationist in regarding associations as the fundamental units of animal learning and memory. They differ over what psychological events control behaviour. For S–R theories, the formation of associations is the critical event. The nature and strength of associations determine subsequent behaviour. Cognitive theories regard the critical events as the retrieval of these associations, analysis of the information embedded in them, and the decision about how to behave based on the information extracted. Retrospective evaluation of available information and decision making are what distinguish cognitive from S–R views of animal learning (Baker & Mercier, 1989). Contemporary work on Pavlovian conditioning clearly shows this cognitive influence, concentrating as it does on 'information', 'temporal and logical relations', 'hierarchical organization' and 'sophisticated representation' (Rescorla, 1988).

It should be stressed that this cognitive view of animal learning and memory, like the behaviourist view that preceded it, is agnostic on the subject of animal thought, animal consciousness, and animal awareness (cf. Griffin, 1984). Discovering that animal learning and memory are more complex than we had previously appreciated leaves us none the wiser about what animals experience. One of the most striking recent findings in the study of human cognition is that a great deal of our own cognitive activity occurs without awareness (Tulving & Schacter, 1990). There is no reason to invoke mentalistic explanations for the behaviour of animals when such explanations offer little help in understanding human behaviour.

Developmental change in cognition

Cognitive development in young animals consists of two very different processes. One is the acquisition of information, which, at least potentially, could occur in the same way as it does in adults. As a result of accumulated experience, naive animals eventually achieve a level of competence, for example in recognizing predators and collecting food, that is typical of adults. Because they are naive, young animals may well do a great deal more of this than adults do. The second process is ontogenetic change in cognitive mechanisms. Young animals may simply not work the way adults do. The development of foraging in birds illustrates this distinction. In general, the foraging success of young birds is much less than that of adults (for review see Wunderle, 1991). Several studies of individually marked birds have shown that improvement in foraging efficiency is rapid, increasing linearly with age (Weathers & Sullivan, 1991) or proportional to the square of age (Heinsohn, 1991). Such improvement need not have a cognitive component. Improvement in sensory or motor capacities could account for the improvement in foraging (see Berridge, Bischof, this volume). But experimental studies, by Beauchamp, Cyr & Houle (1987) and Wunderle & Soto-Martinez (1987), among others (cf. Hogan, Balsam & Silver, this volume), have confirmed that learning plays an important role in the transition to adult feeding behaviour. These two studies, of red-winged blackbirds and bananaquits, respectively, illustrate the distinction in cognitive development referred to earlier. Juvenile red-winged black-birds tend to leave sites after they have located food (a win–shift strategy) and must gradually learn to remain in sites where food does not deplete after a single visit (a win–stay strategy). Adults use the appropriate strategy, depending on whether or not the food resource depletes after a single visit. In contrast, juvenile bananaquits use the strategy that is appropriate to the nature of the food source: adult bananaquits lack this flexibility, and have great difficulty in returning to sites they have just visited. Cognitive development in young red-winged blackbirds, thus, consists of learning, as a result of experience, to adopt the strategy that is appropriate to the nature of the food they are exploiting. Cognitive development in young bananaquits is quite different. Young birds can do something that adults cannot, namely, rapidly adapt their behaviour to the nature of the food source. Young red-winged blackbirds are naive, but can eventually learn about the distribution of food in their environment. Young bananaquits have a cognitive flexibility that adults appear not to possess.

Weanling rat pups provide another example of a change in cognitive organization that is quite distinct from the acquisition of new information. Young rats can learn flavour aversions as early as the first days of life (Martin & Alberts, 1979; Gemberling & Domjan, 1982). They can learn to associate a certain flavour with illness and avoid ingesting anything that carries this flavour. Young rat pups will not, however, form aversions to flavours they experience while nursing (Martin & Alberts 1979; Gubernick & Alberts, 1984). The clearest demonstration of this comes from a series of experiments by Gubernick & Alberts (1984) in which they were able to deliver small quantities of flavoured milk to pups through a cannula that terminated on the tongue. Pups were given approximately 0.2 ml of flavoured milk over a period of about 30 seconds, followed immediately by an injection of lithium chloride. Pups that received their infusion of flavoured milk while attached to the nipple of an anaesthetized female rat formed no aversion to the flavour, while pups that received their milk infusion while lying beside an anaesthetized female, but not attached to the nipple, formed an aversion to the flavour of the infused milk. In neither case did pups receive any milk from the female, because anaesthesia prevented milk let-down. Infusion of flavoured milk followed by injection of saline produced no aversion whether the pups were attached to the nipple or not. These results show that young rat pups are quite capable of acquiring new information, namely that certain flavours predict illness, but are also organized to suppress formation of this association if the flavour that precedes illness is experienced while nursing. Following weaning and experience with solid food, young rat pups do form aversions to flavours experienced while nursing. These results show that young rats that are quite capable of learning that a flavour predicts illness possess an unusual cognitive mechanism, namely the ability to override the formation of such aversions in one highly specific context. Cognitive development in rat pups consists of both the acquisition of information (about flavours and their consequences) and the loss of a cognitive mechanism specific to one phase of early development.

Spatial abilities

Research on spatial abilities has generated a good deal of the current interest in animal cognition. There are a variety of very different but effective mechanisms for orienting in space (Leonard & McNaughton,

1990). Results showing that many animals have extremely good spatial abilities have raised questions about how spatial information is represented by the nervous system. Mechanisms of spatial orientation also undergo ontogenetic change, as shown by recent work with laboratory rats and homing pigeons.

Place learning by rats

Rats placed in a pool of water will learn to swim to a submerged platform and climb onto it to escape (Morris, 1981). The water maze is a useful device for investigating spatial memory because animals must remember the location of the platform in order to find it again. The platform is not visible and there are no odour trails or other beacons in the water to guide the rat. A variety of experiments have shown that rats relocate the submerged platform by remembering its position relative to objects and features outside the pool (Brandeis, Brandys & Yehuda, 1989). Even when the platform protrudes above the surface of the water and rats need remember nothing in order to see the platform and swim to it, they still tend to learn its spatial location.

Rats younger than 20 days of age cannot learn the spatial location of either a submerged or a visible platform. If the platform is visible, they can successfully swim to it, but when tested with the platform removed they show no tendency to return to the platform's former location (Rudy, Stadler, Morris & Albert, 1987). Rats 20 days of age or older can remember the location of both submerged and visible platforms (Rudy et al., 1987; Rudy & Paylor, 1988). The fact that young rats less than 20 days of age can swim to a visible platform but then fail to remember its location is quite informative, because it shows that sensory or motor limitations are not responsible for their failure to learn the platform's location. Some cognitive ability develops around 20 days of age that allows young rats to solve the water maze problem.

It is probably no coincidence that the hippocampus is undergoing rapid maturation at the same time rats acquire the ability to solve the water maze (Altman & Das, 1965; Cotman, Taylor & Lynch, 1973; Crain et al., 1973; Meibach et al., 1981; Pokorny & Yamamoto, 1981). Indeed, Rudy & Paylor's (1988) experiment was designed as a test of a non-spatial hypothesis of hippocampal function, Rawlins' (1985) theory that the hippocampus serves as a temporary memory buffer. Because rats less than 20 days of age have a hippocampus that is still maturing, they may lack a temporary memory buffer, and be unable to integrate the events that precede finding the submerged platform into some representation of the platform's location. This is less likely to be a problem, however, when the platform is continuously visible. Rudy & Paylor (1988) concluded from young rats' inability to remember the location of a successfully located visible platform that they lack the spatial mapping function attributed to the hippocampus (O'Keefe & Nadel, 1976), not a temporary memory buffer (Rawlins, 1985).

Pigeon homing

The hippocampus also plays a role in the development of homing by young pigeons, but in contrast to findings with rats, its role appears to be a transitory one. Pigeons that have experienced homing from distant release sites can return home accurately from novel sites. Bilateral removal of the hippocampus has no effect on homeward orientation at release sites by experienced birds: homeward orientation by hippocampectomized pigeons at release does not differ from the homeward orientation of control birds (Bingman, Bagnoli, Ioalé & Casini, 1984). However, few hippocampectomized birds successfully return home. They become lost within sight of their familiar home loft. The pigeon hippocampus is therefore thought to play a crucial role in the recognition of familiar landmarks in the vicinity of the home loft and to play little role in determining a homeward bearing at a release site (Bingman, 1990). Results with another species of bird, the black-capped chickadee (*Parus atricapillus*), also show that hippocampal lesions produce deficits in the performance of well-learned spatial tasks in a familiar environment (Sherry & Vaccarino, 1989).

The avian hippocampus plays a different role in young inexperienced homing pigeons. Pigeons that underwent bilateral removal of the hippocampus around 4 weeks of age and were then placed in their home loft showed no significant homeward orientation when tested 10 weeks later at three different novel release sites. Control birds that underwent bilateral removal of a comparable amount of neural tissue from a visual forebrain area were significantly homeward oriented at novel release sites, as were unoperated birds (Bingman, Ioalé, Casini & Bagnoli, 1990). This outcome shows that the hippocampus plays an important role in acquisition of the ability to home from distant release sites, while the results described earlier show that it is unnecessary for correct determination of the homeward bearing by experienced birds. The pigeon hippocampus participates in the development of a cognitive ability, which then functions with little or no hippocampal involvement.

Sex differences in spatial abilities

Males and females of a variety of species differ in spatial abilities. In humans, male superiority on tasks requiring mental rotation of objects is well documented (Linn & Petersen, 1985; Halpern, 1986), while female superiority in memory for the spatial locations of objects has been reported (Silverman & Eals, 1992). For voles (*Microtus*), male superiority on laboratory spatial tasks is found in polygamous species, but there is no sex difference on these tasks in a monogamous species (Gaulin & Fitzgerald, 1986, 1989; Gaulin, Fitzgerald & Wartell, 1990; Gaulin & Wartell, 1990). Male home range size is much greater than that of females in polygamous voles, while male and female home ranges are roughly equal in size in monogamous voles. Sex differences in the use of space may lead, in general, to the evolution of sex differences in spatial ability.

Williams, Barnett & Meck (1990) investigated the development of sex differences by manipulating the hormonal status of neonatal rats and observing their performance as adults on the radial-arm maze. Males were either castrated or subjected to sham surgery at 1 day of age. Females received injections of either estradiol benzoate (EB), which causes behaviour to become more male-like, or control injections. Subjects were gonadectomized at puberty if this had not occurred at day 1, to eliminate the effects of adult gonadal secretions, and were tested in the radial-arm maze at 70 days of age.

Overall, control males and EB females behaved alike in acquiring the task more quickly and performing it more accurately than either castrated males or control females. This difference disappeared over the course of further training on the maze. A subsequent experiment showed, remarkably, that control males and EB females used different information to solve the maze than did castrated males and control females. Performance by males and EB females was disrupted when the shape of the test room was changed using curtains, while changing the position of landmarks or removing them altogether had little effect on maze performance. In contrast, control females and neonatally castrated males suffered a decrement in performance when either the shape of the room or the landmarks within it were modified. These results lead to a number of conclusions about the development of sex differences in spatial ability in mammals. First, sex differences favouring males are produced by the neonatal action of the same masculinizing hormones that produce sexual differentiation. Gonadectomy at puberty produces no further effect on spatial abilities. Secondly, males and females may use different information to solve spatial problems like the radial-arm maze, and at least on this task, the use of multiple sources of information by females is correlated with slower acquisition and less accurate performance.

Conclusions

To return to the questions that opened this chapter – do animals have cognitions and do they change ontogenetically? – the answer to both is, yes. The cognitive view of animal learning and behaviour is really a shift in emphasis rather than a radical reformulation of the causal basis of learning and memory. Instead of concentrating on how associations are formed, the cognitive view takes for granted that experience provides animals with a rich representation of their world. The interesting problem, from a cognitive point of view, is to determine what use animals make of this information and what kind of problems they can solve with the information available to them. Thus, animals have cognitions to the extent that they can make adaptive behavioural decisions on the basis of experience.

The causal mechanisms underlying such cognitive activity in animals can clearly be seen to change developmentally. The formation of taste aversions can be suppressed by young rats in a specific context, namely while nursing. The ability to return to a particular spatial location shows a developmental time course, in both young rats and young homing pigeons, and there is evidence for hippocampal involvement in both cases. Sexual differentiation of spatial abilities in rats also shows that the mechanisms of spatial orientation diverge developmentally, under hormonal control. Johnson (this volume) points to comparable changes in cognitive organization in infants, for example from subcortical to cortical mechanisms of face recognition.

Changes in cognitive organization can be separated from the acquisition of information during development. A young rat pup may learn that a certain flavour predicts illness and others do not, and this is readily distinguished from change in the process of acquiring such information. But why does the causal basis of cognition change at all? Undoubtedly, one reason is that adult cognitive organization may require neural organization that is not present in young animals. The effects of developmental change in the hippocampus on spatial orientation in rats is one example of this. The suppression of taste aversions while nursing presents a different picture, however. Gubernick & Alberts (1984) found that young rat pups are capable of forming taste aversions outside a weaning context, and do so. Indeed, developmental change in this case is not the acquisition of a new competence, but the loss of one at around 20 days of age. Thus, a second reason that causal mechanisms of cognition may change during development is, as Alberts & Gubernick (1989) suggest, that the learning and memory capacities of young animals are often adaptations to the changing environment in which development occurs.

Acknowledgements

I would like to thank Steve Mitchell, Hollie Rice, Bill Roberts, and Richard Sorrentino for discussion. Preparation of this chapter was supported by a grant from the Natural Sciences and Engineering Research Council of Canada.

References

Alberts, J. R. & Gubernick, D. J. (1989). Early learning as ontogenetic adaptation for ingestion by rats. *Learning & Motivation*, 15, 334–359.

Altman, J. & Das, G. D. (1965). Autoradiographic and histological evidence of postnatal hippocampal neurogenesis in rats. *Journal of Comparative Neurology*, 124, 319–336.

Baker, A. G. & Mercier, P. (1989). Attention, retrospective processing and cognitive representations. In *Contemporary Learning Theories*, ed. S. B. Klein & R. R. Mowrer, pp. 85–116. Hillsdale, NJ: Lawrence Erlbaum.

Beauchamp, G., Cyr, A. & Houle, C. (1987). Choice behaviour of red-winged blackbirds (*Agelaius phoeniceus*) searching for food: the role of certain variables in stay and shift strategies. *Behavioural Processes*, 15, 259–268.

Bingman, V. P. (1990). Spatial navigation in birds. In *Neurobiology of Comparative Cognition*, ed. R. P. Kesner & D. S. Olton, pp. 423–447. Hillsdale, NJ: Lawrence Erlbaum.

Bingman, V. P., Bagnoli, P., Ioalé, P. & Casini, G. (1984). Homing behavior of pigeons after telencephalic ablations. *Brain, Behaviour and Evolution*, 24, 94–108.

Bingman, V. P., Ioalé, P., Casini, G. & Bagnoli, P. (1990). The avian hippocampus: evidence for a role in the development of the homing pigeon navigational map. *Behavioral Neuroscience*, 104, 906–911.

Brandeis, R., Brandys, Y. & Yehuda, S. (1989). The use of the Morris water maze in the study of memory and learning. *International Journal of Neuroscience*, 48, 29–69.

Cotman, C., Taylor, D. & Lynch, G. (1973). Ultrastructural changes in synapses in the dentate gyrus of the rat during development. *Brain Research*, 63, 205–213.

Crain, B., Cotman, C., Taylor, D. & Lynch, G. (1973). A quantitative electron microscopic study of synaptogenesis in the dentate gyrus of the rat. *Brain Research*, 63, 195–204.

Festinger, L. (1961). The psychological effects of insufficient rewards. *American Psychologist*, 16, 1–11.

Gaulin, S. J. C. & Fitzgerald, R. W. (1986). Sex differences in spatial-learning ability: an evolutionary hypothesis and test. *American Naturalist*, 127, 74–88.

298   David F. Sherry

Gaulin, S. J. C. & Fitzgerald, R. W. (1989). Sexual selection for spatial-learning
    ability. *Animal Behaviour*, 37, 322–331.
Gaulin, S. J. C., Fitzgerald, R. W. & Wartell, M. S. (1990). Sex differences in
    spatial ability and activity in two vole species (*Microtus ochrogaster* and *M.
    pennsylvanicus*). *Journal of Comparative Psychology*, 104, 88–93.
Gaulin, S. J. C. & Wartell, M. S. (1990). Effects of experience and motivation on
    symmetrical-maze performance in the prairie vole (*Microtus ochrogaster*).
    Journal of Comparative Psychology, 104, 183–189.
Gemberling, G. A. & Domjan, M. (1982). Selective associations in one-day-old
    rats: Taste toxicosis and texture-shock associations. *Journal of Comparative
    and Physiological Psychology*, 96, 105–113.
Griffin, D. R. (1984). *Animal Thinking*. Cambridge, Mass.: Harvard University
    Press.
Gubernick, D. J. & Alberts, J. R. (1984). A specialization of taste aversion
    learning during suckling and its weaning-associated transformation.
    Developmental Psychobiology, 17, 613–628.
Halpern, D. F. (1986). *Sex Differences in Cognitive Abilities*. Hillsdale NJ:
    Lawrence Erlbaum.
Heinsohn, R. G. (1991). Slow learning of foraging skills and extended parental
    care in cooperatively breeding white-winged choughs. *American Naturalist*,
    137, 864–881.
Leonard, B. & McNaughton, B. L. (1990). Spatial representation in the rat:
    Conceptual, behavioral, and neurophysiological perspectives. In
    Neurobiology of Comparative Cognition, ed. R. P. Kesner & D. S. Olton,
    pp. 363–422. Hillsdale NJ: Lawrence Erlbaum.
Linn, M. C. & Petersen, A. C. (1985). Emergence and characterization of sex
    differences in spatial ability: A meta-analysis. *Child Development*, 56,
    1479–1489.
Martin, L. T. & Alberts, J. R. (1979). Taste aversions to mother's milk: The
    age-related role of nursing in acquisition and expression of a learned
    association. *Journal of Comparative and Physiological Psychology*, 93,
    430–445.
Meibach, R. C., Ross, D. A., Cox, R. D. & Glick, S. D. (1981). The ontogeny of
    hippocampal energy metabolism. *Brain Research*, 204, 431–435.
Morris, R. G. M. (1981). Spatial localization does not require the presence of
    local cues. *Learning & Motivation*, 12, 239–260.
O'Keefe, J. & Nadel, L. (1976). *The Hippocampus as a Cognitive Map*. Oxford:
    Clarendon Press.
Pokorny, J. & Yamamoto, T. (1981). Postnatal ontogenesis of hippocampal
    CA1 area in rats. I. Development of dendritic arborisation in pyramidal
    neurons. *Brain Research Bulletin*, 7, 113–120.
Rawlins, N. (1985). Associations across time: The hippocampus as a temporary
    memory store. *Behavioral & Brain Sciences*, 8, 479–496.
Rescorla, R. A. (1988) Pavlovian conditioning: It's not what you think it is.
    American Psychologist, 43, 151–160.
Roitblat, H. L. & Von Fersen, L. (1992). Comparative cognition:
    representations and processes in learning and memory. *Annual Review of
    Psychology*, 43, 671–710.
Rudy, J. W. & Paylor, R. (1988). Reducing the temporal demands of the Morris
    place-learning task fails to ameliorate the place-learning impairment of
    preweanling rats. *Psychobiology*, 16, 152–156.
Rudy, J. W., Stadler-Morris, S. & Albert, P. (1987). Ontogeny of spatial

navigation behaviors in the rat: Dissociation of 'proximal'- and 'distal'-cue-based behaviors. *Behavioral Neuroscience*, 101, 62–73.

Sherry, D. F. & Vaccarino, A. L. (1989). The hippocampus and memory for food caches in black-capped chickadees. *Behavioral Neuroscience*, 103, 308–318.

Silverman, I. & Eals, M. (1992) Sex differences in spatial abilities: Evolutionary theory and data. In *The Adapted Mind: Evolutionary Psychology and the Generation of Culture*, ed. J. H. Barkow, L. Cosmides & J. Tooby, pp. 533–549. New York: Oxford University Press.

Terrace, H. S. (1984). Animal cognition. In *Animal Cognition*, ed. H. L. Roitblat, T. G. Bever & H. S. Terrace, pp. 7–28. Hillsdale, NJ: Lawrence Erlbaum.

Tulving, E. & Schacter, D. (1990). Priming and human memory systems. *Science*, 247, 301–306.

Weathers, W. W. & Sullivan, K. A. (1991). Foraging efficiency of parent juncos and their young. *Condor*, 93, 346–353.

Williams, C. L., Barnett, A. M. & Meck, W. H. (1990). Organizational effects of early gonadal secretions on sexual differentiation in spatial memory. *Behavioral Neuroscience*, 104, 84–97.

Wunderle, J. M., Jr. (1991). Age-specific foraging proficiency in birds. In *Current Ornithology*, Vol. 8, ed. D. M. Power, pp. 273–324. New York: Plenum Press.

Wunderle, J. M. & Soto-Martinez, J. (1987). Spatial learning in the nectarivorous bananaquits: juveniles versus adults. *Animal Behaviour*, 35, 652–658.

13

The biological building blocks of spoken language

JOHN L. LOCKE

I still recall a pronouncement from my earliest days of graduate training in linguistics that 'phonemes are the building blocks of language'. I have forgotten who said it – possibly a number of linguists issued similar pronouncements and I am only recollecting a composite. In any case, the statement is *believed* by many individuals who study language. In this chapter I shall show that the phoneme is not the elemental unit best suited for understanding the development of spoken language.

The term phoneme refers to the smallest difference that can differentiate two spoken words. In English, /p/ and /t/ are phonemes because they distinguish 'pie' from 'tie', /s/ and /z/ are phonemes because they distinguish 'hiss' from 'his'. It makes intuitive sense to suppose that these sound units are in fact the building blocks of language because words and grammatical markers are made of such sounds, and phrases and sentences are made of words. But this logic serves us poorly if our intent is to understand what human developments produce linguistic capacity and determine the form of linguistic behaviors. These are ontogenetic questions, and in asking them our concern cannot reside with elemental units of a behavior not yet acquired. Rather, we must concentrate on developmental mechanisms which facilitate or enable behaviors that – as the human child ultimately discovers – are decomposable into those units. From a developmental perspective, then, the phoneme is unavoidably *a posteriori* and therefore incapable of building any of the child's earlier behaviors.

In this chapter, the primary causal mechanisms of spoken language are considered to be found in at least four logically separable but functionally interrelated domains: sensory and perceptual; affective and social; motoric; and conceptual. When these domains are sufficiently operational in the human infant, the behaviors thus produced include words which can represent and communicate experience. With certain futher neurolinguistic developments, early words can be analyzed into their phonemic parts and used to build more language.

Sensory and perceptual processing

From birth, the human infant is pre-eminently a perceptual animal, richly social from its first postnatal moments, oriented to the facial and vocal displays of people who are using language. Whenever the development of behavior among the individual members of a species proceeds along a common path, with a relatively small number of individual variations, development can be said to be canalized (Waddington, 1940, 1975). Canalization provides an internally regulated growth path that can be adhered to as long as the developing animal has access to a supportive environment of the type that usually is available to all members of the species.

I believe there is a species-typical developmental path which leads to spoken language, a canalization for language. We see this when normally developing infants from disparate cultures produce the same error profiles on their way to increasing mastery of language (Locke, 1983). One senses a strong internal regulation of language-learning mechanisms when even severe sensory and neurological disorders fail to produce lasting linguistic deficits (Aram, Ekelman & Whitaker, 1986, 1987). Because linguistic canalization is a defining feature of our species, all human infants have the neural capacity to acquire language despite natural variations in structure, maturational course, and environmental circumstances.

Research in the past several decades tells us that the human infant is not born into 'a blooming, buzzing confusion', as William James once wrote. Indeed, when the human infant is born, it is already slightly acquainted with the vocal sounds it will be hearing for the rest of its life. At one time, 'born with' implied a genetic pre-adaptation which operated with no more than minimal environmental support and required no specific sensory experience. But behavioral embryologists, most notably Gottlieb (1978, 1980), have demonstrated that early postnatal capabilities frequently reflect prenatal experience.

Perceptual learning of vocal patterns begins at a very early age in the human. The fetus is able to hear at about 26 weeks gestational age and probably is able to hear its mother's voice during the final trimester (Birnholz & Benacerraf, 1983). There is evidence that the fetus adapts to tones that are presented to the maternal abdomen, and that the auditory system is sufficiently developed prior to birth for the fetus to discriminate consonant-like sounds (Lecanuet, Granier-Deferre & Busnel, 1989).

DeCasper and Fifer (1980) found that within 3 days of birth, newborns demonstrated a preference for their mother's voice over that of other adult females. They attributed this finding to intrauterine learning because of the favorable signal-to-noise ratios in the uterus for maternal voice (Querleu, Renard & Crepin, 1981; Querleu, Renard & Versyp, 1986). It also is possible that the mother's voice was learned postnatally, for at the age of 16 hours neonates react differently to their own voice than they do to the voice of another neonate (Martin & Clark, 1982). This result suggests that the learning of vocal patterns which index speaker identity occurs extremely rapidly in the human newborn.

Newborns also seem to prefer prosodic patterns that were first heard prenatally (DeCasper & Spence, 1986), and to respond preferentially to the language their mother spoke, presumably also because of intrauterine exposure to its intonation and rhythm (Spence & DeCasper, 1987). All of these findings suggest that the human newborn begins extrauterine life with an auditory pre-adaptation, a strong inclination to track conspecific voice. Infants pay especial attention to a moving voice whose fundamental frequency rises and falls excursively (Fernald, 1985). These prosodic variations characterize positive emotionality; the heavily intoned voice is an affective display. We must presume newborns also track vocal resonance or whatever other information in the voice identifies particular individuals. Such indexical functions also are served by the face: several days after birth, neonates prefer the sight of their mother's face to the face of other women (Bushnell, Sai & Mullin, 1989). Facial familiarity must be brought about by early postnatal learning.

The human infant is strongly attracted to facial as well as vocal movements. Of the moving parts, the eyes are preferred over the mouth; this holds even if the mouth is moving, as in speech (Haith, Bergman & Moore, 1977). The eyes also are an affective organ, routinely conveying information about emotions and social intentions in conjunction with or independent of speech. Where they seemingly are disinterested in a monotonous voice, infants are distressed by an inexpressive or 'poker face' (Cohn & Tronick, 1983).

Vocal prosody and affective facial activity are not included in most formal accounts of spoken language, but their coordinate activities represent an exceedingly active channel between mother and infant. Since the sounds of speech usually are embedded in maternal vocalizations, the auditory pre-adapted infant need not spend a lengthy period searching for the cues to spoken language, and of course would have no reason to do that in any case. When listening to its mother speak, the infant gets vocal information which identifies her and her emotions, but it also is exposed to vocal variations which cue phonetic categories. Research since the early 1970s reveals that infants are aware of these variations, and produce adult-like discrimination curves (Eimas, Siqueland, Jusczyk & Vigorito, 1971; Jusczyk, 1992; see also Harnad, 1987). Even if extensively trained animals respond in superficially similar ways, the human infant's early and untrained phonetic capability bodes well for its learning of speech.

However, it is not entirely clear how this capability is initially applied in real-life circumstances. In infant perception research, subjects typically are exposed to synthetic syllables which are speech-like but carried by a 'lab voice' that is unfamiliar, impersonal, and unemotional. A range of contrasts between phonetic segments, e.g. the consonantal sounds in [ba] and [da], are discriminable when embedded in these unnatural frames. Whether they would be attended to or processed when spoken by a familiar caregiver is unclear.

Fetal exposure explains the human newborn's orientation to the maternal voice. However, the human infant's discrimination of phonetic contrasts, which may be shared broadly with mammals and birds, has no experiential explanation of this sort (see Kuhl, 1991). These cues are unlikely to be heard clearly by the fetus, and there seems to be little effect of postnatal learning by the time of test, often a month or two after birth. Whatever the origins of phonetic perception, at some point in its development the human infant becomes capable of responding both to the indexical and affective components of voice *and* the categorical phonetic elements embedded therein. Henceforth, from sight and sound the infant can tell who is speaking and how they feel about him/her; from audition alone it can know these things and also what specific sounds the speaker is making.

It may seem paradoxical that while hearing is not linked to frequency of vocalization in the young infant, vision is. Studies suggest that the appearance of a friendly face may elicit vocalization (Cohn & Elmore, 1988), and that blind infants vocalize less than sighted ones (Fraiberg, 1979). Vision also may contribute to vocal imitation. When vocalizations are heard but not seen there appears to be less vocal imitation than when they are heard and seen. Dodd (1972) found that among 9- to 12-month-olds, live, face-to-face babbling (e.g. [dadada]) was more likely to elicit vocalization than taped babble (see also Clayton, this volume).

There is evidence of perceptual *learning* during the second 6 months of life but not before. It is primarily during this interval that phonetic category boundaries are re-adjusted in response to differential linguistic stimulation (Eilers, Wilson & Moore, 1977; Werker & Pegg, 1992).

Neural specializations

It is important to know which mechanisms of the adult brain process affect and personally identifying information. We then can ask if these mechanisms develop before, or in effect join or become the neural systems responsible for speech, if indeed they are altogether different mechanisms. This requires that we consider the neurology of personal recognition and emotional interpretation by way of facial and vocal cues.

Behavioral and neurophysiological experiments reveal that species from chicks to primates have an acute awareness of ocular patterns and face movements in conspecifics (Mendelson, Haith & Goldman-Rakic, 1982; Johnson & Morton, 1991; Johnson, Bolhuis, this volume) and process this activity with a dedicated set of 'face neurons' (Desimone, 1991). There are indications that human primates are similarly specialized for face processing (Creutzfeldt, Ojemann & Lettich, 1989a, 1989b) and also may have a similar pre-adaptation to conspecific voice (Holmes, Ojemann, Cawthon & Lettich, 1991).

While logically separable, face and voice processing appear to comprise a single, unified capability for affective communication, with a common neural substrate. The evidence for this is various, and includes the facts that: (1) voice and face memory have similar developmental courses, each gradually improving until adolescence, when they drop simultaneously and co-recover a few years later; (2) human and nonhuman primates use voice and facial expressions interchangeably, to some degree; (3) facial activity can contradict or augment vocal activity, and *viceversa*; (4) experimental evidence reveals that the same cerebral hemisphere – the right – is disproportionately responsible for facial and vocal affect; and (5) unilateral right hemisphere damage frequently disrupts both face and voice processing, and impairs both facial and vocal encoding of socially relevant information.

Affective responding and social interaction

Attachment

Because the human infant is born helpless, and is very much dependent upon its mother, it is in need of maternal attachment (Bowlby, 1969). I raise this issue here because I believe the primary points of articulation between mother and infant are facial and vocal. As a general rule, attached infants feel more secure when their mother is around, and they can know if she is nearby by tuning in to self-identifying vocal information. By tracking the movements of the face and voice, the infant also can infer the emotional state of the mother, including any desire she might have to act lovingly and nurturantly. For reasons closely related to survival, then, the human infant pays especial attention to a complex of cues which happens to contain all the information a parent can supply about spoken language. Why this biologically obligatory channel just happens to contain all the cues to spoken language is a matter on which I have speculated elsewhere (Locke, 1993a).

Vocal turn taking

Infants, even deaf ones, vocalize freely in the first months of life. Initially, they vocalize in distress, but increasingly, over the first year of life, infants vocalize in times of comfort. To hear its mother's voice, the loquacious infant must turn off its own voice, and at 3 to 4 months infants generally begin to suppress vocalization when their mother is talking, and start up when she is silent (Ginsburg & Kilbourne, 1988). This tendency, in coordination with similar dispositions in their mother, produces vocal alternations. Vocal turn taking thus gives the superficial appearance of a conversation, even though the infant's own vocalizations are devoid of linguistic content. Such 'conversations' maximize infants' opportunity to hear and respond to maternal speech, and provide mothers with an opportunity to hear and respond to the sounds of their infant. Success at this level is correlated with success at vocal imitation (Pawlby, 1977).

Eye gaze

The eyes play an important role during the learning of an initial lexicon. This value is seen primarily in joint attention. Joint attention refers to those moments in which infant and caregiver are focused on the same objects or events. Mothers watch their infants' eyes closely to see what they are looking at, and then name or otherwise comment on that material (Collis,

1977). Utterances by either party during joint attention episodes frequently are interpreted as comments about the objects or events that are attended jointly. This greatly narrows the semantic field, and allows infants to guess at their mothers' intended referrent and *vice versa*.

Vocal accommodation

Whether they intend to or not, parties that speak or vocalize together tend to take on some of the vocal attributes of their interlocutors. There is evidence of such vocal accommodation in nonhuman primates and in human children (Maurus, Barclay & Streit, 1988; Street, Street & van Kleek, 1983). This practice is developmentally significant in the infant, for even if accommodation is motivated by a generalized desire for intimacy or acceptance, as seems likely, it can produce a superficially linguistic result to which others can react. It may be responsible for the earliest words and phrases of children, which are generally considered to be formulaic, i.e. perceived and reproduced as a whole, as gestalts (Locke, 1993a).

Development of motor behavior
Babbling

During the first 6 months of life, there may be numerous occasions on which the infant communicates by way of voice, but in almost every instance the communication is affective and implemented by way of vocal resonance or frequency. The possibility for differentiation of this vocal signalling system into a *bona fide* sound system becomes evident when babbling begins, typically around 7 months (Koopmans-van Beinum & van der Stelt, 1986). At this point, infants develop the ability to chop up the smooth flow of voice into syllable-like pulses. Babbling refers to the production of these syllabic patterns, which usually are of consonant–vowel shape (e.g. [da]), and often produced reduplicatively (e.g. [dadada]). Initially, this behavior may be produced by oscillations of the mandible with relatively passive positioning of the lips or tongue.

The specific sounds – the particular movement patterns which produce the sounds, really – are not obviously inspired by listening to others talk (Locke, 1983), though the readiness to indulge in this infant version of speech may be. Rather, the inclination to make certain vocal movements seems to occur spontaneously, and in terms of phonetic features tends to be largely the same from one infant to the next (Locke, 1983). As a number of perfectly soundless motor functions, e.g. crawling, appear at about the same time as babbling (van der Stelt & Koopmans-van Beinum, 1986), the coordinative aspects of reduplicative syllabic activity may be due more to motoric than to auditory factors. Nevertheless, this activity is presumably not kinesthetically so enjoyable that it will be done with the same frequency by infants who are aphonic or deaf, at least over the first year of life.

Initially, babbling has a small repertoire of oral movements, as far as the ear can tell. The most common point of contact in babbling is between the apex of the tongue and the anterior portion of the hard palate. Produced with continuous voicing this yields [dada] if orally released or [nana] if nasally emitted. In time, other movements and points of contact develop, as does coordination of laryngeal and oral activity, and a working repertoire of nearly a dozen sounds may be heard consistently by the end of the first year. Significantly, these sounds are available to the infant for intentional use; studies of the phonetic composition of early words reveal that children draw heavily upon the sounds used in babbling (Locke, 1983). It becomes important, then, to ask what motivates these particular patterns of movement, these articulatory building blocks.

Modal action patterns

Elsewhere (Locke & Pearson, 1992), I have described the infant's stereotyped oral movements as phonetic action patterns with properties not unlike Barlow's (1977) modal action patterns. By this I mean that certain action patterns are available to the human infant by virtue of it being human, with no more than a minimal role for early experience. The basis for favoring particular movements is to be found in theories of vocal tract physiology (Locke, 1983). Like other modal action patterns, the infant's phonetic action patterns are relatively fixed. They tend to occur across infants, regardless of specific phonetic experience. The phonetic action patterns of infants also resemble closed behavioral systems (Mayr, 1976), for they tend to be relatively unmodified by environmental stimulation.

We might illustrate this by looking at several real cases. The [m] sound is relatively common in infant vocalization, the American English [r] is extremely rare. When we look at archival data on established languages, we see that the /m/ phoneme occurs in practically every language in the world, the American English [r] in very few languages. When one counts up the variants of these two sounds, on frequency grounds alone one ought to find nearly 20 times more /m/ than /r/ variants, but in fact one finds more of the latter. I therefore take late appearance in development to mean that an oral movement pattern is less fixed and more susceptible to ambient experience.

No other primate does anything quite like babbling. The lip-smacking of monkeys is not close, really, as the sounds seemingly are produced without laryngeal support, and the activity tends to occur only as a display in certain social contexts (Redican, 1975). Nonhuman primates also give little evidence of vocal control in experiments or seminaturalistic observation (Myers, 1976), and indeed are far less inclined than humans to vocalize at all (Gardner, Gardner & Drumm, 1989). This is a major difference between human and nonhuman primates. However, the subsong of many bird species (see DeVoogd, this volume) is quite analogous to babbling. It is likely that subsong plays a similar role in the development of adult song as babbling plays in the development of speech.

This bridge between babbling and speech is supported by the phonetic topography of languages themselves. If an infant says [dada], someone is likely to think that it said 'dada' or even 'daddy'. That is to say, the infant can get phonological and lexical credit for its phonetic action patterns. What explains this fortuitous turn of events? How can the infant learn to speak without attending to, storing, retrieving and executing phonetic representations? To answer this we must acknowledge the presence, in most lexicons, of *baby words* such as 'bye-bye', 'papa', 'mommy', 'pee-pee', 'kaka' and other words that are made up of the consonant–vowel reduplications that occur in babbling (Ferguson, 1964). These items reflect a lexicalization of sound patterns which are natural to the infant, i.e. not obviously inspired by exposure to the specific stimulation of others. We must assume such 'user- friendly' words are easily learned or, better yet, appropriated by the infant with a vocal tract whose functions are immature but whose output is predictable.

Babbling also may facilitate development of a speech-learning system. Even prior to the onset of babbling, infants reveal an awareness of visual-motor and auditory correspondences associated with the act of speaking (Dodd, 1979; Kuhl & Meltzoff, 1982). This suggests that elements of the monitoring system that will be used to learn speech already are in place prior to a great deal of articulatory self-production. But with babbling experience, I would hypothesize that the infant's vocal monitoring system elaborates and becomes more effective. Infants who do not babble may be in double jeopardy, then, because they are not so inclined, and do not receive the beneficial effects of articulatory–auditory experience. Later, we will look at the language development of children with delayed babbling.

Conceptual developments

For several decades, the mere mention of conceptual development implied an interest in Piaget's notions on the emergence of basic concepts of object constancy, symbolic behavior, and so on. And, indeed, various observers have reported correlations between children's Piagetian stage and their linguistic stage. Some nonhuman primates go through the same stages (Chevalier-Skolnikoff, 1989), but obviously do not have language, so conceptual development of a Piagetian sort does not ensure linguistic capacity.

Nonlinguistic categorization

To show that infants put objects in the same category, it is necessary to show that the items are discriminable. In some studies that have done this, certain types of low-level categorization were performed by infants as young as 4 months of age. Using exemplars of the category bird, Roberts & Horowitz (1986) obtained evidence for categorization beginning between 7 and 9 months. Using drawings of faces, Sherman (1985) found that at 10 months – one to several months before the appearance of words, typically – infants demonstrate the ability to abstract category-level information from their perceptual experience with the objects that they discriminate reliably.

When a child names a novel thing, it can be argued that it has figured out what the thing is, and that the emergence of expressive vocabulary reflects the growth of the ability to categorize (Reznick & Goldfield, unpublished). Reznick (1989) points out that there are at least three levels of categorization (also see Markman references below). In level one categorization, the members of a category are perceptually indiscriminable, as in the case of elements that are categorically perceived (e.g. stop consonants). In level two categorization, the elements share a perceptual feature that causes them to be seen as members of a set, but they are discriminable from one another (e.g. persons). In level three categories, the members are not necessarily similar perceptually but by way of their similar function, cultural value or some other association (e.g. household objects; vehicles).

Reznick believes that the development of categorization ability is difficult to study because some categorizations are modality sensitive, and there are no clear-cut procedures for measuring the relevant behaviors. Nevertheless, he reviews literature suggesting that level one and two categorizations emerge under a variety of measurement conditions by the middle of the first year. Some level three categorizations are evident by the end of the first year of life. This would suggest that categorization ability is functioning by the time, approximately 12 to 18 months, when lexical knowledge and use are first revealed. Categorization ability is almost certainly precursive and prerequisite to the ability to name and to use the names of things creatively.

Taxonomic sensibilities

In the last few years, some of these issues have been revisited in refreshingly different contexts. One concept that shows great promise in developmental linguistics is a taxonomic one. When we call animals of various sizes and shapes 'dog', we treat them as members of the class {DOG}. In effect, we classify them.

Infants tend to be very sensitive to the things objects are used for or associated with. In a series of experiments by Ellen Markman and her colleagues at Stanford (Markman & Hutchinson, 1984; Markman, 1991, submitted), children were shown pictures and asked to indicate to which other pictures the target photos were thematically or taxonomically related. For example, the child might be handed a picture of a dog and asked to 'find another one that is the same kind of thing' (or merely to 'find another one') from a set containing a cat (the taxonomic alternative) and a bone (the thematic choice).

Typically, children as young as 18 months chose the thematic alternative on this task which is oriented around objects. But Markman has used an interesting control which shifts the focus to (nonsense) *words*. For example, the experimenter will say 'See this? It's a kind of dax. Can you find another kind of dax?' Just hearing this new term in reference to the same object greatly increases the number of taxonomic responses. It is as though children know intuitively that words refer to collections of similar objects, i.e. to taxonomies. Merely naming an object seems to weaken the thematic associations the child normally would call to mind for that object, and for its name. This makes it possible, Markman noted, for words to achieve their fullest flexibility and combinatorial power. For as the names of categories of similar objects, words stand free of associative baggage, eligible to be linked with other words (taxonomic labels) to yield the desired associations, and only those associations.

These findings have been replicated by Waxman & Kosowski (1990), who found additionally that children's taxonomic bias may be associated solely with nouns; no such effect was observed for adjectives. They also found that children as young as 2 years have some sense of subordinate–supraordinate relations, revealing in their first and second choices for novel nouns the awareness that perceptually disparate objects such as bird and mouse belong to a single (animal) class.

Markman and her colleagues also have determined that when an unfamiliar object has component parts, any name that is supplied is taken by the child to identify the whole object and not the parts. When the object and its name are familiar, however, and children are supplied with a novel word, they treat the new term as the name of one of the object's parts. This, which Markman calls the *whole object* assumption, constrains the range of possible meanings for a new word.

Semantic constraints are offered also by the *mutual exclusivity* assumption. Evidence for such an assumption usually is produced when children see a familiar and an unfamiliar object and hear an unfamiliar word. Almost invariably, they select the unfamiliar object, rather than assume that the novel word might be another name for the familiar thing, or the name of an attribute of the object that they know.

Other minds

Linguistic progress requires development of articulatory abilities, massive storage of lexical items and phrases, and grammatical analyses. To plunge deeply and creatively into language, infants must have incentives. One concept that is likely to be important in this regard is the belief that we all have our own feelings, thoughts, and knowledge. For example, some researchers have looked at the emergence of empathy. Others have traced changes in the capacity for pretend play and acts of deception (Sodian, Taylor, Harris & Perner, 1991). Still others have looked at the child's emergent awareness of 'other minds' – the fact that other people have different knowledge and thoughts than the child, and that it thinks that they do not (Leslie, 1987).

One way in which this appreciation of other minds has been demonstrated is with a simple game. In full view of the child, the researcher marches onto a table a girl doll who is carrying a ball. The ball is put in a cup. The doll is then made to march off the table, and while it is out of sight the investigator takes the ball out of the cup and puts it in its pocket. Then the doll is brought back to the table and the investigator asks the child 'Where will she look for the ball?' The normally developing child has the girl look in the cup, which is where they know she last saw it and now will think it is. Even children with Down's syndrome behave like this, but children with the primary affective disorder of autism have the doll look in the researcher's pocket (Baron-Cohen, Leslie & Frith, 1985).

On logical grounds, the concept of other minds is critical to language development. It seems unlikely that the child's initial lexicon would expand if it did not feel a strong urge to communicate, and the mere idea of communication makes little sense unless individuals assume that they know things that other people do not, and *vice versa*. In other words, the drive to communicate expands referential capacity, a point that has been argued more extensively elsewhere. Interestingly, the theory of other minds is one of the few concepts that have been linked to language that nonhuman primates do not share with us (Premack & Woodruff, 1978).

Neurolinguistic capabilities

The biological building blocks I have identified above all work to create the social capability for interpersonal communication, the perceptuomotor capability for vocal communication, and the cognitive capability for abstract referential communication. There are neural specializations subserving these functions (Locke, 1992), and in some cases what appears to be a unique phylogenetic history. However, nonhuman primates seem to share some level of function in several of these domains and, as we know, they have some communicative and referential capability. Where the discontinuity arises, and where spoken language separates itself from other systems for communication and representation, is linguistic grammar.

The pregrammatical child is not mute. The 18- to 24-month-old infant uses single words and stereotyped phrases to refer to things in restricted social contexts. Very few of its utterances are creative and none of them is rule governed. But the child is a talker, and it stands at the threshold of grammatical language.

Grammar is not to be confused with syntax. As Liberman (1970) pointed out years ago, language has several grammars. We are all used to thinking that sentences are constructed by various ordering rules but so are words and syllables. The phonology of a language is a set of organizational principles which tell speakers, among other things, which sequences of sounds are allowable and which are not. Morphological principles tell us about various 'markers'. For example, if we are going to negate, we must add a prefix like 'im-' to words such as 'permissible' but 'in-' to words having the form of 'considerate'. To pluralize, we add /s/ to 'lock' and /z/ to 'key'. Phrases such as the boy pushed the girl and the boy was pushed by the girl are comprehensible only by those who understand the rules of syntax.

Though we are used to being told that phonemes are the building blocks of words, in fact words come to the child first. Systems that develop are systems that differentiate, and phonology is no exception. When the child with a small lexicon discovers the phonemic principle, it automatically discovers the combinatorial privilege of phonemes. What could only have been stored as a series of holistic templates can now be organized by individual sounds and stored more efficiently.

None of this can be explained by reference to social, perceptual, motoric or conceptual advances of any kind that we currently know about. The advances mentioned above are all necessary for spoken language, but they are insufficient for grammatical functioning. Some other capability is needed, but it is unclear what, if anything, could be a building block for grammar. For this reason, among others, grammatical capability has been conceptualized as modular, an encapsulated domain of information processing with its own neural mechanisms (Fodor, 1983; Liberman & Mattingly, 1991). This module is analytical and computational, but it is incapable of gathering its own material. To assume this would be to assume that Nature created massively redundant informational systems which are not logically necessary and do not demonstrably exist. Rather, it makes more sense – and is more parsimonious – to assume that facial and vocal processing systems, transmitting affective information in the context of mother–infant interaction, initially, attract children to utterances which are replicable in the absence of grammatical capability.

My argument is that speech is a facial and vocal display (Locke, 1993a, 1993b, in press) that is interpreted by social, perceptuomotor and conceptual developments, which supply the foundation for reasonably complex forms of communication and representation. In learning language, utterance material is sollicited for and collected by neural systems that are specialized for such socially cognitive operations. This material is then fed to a grammatical module, which analyzes and imputes organization to incoming utterances. The set of rules that is acquired in this fashion is then available for use in decoding the utterances of others and in the encoding of one's own utterances. But the other systems must precede the induction and application of grammatical principles (Locke, 1992).

Building blocks in the evolution of linguistic capacity

How did our two cerebral hemispheres come to divide up language as they do? While macaques process certain types of species-specific information on the left side of the brain, they appear to process affective and indexical vocal information on both sides of the brain (Petersen *et al.*, 1984), and it is speculated that in their efforts to identify their fellow hominid, and to infer its motives, our prehistoric ancestors relied on both hemispheres, too. In hominids, both hemispheres may have processed vocal information of all types, before there was anything as complex as present-day spoken language. We may speculate that when our ancestors began to interrupt the voice with their articulators, their vocal repertoire took on a segmental content whose mode of production required precise timing and co-ordination. These unfamiliar, transient patterns with little affective or indexical value naturally were sent to a hemisphere, the left, which for other reasons had become better able to coordinate precise movements in any reproductions (Calvin, 1983; Ojemann, 1984; MacNeilage, 1986), and which ultimately became analytical. The competition drove out vocal affect, leaving the right hemisphere to control this important communicative function. In ontogeny, correspondingly, our expectation is that this right hemisphere specialization for vocal affect would develop before the left hemisphere takes on the responsibility for consonants.

The ontogenetic sequence seems not to be dissimilar. The young infant has mechanisms on both sides of the brain that participate in affective communication. The left hemisphere assumes motor and perceptual control of activity of the type associated with grammatical language. The neural machinery needed to process information which carries spoken language is active from the start and later becomes linguistic. It remains to be seen if there are other systems of the brain which are linguistic from the start, merely waiting to become active.

When the blocks fall: developmental language disorders

If the biological building blocks I have identified above are, in fact, critical to language, we should expect to find language disorders when the preconditions for language are not met, i.e. when the building blocks fall.

Affective disorders

Populations with a high incidence of affective disorders, such as individuals with Down's syndrome and autism, tend also to have an abnormally high rate of developmental language disorder. Thompson, Cicchetti, Lamb & Malkin (1985) evaluated facial and vocal affect in Down's syndrome infants at 19 months of age and normally developing control infants at just over 12 and 19 months. Infants were submitted to the Strange Situation, a procedure consisting of seven 3-minute episodes in which the infant is alone, with its mother, with a stranger, or with its mother and a stranger.

During these episodes, observers classified facial and vocal displays of emotion. Analyses of peak intensity and emotional range revealed that the Down's syndrome infants showed consistently less facial and vocal affect than both groups of control subjects during all except one episode of the Strange Situation. They were less distressed when their mother left the room and they regained their composure sooner when she returned. Down's syndrome infants were slower to show even the attenuated changes they did reveal, and they exhibited a restricted range of emotional responding. Overall, their emotional reactions were subdued, their socioemotional variability flat.

There was little change in vocal or facial affect with age. The infants with Down's syndrome were consistently less expressive in both modes. Since the scales used to rate vocal and facial affect were discommensurate, the useful comparison here is the vocal–facial relationship across the three groups. Except when the infant was with mother, and with its mother and a stranger, where there may have been more facial expressivity, there were similar scale values for vocal and facial affect among the normally developing infants. The Down's syndrome infants, by comparison, were less overtly affective in the vocal than in the facial mode.

In their study of elicited humor responses, Cicchetti & Sroufe (1976) found that Down's syndrome infants were significantly less likely to laugh in humorous situations than infants who were developing normally. The four most hypotonic Down's syndrome infants in their study *never laughed before the age of* 13 *months* and rarely laughed then. They assumed that hypotonic infants have difficulty 'processing incongruity fast enough to generate the tension required for laughter.' (p. 923). This difficulty was thought to be the joint result of biochemical and cognitive factors.

Clinically, it has long been noted that autistic children look less at other people's faces than do normally developing children (Kanner, 1943). There is a limited amount of support for this observation in the scientific literature. For example, Hutt & Ounsted (1966) found that autistic children spent significantly less time than normal controls looking at drawings of human faces, but about the same amount of time looking at a facial contour, a monkey face and a dog face. Other studies have explored the facial looking patterns of autistic children, but as Weeks & Hobson (1987) observe, the artificiality or other weaknesses of the early face studies makes them 'relatively unenlightening'. What one really would like to see, of course, is quantified reactions to live displays of facial emotion, and here the literature is sparse.

In any case, if we assume for the moment that clinical impressions are correct, and that autistic children spend little time looking at faces, are they indifferent to them or do autists actively avoid looking at faces? This is an important distinction, for if autistic children are indifferent – giving faces the same attention as inanimate objects – then for them faces lack a species-chacteristic appeal that normally is observed from birth, when faces are preferred over other visual patterns (Goren, Sarty & Wu, 1975). Weeks & Hobson (1987) seem to lean toward this hypothesis, suggesting that autistic children lack 'a biologically based attentiveness and emotional responsiveness to certain of the bodily features of others, including features of emotional expression.' (p. 148).

Langdell (1978) made some interesting comparisons of normal children's performance with the processing patterns of autistic and retarded children. He began by photographing the face of 10- and 14-year-old autistic children. He also took black and white pictures of two groups of normally developing children, one matched for chronological age and the other for mental age, as well as retarded subjects of similar mental age scores. These photos were shown to the children in each group for identification. Pictures were presented either in the normal upright fashion, or were inverted or masked in seven different ways. The masks revealed the nose or eyes only; everything below the mouth, nose, or eyes; or everything above the eyebrows or eyes.

Consistent with other research, Langdell found that normal and retarded subjects were better able to recognize their friends from the upper half of facial displays than the bottom half. The two groups of autistic subjects differed somewhat from each other, but on the whole they performed in precisely the opposite fashion from the normal children, recognizing far more friends from *the lower half* of their photos. Langdell's data indicate that normal subjects made half as many errors when scanning faces that were revealed from the eyes up as when they saw faces that were exposed from the eyes down. Autistic children, on the other hand, made three times as many errors on the 'eyes up' photos as on the 'eyes down' pictures.

It is not obvious how these data should be explained. Langdell's own speculations embraced the prospect that autistic children do not view the face as a social stimulus, and therefore would not be particularly drawn to (socially expressive) eyes. He also wondered if autistic children might focus upon the mouth area in an attempt to compensate for unusual difficulties in comprehending the auditory components of speech.

Motor speech delay

If babbling is related to speech developmentally, one might suppose that delays or deviations in babbling foretell delays or deviations in speech. Stoel-Gammon (1989) found that nested in a larger sample of 34 children there were two whose early phonetic development was atypical. One child produced few canonical babbles from 9 to 21 months. The other had an unusual pattern of sound preferences in its babbling. At 24 months, the words of both subjects were produced with a more limited phonetic repertoire and with simpler syllable shapes, compared to peers. With a somewhat different population and mode of analysis, compatible findings were provided by Whitehurst *et al.* (1991), who found that the single strongest predictor of language was the ratio of consonants to vowels in an earlier sampling.

Jensen *et al.* (1988) longitudinally followed some infants who were, or were not 'at risk' for developmental delay (based on Apgar scores, birthweight, and presence or absence of neonatal cerebral symptoms). During the first year, subjects who were judged to be at risk produced significantly fewer consonant-like sounds and reduplicated syllables than the normal children. Some 5 years later, a much higher proportion of the children at risk also scored below age level on a language test.

Reduced turn taking

If vocal turn taking facilitates language development, one might expect a later onset or reduced occurrence in children with delayed language. In this regard, it is interesting that two studies have found that Down's syndrome infants are slow to engage in vocal turn taking. Berger & Cunningham (1983) placed normally developing and Down's syndrome infants, with their mothers, in two situations. In the silent condition, the mother was to sit silently with her infant. In the talking condition, mothers were to talk as they normally would. When the mother was silent, both groups vocalized with increasing frequency over age. When the mother talked, normally developing infants vocalized with increasing frequency up to the age of 13 to 16 weeks, when their vocal frequencies declined. However, the retarded infants continued to vocalize past that age. The authors thought this difference might be due to a reduced interest in the mother's speech by infants with Down's syndrome.

Jones (1977) also studied Down's syndrome and normally developing infants. Her subjects were 8 to 19 months old at the beginning of the study. In her coding scheme maternal and infant vocalizations were considered 'clashes' if separated by less than a second. It was found that Down's syndrome infants were involved in more of these vocal clashes with their mothers than the normally developing infants. Because most of the clashes occurred after the mother had begun speaking, it appeared to Jones that the majority were the 'child's fault'.

Concluding comments

Phonemes may be the linguistic building blocks used to construct spoken language, but the ontogenetic material used to produce linguistic capacity is of a different type. In the first instance, the biological building blocks include perceptual developments which range from species-specific biases and fetal experience to infant learning of ambient vocal patterns. There appears to be a specialized neural system, already active at birth, which orients attention to the human face and voice. This specialization is revealed in a parallel developmental course for indexical face and voice processing, as well as similarity in their phylogenetic history, response to focal lesions, natural usage patterns of competent speakers, and a variety of psychophysiological measures.

The human infant is dependent upon its mother for nurturance and protection, a dependence which fosters emotional attachment. Attached parties are joined at particular points, which include the face and voice. Since the cues to spoken language are embedded in movements of the face and voice, emotional dependency may be indirectly responsible for the early learning of speech-like material. Arising from maternal–infant interactions are the tendency to vocally alternate or take turns, to use eye gaze patterns as indices of attention – permitting infants to associate verbal labels with objects – and to accommodate to the vocal properties of the partner.

Initially, the infant accommodates to the vocal patterns of its home, probably unconsciously and perhaps contagiously. The act of vocal accommodation represents a subset of the larger constellation of verbal and motor activities which are observed by the infant and incorporated into its repertoire. Vocal accommodation is a broad net, and it brings in a number of linguistically irrelevant vocal behaviors along with real language material. As a consequence, infants' speaking behaviors probably resemble their parents' in matters of rate, tone and style, as well as lexical substance.

Clearly, a vocally accommodating infant who is aware of referential eye movements is in good shape as it approaches the age at which speaking would be expected, but it also must possess a degree of vocal–motor control. The onset of babbling reveals species-characteristic vocal action patterns which also are prominent in child speech and, universally, in the formal properties of linguistic sound systems. With increasing control over these phonetic action patterns, the human infant is able to achieve a number of word-like utterances, and at this point frequently is credited with linguistic capability.

However, truly linguistic behavior must await the development of a strong drive to communicate, discovery of taxonomic principles, and the emergence of grammatical capability. The urge to communicate logically rests on the presumption that one has information that others lack, and *vice versa*. The cognitive substrate for this presumption is revealed in empathy, pretend play and deception, and more formally in experimental tasks. Emergence of the taxonomic principle is revealed when very young children begin to treat words as the names of classes of like phenomena. Grammar appears to be a heavily encapsulated or modular function, not closely correlated with other cognitive capabilities and lacking in non-human primates.

If the behavioral advances identified above are functionally linked to language, it follows that difficulties at early levels will be correlated with later problems. Studies of children with primary affective disorders, early deficits or deviations in vocal–motor development, and in vocal turn taking, tend to bear this out, though a great deal more investigation is needed.

Note

Most of the topics discussed in this chapter are dealt with more extensively in Locke (1993a).

References

Aram, D. M., Ekelman, B. L. & Whitaker, H. A. (1986). Lexical retrieval in left and right brain lesioned children. *Brain and Language*, 27, 75–100.

Aram, D. M., Ekelman, B. L. & Whitaker, H. A. (1987). Spoken syntax in children with acquired unilateral hemisphere lesions. *Brain and Language*, 31, 61–87.

Barlow, G. W. (1977). Modal action patterns. In *How Animals Communicate*, ed. T. A. Sebeok, pp. 98–134. Bloomington: Indiana University Press.

Baron-Cohen, S., Leslie, A. M. & Frith, U. (1985. Does the autistic child have a 'theory of mind'? *Cognition*, 21, 37–46.

Berger, J. & Cunningham, C. C. (1983). Development of early vocal behaviors and interactions in Down's syndrome and nonhandicapped infant–mother pairs. *Developmental Psychology*, 19, 322–331.

Birnholz, J. C. & Benacerraf, B. R. (1983). The development of human fetal hearing. *Science*, 222, 516–518.

Bowlby, J. (1969). *Attachment and Loss*. Vol. 1 *Attachment*. New York: Basic Books.

Bushnell, I. W. R., Sai, F. & Mullin, J. T. (1989). Neonatal recognition of the mother's face. *British Journal of Developmental Psychology*, 7, 3–15.

Calvin, W. H. (1983). Timing sequencers as a foundation for language. *Behavioral and Brain Sciences*, 2, 210–211.

Chevalier-Skolnikoff, S. (1989). Spontaneous tool use and sensorimotor intelligence in *Cebus* compared with other monkeys and apes. *Behavioral and Brain Sciences*, 12, 561–627.

Cicchetti, D. & Sroufe, L. A. (1976). The relationship between affective and cognitive development in Down's syndrome infants. *Child Development*, 47, 920–929.

Cohn, J. F. & Elmore, M. (1988). Effect of contingent changes in mothers' affective expression on the organization of behavior in 3-month-old infants. *Infant Behavior and Development*, 11, 493–505.

Cohn, J. F. & Tronick, E. Z. (1983). Three-month-old infants' reaction to simulated maternal depression. *Child Development*, 54, 185–193.

Collis, G. M. (1977). Visual co-orientation and maternal speech. In *Studies in Mother–Infant Interaction*, ed. H. R. Schaffer, pp. 355–375. New York: Academic Press.

Creutzfeldt, O., Ojemann, G. & Lettich, E. (1989a). Neuronal activity in the human lateral temporal lobe. I. Responses to speech. *Experimental Brain Research*, 77, 451–475.

Creutzfeldt, O., Ojemann, G. & Lettich, E. (1989b). Neuronal activity in the human lateral temporal lobe. II. Responses to the subject's own voice. *Experimental Brain Research*, 77, 476–489.

DeCasper, A. & Fifer, W. P. (1980). On human bonding: Newborns prefer their mothers' voices. *Science*, 208, 1174–1176.

DeCasper, A. & Spence, M. (1986). Prenatal maternal speech influences newborns' perception of speech sounds. *Infant Behavior and Development*, 9, 133–150.

Desimone, R. (1991). Face-selective cells in the temporal cortex of monkeys. *Journal of Cognitive Neuroscience*, 3, 1–8.

Dodd, B. (1972). Effects of social and vocal stimulation on infant babbling. *Developmental Psychology*, 7, 80–83.

Dodd, B. (1979). Lip reading in infants: Attention to speech presented in- and out-of-synchrony. *Cognitive Psychology*, 11, 478–484.

Eilers, R. E., Wilson, W. R. & Moore, J. M. (1977). Developmental changes in speech discrimination in infants. *Journal of Speech and Hearing Research*, 20, 766–780.

Eimas, P. D., Siqueland, E. R., Jusczyk, P. & Vigorito, J. (1971). Speech perception in infants. *Science*, 171, 303–306.

Ferguson, C. A. (1964). Baby talk in six languages. *American Anthropologist*, 66, 103–114.

Fernald, A. (1985). Four-month-old infants prefer to listen to motherese. *Infant Behavior and Development*, 8, 181–195.

Fodor, J. (1983). *Modularity of Mind*. Cambridge, Mass.: MIT Press.

Fraiberg, S. (1979). Blind infants and their mothers: An examination of the sign system. In *Before Speech: The Beginning Of Interpersonal Communication*, ed. M. Bullowa, pp. 149–169. New York: Cambridge University Press.

Gardner, R. A., Gardner, B. T. & Drumm, P. (1989). Voiced and signed responses of cross-fostered chimpanzees. In *Teaching Sign Language to Chimpanzees*, ed. R. A. Gardner, B. T. Gardner & T. E. Van Cantfort, pp. 29–54. Albany, NY: State University of New York Press.

Ginsburg, G. P. & Kilbourne, B. K. (1988). Emergence of vocal alternation in mother–infant interchanges. *Journal of Child Language*, 15, 221–235.

Goren, C. C., Sarty, M. & Wu, P. Y. K. (1975). Visual following and pattern discrimination of face-like stimuli by newborn infants. *Pediatrics*, 56, 544–549.

Gottlieb, G. (1978). Development of species identification in ducklings: IV. Change in species-specific perception caused by auditory deprivation. *Journal of Comparative and Physiological Psychology*, 92, 375–387.

Gottlieb, G. (1980). Development of species identification in ducklings: VI. Specific embryonic experience required to maintain species-typical perception in Peking ducklings. *Journal of Comparative and Physiological Psychology*, 94, 579–587.

Haith, M. M., Bergman, T. & Moore, M. J. (1977). Eye contact and face scanning in early infancy. *Science*, 198, 853–855.

Harnad, S. (ed.) (1987). *Categorical Perception*. Cambridge: Cambridge University Press.

Holmes, M. D., Ojemann, G. A., Cawthon, D. F. & Lettich, E. (1991). Neuronal activity in nondominant human lateral temporal cortex related to short term spatial memory and visuospatial recognition. *Society for Neuroscience Abstracts*, 17, 476.

Hutt, C. & Ounsted, C. (1966). The biological significance of gaze aversion with particular reference to the syndrome of infantile autism. *Behavioral Science*, 11, 346–356.

Jensen, T. S., Boggild-Andersen, B., Schmidt, J., Ankerhus, J. & Hansen, E. (1988). Perinatal risk factors and first-year vocalizations: Influence on preschool language and motor performance. *Developmental Medicine and Child Neurology*, 30, 153–161.

Johnson, M. H. & Morton, J. (1991). *Biology and Cognitive Development: The Case of Face Recognition*. Oxford: Blackwell.

Jones, O. H. M. (1977). Mother–child communication with pre-linguistic Down's syndrome and normal infants. In *Studies in Mother–Infant Interaction*, ed. H. R. Schaffer, pp. 379–401. New York: Academic Press.

Jusczyk, P. W. (1992). Developing phonological categories from the speech signal. In *Phonological Development: Models, Research & Application*, ed. C. Ferguson, L. Menn & C. Stoel-Gammon, pp. 17–64. Parkton, Maryland: York Press.

Kanner, L. (1943). Autistic disturbances of affective contact. *Nervous Child*, 2, 217–250.

Koopmans-van Beinum, F. J. & van der Stelt, J. M. (1986). Early stages in the development of speech movements. In *Precursors of Early Speech*, ed. B. Lindblom & R. Zetterstrom, pp. 37–50. New York: Stockton Press.

Kuhl, P. J. (1991). Perception, cognition, and the ontogenetic and phylogenetic emergence of human speech. In *Plasticity of Development*, ed. S. E. Branth, W. S. Hall & R. J. Dooling, pp. 73–106. Cambridge, Mass.: MIT Press.

Kuhl, P. K. & Meltzoff, A. N. (1982). The bimodal perception of speech in infancy. *Science*, 218, 1138–1141.

Langdell, T. (1978). Recognition of faces: An approach to the study of autism. *Journal of Child Psychology and Child Psychiatry*, 19, 255–268.

Lecanuet, J.-P., Granier-Deferre, C. & Busnel, M.-C. (1989). Differential fetal auditory reactiveness as a function of stimulus characteristics and state. *Seminars in Perinatology*, 13, 421–429.

Leslie, A. M. (1987). Pretense and representation: The origins of 'theory of mind'. *Psychological Review*, 94, 412–426.

Liberman, A. M. (1970). The grammars of speech and language. *Cognitive Psychology*, 1, 301–323.

Liberman, A. M. & Mattingly, I. M. (1991). A specialization for speech perception. *Science*, 243, 489–494.

Locke, J. L. (1983). *Phonological Acquisition and Change*. New York: Academic Press.

Locke, J. L. (1992). Neural specializations for language: A developmental perspective. *Seminars in the Neurosciences*, 4, 425–431.

Locke, J. L. (1993a). *The Child's Path to Spoken Language*. Cambridge, Mass.: Harvard University Press.

Locke, J. L. (1993b). The role of the face in vocal learning and the development of spoken language. In *Developmental Neurocognition: Speech and Face Processing in the First Year of Life*, ed. B. de Boysson-Bardies, S. de Schonen, P. Jusczyk, P. MacNeilage & J. Morton, pp. 317–328. Dordrecht: Kluwer Academic Publishers.

Locke, J. L. (in press). Development of the capacity for spoken language. In *Handbook of Child Language*, ed. P. Fletcher & B. MacWhinney. Oxford: Blackwell.

Locke, J. L. & Pearson, D. M. (1992). Vocal learning and the emergence of phonological capacity: A neurobiological approach. In *Phonological Development: Models, Research, and Application*, ed. C. Ferguson, L. Menn, & C. Stoel-Gammon, pp. 91–130. Parkton, Maryland: York Press.

MacNeilage, P. F. (1986). Bimanual coordination and the beginnings of speech. In *Precursors of Early Speech*, ed. B. Lindblom & R. Zetterstrom, pp. 189–201. New York: Stockton Press.

Markman, E. M. (1991). The whole object, taxonomic, and mutual exclusivity assumptions as initial constraints on word meanings. In *Perspectives on Language and Cognition: Interrelations in Development*, ed. J. P. Byrnes & S. A. Gelman, pp. 72–106. New York: Cambridge University Press.

Markman, E. M. & Hutchinson, J. E. (1984). Children's sensitivity to constraints on word meaning: Taxonomic vs. thematic relations. *Cognitive Psychology*, 16, 1–27.

Martin, G. B. & Clark, R. D. (1982). Distress crying in neonates: Species and peer specificity. *Developmental Psychology*, 18, 3–9.

Maurus, M., Barclay, D. & Streit, K.-M. (1988). Acoustic patterns common to human communication and communication between monkeys. *Language and Communication*, 8, 87–94.

Mayr, E. (1976). Behavior programs and evolutionary strategies. In *Evolution and the Diversity of Life: Selected Essays*, ed. E. Mayr, pp. 694–711. Cambridge, Mass.: Harvard University Press.

Mendelson, M. J., Haith, M. M. & Goldman-Rakic, P. S. (1982). Face scanning and responsiveness to social cues in infant rhesus monkeys. *Developmental Psychology*, 18, 222–228.

Myers, R. E. (1976). Comparative neurology of vocalization and speech: Proof of a dichotomy. *Annals of the New York Academy of Sciences*, 280, 745–757.

Ojemann, G. A. (1984). Common cortical and thalamic mechanisms for language and motor functions. *American Journal of Physiology*, 246, R901–R903.

Pawlby, S. J. (1977). Imitative interaction. In *Studies in Mother–Infant Interaction*, ed. H. R. Schaffer, pp. 203–224. New York: Academic Press.

Petersen, M. R., Zoloth, S. R., Beecher, M. D., Green, S., Marler, P. R., Moody, D. B., & Stebbins, W. C. (1984). Neural lateralization of vocalizations by Japanese macaques: Communicative significance is more important than acoustic structure. *Behavioral Neuroscience*, 98, 779–790.

Premack, D. & Woodruff, G. (1978). Does the chimpanzee have a theory of mind? *Behavioral and Brain Sciences*, 1, 515–526.

Querleu, D., Renard, X. & Crepin, G. (1981). Perception auditive et reactivité foetale aux stimulations sonores. *Journal de Gynecologie, Obstetrique et Biologie de la Reproduction*, 10, 307–314.

Querleu, D., Renard, X. & Versyp, F. (1986). Vie sensorielle du foetus. In *L'Environnement de la Naissance*, ed. G. Levy & M. Tournaire, pp. 15–46. Paris: Vigot.

Redican, W. K. (1975). Facial expressions in nonhuman primates. In *Primate Behavior: Developments in Field and Laboratory Research*, ed. L. A. Rosenblum, pp. 103–194. New York: Academic Press.

Reznick, J. S. (1989). Research on infant categorization. *Seminars in Perinatology*, 13, 458–466.

Roberts, K. & Horowitz, F. D. (1986). Basic level categorization in seven- and nine-month-old infants. *Journal of Child Development*, 13, 191–208.

Sherman, T. (1985). Categorization skills in infants. *Child Development*, 56, 1561–1573.

Sodian, B., Taylor, C., Harris, L. & Perner, J. (1991). Early deception and the child's theory of mind: False trails and genuine markers. *Child Development*, 62, 468–483.

Spence, M. J. & DeCasper, A. J. (1987). Prenatal experience with low-frequency maternal-voice sounds influence neonatal perception of maternal voice samples. *Infant Behavior and Development*, 10, 133–142.

Stoel-Gammon, C. (1989). Prespeech and early speech development of two late talkers. *First Language*, 9, 207–224.

Street, R. L., Street, N. J. & van Kleek, A. (1983). Speech convergence among talkative and reticent three-year-olds. *Language Sciences*, 5, 79–96.

Thompson, R., Cicchetti, D., Lamb, M. & Malkin, C. (1985). The emotional responses of Down syndrome and normal infants in the Strange Situation: The organization of affective behavior in infants. *Developmental Psychology*, 21, 828–841.

Van der Stelt, J. M. & Koopmans-van Beinum, F. J. (1986). The onset of babbling related to gross motor development. In *Precursors of Early Speech*, ed. B. Lindblom & R. Zetterstrom, pp. 163–173. New York: Stockton Press.

Waddington, C. H. (1940). *Organisers and Genes*. Cambridge: Cambridge University Press .

Waddington, C. H. (1975). *The Evolution of an Evolutionist*. Ithaca, New York: Cornell University Press .

Waxman, S. R. & Kosowski, T. D. (1990). Nouns mark category relations:

Toddlers' and Preschoolers' word-learning biases. *Child Development*, 61, 1461–1473.

Weeks, S. J. & Hobson, R. (1987). The salience of facial expression for autistic children. *Journal of Child Psychology and Psychiatry*, 28, 137–152.

Werker, J. F. & Pegg, J. E. (1992). Are changes in infant speech perception related to the emergence of a honological system? In *Phonological Development: Models, Research, and Application*, ed. C. Ferguson, L. Menn & C. Stoel-Gammon, pp. 285–311. Parkton, Maryland: York Press.

Whitehurst, G. J., Smith, M., Fischel, J. E., Arnold, D. S. & Lonigan, C. J. (1991). The continuity of babble and speech in children with specific language delay. *Journal of Speech and Hearing Research*, 34, 1121–1129.

Part five

Learning and development

14

Behavioral change as a result of experience: toward principles of learning and development

PETER D. BALSAM AND RAE SILVER

The study of the role of experience during ontogeny provides an opportunity for testing the generality of existing principles of learning and may lead to the discovery of new ones. Developmental psychologists and learning theorists share an interest in how experience produces behavioral change, yet differences in the set of problems, methods, and levels of conceptualization traditionally employed have led to largely nonoverlapping areas of research (cf. Shettleworth, this volume). Nevertheless, the relationship between learning theory and behavioral development has been studied within both traditions. At one extreme is the notion that much of development can be explained by learning principles (Baer & Wright, 1975; Bijou & Baer, 1961; Gewirtz & Boyd, 1977; Watson, 1930; Holt, 1931; Skinner, 1953; Staats, 1975). At the other extreme is the suggestion that learning principles have contributed little to understanding behavioral development (e.g. Gottlieb, 1983).

There are two strategies for exploring the contributions of learning to behavioral development. One approach, not taken in this chapter, is to study animals of different ages in traditional learning experiments (Amsel, 1979; Spear, 1978; Johanson & Terry, 1988; Spear & Rudy, 1991). In so far as age-related differences in performance on a learning task can be attributed to changes in process, this strategy reveals the range of processes whereby experience might induce changes in behavior during development. For example, Rudy and his colleagues (Rudy, 1992) have suggested that young animals can learn about single events before they can learn about relationships between events during ontogeny.

In a second approach the role of learning in the development of naturally occurring behavior is analyzed (Hogan, 1973, 1977, 1984; Galef, 1977). Learning principles that apply during normal development may emerge through the analysis of the role of experience in the ontogeny of specific responses. The starting point of this approach is to identify a functionally significant behavior such as eating (Hogan, 1973, 1977, 1984, this volume; Galef, 1977) or breeding (Lehrman, 1955, 1962, 1970a) and then to analyze the mechanisms whereby experience influences the development of this behavior.

This chapter explores the adequacy of contemporary learning theory in dealing with transitions in development. Differences between 'learning' and 'developmental' approaches are then compared as a way of specifying how juvenile and adult learning might differ. The analysis describes aspects of development that can be accounted for by learning theory, and specifies areas in the study of learning which must be developed in order to achieve common principles for describing the role of experience in development and in adult behavior.

Behavioral transitions

How new functions and abilities emerge is the *raison d'être* of the developmentalist. On the one hand there is a concern with continuity, or how new responses and abilities derive from the behavior at previous stages. On the other hand, theorists struggle with the problem of discontinuity in development; somehow new responses or abilities emerge that apparently are not derived from previous stages (see Berridge, this volume). According to Gottlieb (1983, p. 4), the problem of discontinuity is much more important than that of continuity, yet experimental strategies for the analysis of continuities are more accessible. The analysis of transitions in behavior is one of the potential areas where learning principles may contribute to an understanding of development.

The concepts of continuity and discontinuity in development have two meanings, not always specified. On the one hand, there is the problem of continuity and discontinuity in the *stimuli* that evoke a given response. For example, during development the control of behavior changes through a progression of different sense modalities (Gottlieb, 1983). On the other hand, there is the problem of *response* continuity/discontinuity, or homotypic versus heterotypic response transitions in development (Kagan, 1971; Lewis & Starr, 1979; Werner, 1957). The question here is whether the form of a developmental precursor shares elements in common with the developmental successor of the response. For example, one might study how adult feeding emerges from suckling in mammals or how the adult feeding behavior becomes more skilled. By distinguishing between response and stimulus transitions, the analysis of discontinuity in development is made accessible for analysis (cf. Hogan, 1988).

Stimulus transitions

Embedded in the problem of continuity/discontinuity is the notion of transition from one source of control to another. Two kinds of transitions of stimulus control occur in development: the sensory modality controlling a given response changes with time, and the range of effective stimuli narrows. Learning theory has traditionally been concerned with stimulus control and has a great deal to contribute to this aspect of the study of transitions in development.

Sensory transition

It is often the case that the sensory dimensions that control a particular response will change during development. A classic example of change in controlling modality is illustrated in Tinbergen & Perdeck's (1950) studies of the development of the gaping response in young blackbirds. Upon hatching, the nestlings are blind. The response is released by vibrotactile stimulation produced by the parents landing on the nest. Later, it is released by the visual cue presented by the sight of the parent.

Developmental changes in the controlling sense modalities have also been described in a number of other behavioral systems including huddling (Alberts, 1978; Alberts & Brunjes, 1978) and feeding (Johanson & Hall, 1981) in rat pups and home orientation in kittens (Freeman & Rosenblatt, 1978; Rosenblatt, 1979, 1983). For example, the newborn rat and kitten's earliest responses (huddling and home orientation) are released by thermal and tactile stimuli. Later in development, olfactory and then visually released responses appear. It has been suggested further that these transitions of control from one sense modality to another follow a relatively stable order during ontogeny (Gottlieb, 1983) and are attributable to sensorimotor and neural maturation (Rosenblatt, 1979).

Although a stimulus control perspective (Terrace, 1966; Mackintosh, 1977; Balsam, 1988) has not been applied to the analysis of behavioral transitions in development, the problem in developmental and learning analyses seems comparable. Learning theory attempts to account for the transfer of control from one stimulus dimension to another. Applied to the foregoing developmental example, the original modality continues to be functional, yet a new modality comes to control the response. In learning paradigms such transfer is thought to take place when there is a change in the relative validity of cues (Balsam, 1988). The study of how one element of a compound cue can overshadow learning of other aspects of the stimulus situation or how prior learning about a cue can block learning about new cues is an active area of investigation (see Klein & Mowrer, 1989, for several approaches). Such processes may play an important role during ontogeny (cf. Bolhuis, de Vos & Kruijt, 1990).

In developmental transitions, the earlier maturing sensory modality may provide cues that are less predictive of outcomes than are cues provided by later maturing sense modalities. This seems plausible as the earliest sensory systems to mature in mammals (thermal, tactile, olfactory) tend to provide temporally and/or spatially diffuse information, while later maturing sensory systems (auditory, visual) tend to provide information that is more localized. Hence, the overshadowing of a temporally remote cue by a temporally proximal cue (Kehoe, Gibbs, Garcia & Gormezano, 1979) may contribute to the transition in sensory control. These transitions, furthermore, may be facilitated by the formation of inter-element associations and/or potentiation when information is provided through more than one sense modality (Durlach & Rescorla, 1980; Lett, 1980; cf. Bolhuis & van Kampen, 1992). In such situations, learning about one cue facilitates the acquisition of control by added cues. Stimulus control principles can reveal the continuity between apparently discontinuous developmental stages of sensory control and leads to strong predictions about the necessary conditions for the occurrence of stimulus transitions. If indeed new cues come to control behavior on the basis of their predictive validity, then one should be able to alter the normal sequence of sensory transitions by degrading the predictive contingencies that characteristically occur. Thus in the example described above, if the sight of the parent did not predict food for the young blackbird or if absence of the parent did not predict the absence of food (Rescorla, 1968), the sight of the parent might not continue to elicit gaping (Allesandro et al., 1989).

Stimulus specificity

Many transitions in development involve the acquisition of appropriate stimulus control because the young of many species must learn where and when to perform a particular response. Individual experience has some influence on the specific situations that come to control behavior, including behavior such as feeding (Hogan, 1977; Galef, 1992; Balsam, Deich & Hirose, 1992), social interactions (Bolhuis, 1991), reproductive behavior (Domjan, 1992), and aggressive/defensive reactions (Kuo, 1967). One mechanism in the acquisition of stimulus specificity may be that underlying

Pavlovian conditioning. Young animals learn which specific situations are correlated with specific outcomes.

For example, the young of species that have variable food sources are likely to have to learn about the specific qualities of food during their transition to independent feeding (Balsam, Deich & Hirose, 1992). Associative learning probably plays a very important role in this process (Hogan, 1977; Hogan-Warburg & Hogan, 1981). This is illustrated in our own work. Ring dove squab are initially fed by parental regurgitation, and learn to peck at seed during the third week after hatching. Squab must learn to associate seed with positive ingestional consequences for the normal development of pecking (Graf, Balsam & Silver, 1985; Balsam, Graf & Silver, 1992). In these experiments, parents and squab were fed seed ground into a fine powder. Beginning on day 14 after hatching and on all subsequent days, squab were put into a test chamber with seed on the floor for 20 minutes. We found that powder-reared squab pecked very little during these test sessions as compared to seed-reared subjects, indicating that direct experience with seed is important for development of the adult response.

We also investigated normal development of pecking by analyzing whether experience with seed followed by feeding was sufficient to increase squab pecking (Balsam, Graf & Silver, 1992). Squab were reared on the powdered-seed diet, but given different types of experience with seed and feeding. There was a low level of pecking at seed even in subjects that had no prior experience with grain. Significantly more pecking occurred when visual exposure to seed was followed by immediate feeding, but pecking did not increase above control levels if feeding was delayed for an hour after exposure. We interpret this result to mean that the Pavlovian contingency is sufficient for the squab to learn what items to peck. There is a tendency for the squab to peck at grain-like objects, but this tendency is weak and it habituates rapidly unless seed is paired with positive ingestional consequences. The pairing of the sight of grain with food selectively increases the tendency of the squab to peck at seed. We cannot say whether the enhanced pecking is the result of a general motivational effect, an enhanced attention to seed, and/or a change in the incentive properties of the seed. Nevertheless, it is clear that the Pavlovian contingency directs the pecking at appropriate targets.

Lastly, stimulus control may be induced through observational learning. Young animals may learn when and where to perform a particular response by observing the behavior of adults. Specific choices of food items are influenced by choices of conspecifics (Turner, 1964; Epstein, 1984;

Suboski & Bartashunas, 1984; Biederman & Vanayan, 1984). It seems likely that such influences would be effective during ontogeny as well as in adulthood.

The effectiveness of observational learning as a mechanism whereby animals learn about new controlling stimuli may also depend on individual experience. In many species, young feed with their parents (Galef, 1992), and the feeding behavior of the parent becomes a good signal for predicting which objects in the environment will be associated with positive ingestional consequences (see Balsam, Deich & Hirose, 1992, for a more detailed discussion).

Stimulus differentiation

During ontogeny, the class of stimuli that are effective in eliciting a response from the young animal becomes defined more narrowly (cf. Bolhuis, 1991). For example, Hailman (1967) showed that 1–2-day-old laughing gulls beg equally at almost any pointed object that moves in front of them. As the chicks grow older, stimuli that evoke the begging response become progressively more limited. First, models that do not resemble laughing gulls come to be rejected. Finally, only models that resemble their parents are capable of evoking the begging response of the young. It seems likely that the 'sharpening' of stimulus control with extended training (Hearst & Koresko, 1968), and/or differential reinforcement (Jenkins & Harrison, 1960; Spear, Kucharski & Miller, 1989) contribute to the narrowing in the range of stimuli which evoke begging.

Response transitions

There are two ways in which the dominant response that occurs in a particular situation can change. The first type of response transition, *response selection*, occurs when an animal switches from one existing response form to another to fulfill the same function. For example, in ontogeny, when mammals go from suckling to eating there is a period of time when both responses are available and effective at getting food. The question then is why one response is selected over another. Learning psychology has much to contribute to this analysis since much is known about the determinants of choice and, in particular, how the success or failure of a response affects its future occurrence.

The second aspect of the response transition problem is the question of *response induction*. This aspect is concerned with how new response forms come about in the first place. This is a difficult problem because of the apparent discontinuity in response forms. The precursor of a given response may be very different from its successor. An extreme example of this is the analysis of behavior which is expressed correctly the first time it occurs in ontogeny (i.e. it appears to be discontinuous with earlier forms). The induction of new response forms has been little studied in traditional learning psychology. We describe below the beginnings of such an analysis and its implications for conceptualizing developmental changes (see also Berridge, this volume).

Response selection

Global transitions Generally, as an animal gets older there is a transition period in which immature forms of behavior decline and adult forms increase (see Groothuis, this volume). During the transition, both forms of behavior may be functional and the rate of transition may be influenced greatly by individual experience. For example, there may be considerable variation in the environments in which the young of altricial species develop adult feeding behaviors. The timing and speed of the transition may depend on the availability of adult foods as well as on the availability and quality of juvenile food sources, including food provided by the parent. This view implies that transitions depend on the success or failure of the alternative feeding behaviors. In our analysis of the ontogeny of feeding in ring doves we have measured the efficiency of both the juvenile and adult forms of feeding during the transition from begging for parental food to pecking for seed. As Figure 14.1 shows, begging becomes less likely to result in parental feeding while pecking becomes more likely to result in the ingestion of seed. This pattern gives credibility to the hypothesis that the transition to pecking is modulated by the relative success that each individual has at obtaining food in these two ways.

In this view, a reduction in the amount of food supplied by the parents should be an important impetus for the initiation of independent feeding. Consistent with this possibility, a decreasing tendency to feed young as they mature has been documented in macaques (Kaufman & Rosenblum, 1969), domestic cats (Martin, 1986), white£tailed and fallow deer (Gauthier & Barrette, 1985), rats (Reisbick, Rosenblatt & Mayer, 1973; see also Fleming & Blass, this volume), and ring doves (Wortis, 1969; Hirose & Balsam, submitted). In northern wheatears, Moreno (1984) found that sharply decreased rates of parental feeding coincided with a marked increase in foraging by young. These young also seemed to shift between dependent and independent feeding, depending on the payoffs for these behaviors at different times of day. Young spotted flycatchers also showed

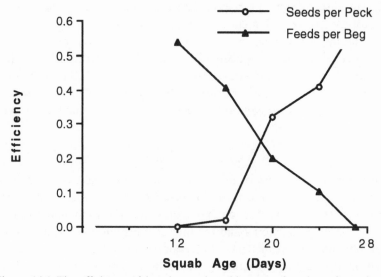

Figure 14.1 The efficiency of begging and pecking as a function of squab age. Begging efficiency is the number of 10-second intervals during which squab were fed by parents, divided by the number of intervals with begging during the first 60 minutes after lights were turned on each morning. Pecking efficiency was measured by dividing the number of seeds eaten by the number of pecking movements during a 20-minute test session.

a sudden increase in rate of prey capture attempts at the time when self-feeding became more efficient than begging for food from their parents (Davies, 1976). Furthermore, Davies (1978) was able to manipulate the age at which hand-reared great tits increased their rate of self-feeding by varying the age at which he increased the difficulty of obtaining food by begging. Similarly, in rats, variation in dam's milk supply affects the age of weaning (Stern & Rogers, 1988).

A second implication of this view is that the rate of transition will also depend on the quality of food available for independent feeding. Again, there is considerable evidence that the transition to independent feeding is modulated by this source of reward. In years of famine, varied tits remain dependent on their parents longer than they do in years of plenty (Higuchi & Momose, 1981). Experimental variations of the quality of food available for independent feeding show that rat pups' transition to independent feeding is modulated by food quality (Cramer, Thiels & Alberts, 1990). In sum, these studies illustrate that operant contingencies arrange differential rewards for the adult and juvenile forms of behavior and play a significant role in global response transitions.

Response differentiation A second way in which behavior changes during development is that it becomes more organized and skilled. One can find examples of this in all domains of behavioral development (e.g. Kruijt, 1964; Groothuis, Hogan, this volume). The increased efficiency of ring dove feeding on seed (see Figure 14.1) provides a good example of response differentiation. Successful pecking requires the squab to move its body to the proper location, execute an accurate head thrust, and to perform a gaping movement in which the opening and closing of the beak is appropriate for grasping the target seed. Additionally, the body positioning, thrust, and gape components must all be coordinated properly for a peck to result in the successful ingestion of food. This response is highly stereotyped in adult birds. We have studied the ontogeny of the topography of this response by using a system which permits continuous monitoring of avian gape (Deich & Balsam, 1993, in press). We have found that squab's early pecking is highly variable. Many food-directed thrusts have no gaping. Pecks that have an accompanying gape vary greatly and bear very little resemblance to those of the adult. Pavlovian pairings of the sight of seed with positive ingestional consequences do not induce the adult form. Squab must have successful experience with handling and ingesting seed for the adult topography to emerge. Furthermore, the gape of a different member of the *Columbidae*, the pigeon, has been shown to be susceptible to operant control (Deich, Allan & Zeigler, 1988). We believe that response shaping occurs through feedback from successful and unsuccessful pecks such that the gape component and its coordination with the thrust component become the more effective adult form. Operant conditioning is proposed as the underlying operative process because neither maturation nor extensive Pavlovian training is sufficient to produce the stereotyped adult response. Lastly, the movement from the multiple and variable gape forms early in ontogeny to the stereotyped adult form is consistent with the hypothesis of operant learning.

Mechanisms of response selection Note that we cannot say if it is the absolute level of success for a given behavior or its success relative to other behaviors that determines the time course of the transition to adult responses. However, on the basis of a considerable literature on the determinants of choice (Williams, 1988), we expect that the relative payoff is the controlling variable. In any case, one likely role of individual reinforcement history is to modulate the overall rate and timing of developmental transitions.

An unspecified aspect of our analysis is the nature of the process or processes that underlie response selection. The process whereby reinforcers increase the frequency of behavior has been conceptualized in two general ways (cf. Staddon & Simmelhag, 1971; Staddon, 1983). In one, competition between responses occurs such that when one is strengthened, alternative responses become weaker. In the other, a set of responses is present initially, and those which are not strengthened are lost to future actualization. It is possible to conceive of shaping as the loss of response forms that are not followed by reward. Whether shaping involves response strengthening and/or the loss of unselected responses has not been resolved empirically.

Response induction

Lehrman (1955) showed that exogenously administered prolactin induces feeding of squab in parent doves with previous breeding experience, but not in reproductively inexperienced animals. This finding raises the question as to how doves breeding for the first time can respond appropriately. According to Lehrman (1964):

The answer lies in the fact that we must not think of 'experience' as being limited to the effects of having performed the same behavior before. When a dove builds a nest and thus becomes attached to the nesting site so that she spends most of her time there, she is acquiring 'experience', through which she becomes oriented to the nest in such a way that (a) the nest is the place where she is most likely to lay the egg, and (b) she is bound to come into contact with the egg after it appears even without having any intention to sit on it. Similarly, when doves sit on the eggs, they become attached to the nest-site even more, and they are there when the eggs hatch, so that they come into contact with the newly hatched squab. This means that experience is playing a role in the succession from stage to stage even during the first breeding cycle.

This example illustrates that the response made to a stimulus at one point in time is not necessarily the response which later emerges as learned; in this case, the developmental precursor is not necessarily topographically related to the response which emerges. This apparent discontinuity in response form may derive from the operation of a number of learning mechanisms. Below we describe how sensitization, habituation, Pavlovian conditioning, differential reinforcement, and observational learning might potentially underlie these saltatory changes in response form.

Sensitization When sensitization occurs, the presentation of a stimulus, particularly a strong one, results in the enhanced ability of the cue to elicit a new response. Sometimes, the potentiated response is a stronger version of the elicited response so there is apparent continuity. Lipsitt & Kaye

(1965) found that repeated nipple presentation strengthened the sucking reflex in human infants. However, the potentiated response need not be similar to the one originally elicited by a cue. For example, in the sea slug, *Aplysia*, elicitation of the gill withdrawal response by mechanical stimulation of the gill makes the defensive inking response more likely. If a strong stimulus is presented periodically, so-called schedule-induced behaviors may be potentiated (Staddon, 1977; Wetherington, 1979). The behavior which tends to occur following reinforcer presentation depends on the environmental supports available during these periods. For example, following periodic food presentation, rats will drink excessively, chew on wood chips, run in a wheel, or be aggressive toward a conspecific depending on which stimuli are present (cf. Staddon, 1977). Like sensitization, the reactivity to a cue is potentiated by the presentation of a strong stimulus. However, multiple reinforcer presentations are required to produce this effect, and the strength of the potentiation is dependent on the interval between reinforcer presentations. Perhaps general arousal produces the enhanced reactivity (Killeen, 1984). Again, the induced responses may bear no resemblance to the responses previously observed in that situation.

Habituation When habituation occurs, the repeated presentation of a stimulus may result in a diminution of the response controlled by that cue. If the response has multiple components, such as aggressive or courtship displays, repeated presentations of a releasing stimulus may cause different components to habituate at different rates. Such differential habituation could result in significant changes in the organization of response components. Alternatively, the habituation of one response may disinhibit other weaker response tendencies. Fear induced by separation from parents may inhibit feeding. The apparent abrupt onset of independent feeding in some altricial species may be a concomitant of habituation of the fear response. As Galef (1977) has shown, rat pups will generally avoid new foods. However, feeding responses will emerge if pups are exposed to the odor of specific foods. Interestingly, in rats older than 21 days this odor exposure must occur in a social context (Galef & Kennett, 1987). In both cases the emergence of the feeding response depends on exposure to the odor. Thus, the first occurrence of independent feeding may depend on habituation of the neophobic response.

Pavlovian conditioning Pavlovian conditioning may produce a discontinuous change in response forms. When a cue is paired with a primary reinforcer an entire behavior system is conditioned (Hogan, 1974, 1988; Timberlake & Lucas, 1989). Thus, when a conditioned stimulus is paired with a noxious stimulus the whole defensive system is conditioned; when paired with a sexual outcome the reproductive system is conditioned; and when paired with food an entire feeding system is conditioned.

In the prototypical Pavlovian learning experiment the conditioned response (CR) is similar to the unconditioned response (UR). However, cases in which CR and UR resemble each other are common to the psychology laboratory but probably fairly uncommon in nature. Even in laboratory studies of Pavlovian conditioning it is clear that the form of the CR is not invariably similar to that of the UR (Holland, 1984a). For example, if a brief keylight is paired with grain, pigeons will come to peck at the keylight with a peck topography that is quite similar to the feeding topography (Jenkins & Moore, 1973; LaMon & Zeigler, 1988). However, if general contextual cues signal food presentation, the CR is an increase in locomotor activity (Mustaca, Gabelli, Papini & Balsam, 1991). CR topographies are influenced by the reinforcer type (Pavlov, 1927; Jenkins & Moore, 1973), reinforcer frequency (Innis, Simmelhag-Grant & Staddon, 1983), antecedent stimulus type (Timberlake & Grant, 1975; Holland, 1977), stimulus duration (Holland, 1980), as well as by properties of the learning and test context (Balsam, 1985). It is clear that the response that emerges as learned does not necessarily resemble any previously observed response to the unconditioned stimulus or any previously experienced conditioned stimulus.

The implication of these findings for an ontogenetic analysis is that as perceptual and response systems develop, the expression of underlying associations may change. For example, Johanson, Hall & Polefrone (1984) found that the conditioned response evoked by odors paired with milk infusion changed as rat pups got older. The youngest pups showed no CR when trained, but a CR emerged when they were tested at an older age. Furthermore, the CR of older pups differed from the CRs of younger pups. In this way, Pavlovian conditioning may mediate some discontinuities in response form during ontogeny.

Chaining Response forms can change in an abrupt way during ontogeny through chaining. The development of one response may bring the young animal in contact with stimuli that elicit new responses. The new component of this sequence appears to develop discontinuously from its predecessor. We suggest that this mechanism may be very important in the ontogeny of the adult pecking response in doves. The earliest form of thrusting to get food occurs when squab beg for food. However, these thrusting movements are not accompanied by any gapes (Deich & Balsam, in press). Once squab start thrusting at grain, gapes occur. As mentioned above, the thrusting component of this peck is induced by Pavlovian contingencies. Deich and Balsam (1993) found that gapes are four times more likely to occur in proximity to a head thrust at seed than they are when the birds are not pecking. Though we have not identified the specific eliciting stimulus for gaping, it seems plausible that the visual, kinesthetic or vestibular cues produced by movement of a squab toward the substrate induce gaping. Thus the saltatory emergence of gaping while pecking in the ring dove is likely to be the result of a two-step process. Pavlovian associations between seed and ingestional consequences induce the thrusting movement toward the ground. This movement brings the squab into contact with the stimuli that elicit gaping.

Induction by differential reinforcement

(1) *Spatial and temporal variation.* Abrupt response transitions occur in development through the process of response shaping. In the classical analysis of shaping based on Skinner's (1938) characterization of this process, one reinforces closer and closer approximations to the target response. After an approximation is established reinforcement is withheld. This extinction induces quantitative variation in the behavior and a closer approximation is selected for reinforcement. In this way behavior is 'molded like a sculptor molds clay' (Skinner, 1953). There is strong evidence that the spatial and temporal distribution of behavior may be molded this way (see Stokes & Balsam, 1991, for a review). During development it seems likely that the spatial and temporal behavioral variation needed for reinforcement to fine tune responses occurs when the behavior of the juvenile fails (Deich & Balsam, 1993) or generates aversive consequences (West, King & Arberg, 1988).

(2) *Variation in component organization.* While behavior may be refined along continuous dimensions, its initial occurrence in development is not often seen as being molded from precursors. This is so because behavior is organized into discrete components (Timberlake, 1983; Timberlake & Lucas, 1989). Support for this point of view comes from developmental and neurophysiological analyses of many types of motor behavior (Fentress, 1983;

Gallistel, 1980; Golani & Fentress, 1985; Berridge, this volume) including operants (Teitelbaum, 1977). Additionally, studies of shaping and response induction show that reinforcement or extinction of one response may increase the frequency of topographically unrelated responses. For example, reinforcing novel behavior can result in the production of responses that are topographically unrelated to previously reinforced responses (Pryor, Haag & O'Reilly, 1969; Goetz & Baer, 1973; Holman, Goetz & Baer, 1977). Similarly, reinforcing one member of a positively covarying set of responses may increase the frequency of other members with quite different topographies, while extinction of one member of a set of negatively-covarying responses may increase the frequency of topographically dissimilar behavior (Hogan, 1964; Kara & Wahler, 1977; Russo, Cataldo & Cushing, 1981; Wahler & Fox, 1981; Kazdin, 1982). Thus a complete understanding of the role of reinforcement in producing new response forms requires analysis of how new components are induced as well as how they are selected and reorganized by reinforcement.

The central problem for understanding response induction is to identify the linkages between topographically different response forms. There is no adequate set of principles in learning theory for describing this aspect of response shaping, although some have grappled with the problem (Moore, 1973, Segal, 1972; Timberlake & Lucas, 1989). What is lacking for understanding the emergence of a response through shaping is a set of rules which describe how response forms are determined. Two kinds of information seem necessary to develop these rules.

First, motivational and environmental determinants of response form must be studied, and second, the organization of response forms must be described. As described above, we already know some of the environmental factors that determine response form. Topographies are influenced by the reinforcer type and frequency, antecedent stimulus type and duration, and by properties of the learning and test context.

Less is known about the role of response organization in the determination of response form. Here the question is one of understanding the structural and functional linkages between response topographies. Specifically, how does the induction of one response form lead to the induction of other, different forms. Although these linkages have not been described previously in learning paradigms, the techniques for doing so have been developed in the ethological literature on the study of behavior system organization (Tinbergen, 1951; von Holst & von St Paul, 1963; Baerends, 1976; Hogan, 1988) and of species-typical behavior transitions (Slater, 1973; Colgan, 1978; Fentress, 1983).

Stokes & Balsam (1991, submitted) have begun a similar analysis in the domain of adult learning by studying how rats acquire the barpress response through shaping. In these studies the rat's behavior in an operant chamber is classified into discrete categories. The categories of interacting with the bar include touching the bar with the right or left paw, grasping it with both paws, nosing it, biting it, and so on. Bar-directed behaviors were observed in a group of four rats under a number of conditions. One observation occurred during a session in which subjects were water deprived and given dipper training in which the dipper was presented regardless of what the rat was doing on a variable time 30-second schedule. The rats were then shaped to press the bar and earned 100 reinforcers. The results showed that no new components emerged during shaping. The effect of shaping and reinforcement of particular topographies is not to create new components but merely to modulate the frequencies of components that had been induced by the dipper-training procedure. Reinforcement also acts to organize these components into highly structured sequences.

While this type of study represents a descriptive starting point for analyzing the role of reinforcement in inducing new sequences, general principles of induction have not yet emerged. Both from the point of view of studying adult learning and of analyzing ontogenetic change, it is important to work out how reinforcement induces new component organizations. The lack of principles for specifying how environmental constraints and structural and functional linkages between response components interact to produce new behavioral variants is a serious impediment to understanding the role of experience in both adults and juveniles.

Imitation Imitation also provides a means for acquiring new response forms in ontogeny. Animals of many species are capable of imitation (Whiten & Ham, 1992). Though there are few examples of imitation that are unassailably instances of copying the behavior of a model (Galef, 1990), there are some seemingly well-controlled demonstrations of this phenomenon (Heyes & Dawson, 1990; Palameta & Lefebvre, 1985). It therefore seems quite likely that new responses might be induced in the young by observing adults (see also Groothuis, this volume).

Thus, specification of the rules for induction of response topography is crucial to a complete analysis of the role of experience in behavior change. The problem of understanding response induction in adults is identical to this aspect of the developmentalists' problem of studying the ontogeny of responses which do (continuity) and do not (discontinuity) share structural properties with their precursors. Within the limits of current knowledge, the analysis of the role of established learning processes in the ontogeny of behavior provides a powerful conceptual framework for studying how new response forms emerge in development.

Concomitant stimulus and response transitions

At this point it is probably clear that although we have separated our discussions of stimulus and response transitions during ontogeny, both transitions frequently occur together. For example, we have described this kind of change in our studies of the transition from dependent to independent feeding in ring dove squab (Balsam, Deich & Hirose, 1992). Initially, the squab obtain food when the parents grasp their bills and regurgitate crop milk into their gaping beaks. The gaping response is initially under vestibular or proprioceptive control. In the first few days of life the gaping response comes to be elicited by tactile stimulation of the beak. By the second week after hatching, the sight of the parents elicits a begging response which results in parental feeding of the young. During this time period the begging response becomes more elaborate and directed at the parents. By about 12 days of age, the sight of the parent elicits vigorous approach, begging, and squealing from the young squab. Once they leave the nest, the young continue to beg from the parents, but the parents increasingly refuse to feed them (Hirose & Balsam, submitted; Wortis, 1969). Finally, the young peck on their own, and by day 21 they can feed independently.

The present view posits two steps in the acquisition of the adult consummatory response. First, the occurrence and direction of the thrust component are determined on the basis of Pavlovian pairings of the sight of grain with feeding, and second, the gape component and its coordination with the thrust component are moved toward the more effective adult form by a process of response shaping. Further, we suggest that the simultaneous decline in parental feeding and increase in pecking efficiency allows for this transition in foraging strategy. This analysis of the acquisition of pecking by doves serves as a model of how learning theory can contribute to the analysis of developmental transitions. To the extent that behavioral development involves changes both in stimulus control and response form, learning theory can provide examples of the experimental and conceptual analysis of that distinction.

Development versus learning theory

In the preceding section we have suggested that learning principles can be applied directly to the analysis of developmental transitions. The success of such an analysis depends on the correspondence between the conceptualizations of learning theory and those of developmental change. If learning mechanisms engaged during development represent specialized adaptations, then they may share little in common with adult learning. However, as Shettleworth (this volume) points out, specialization may occur in the necessary conditions for learning, including parametric differences in learning sensitivities, the contents of learning, or in how learning is translated into performance. Because the developmental and learning traditions for analyzing the role of experience in behavioral change have remained so distinct, one must entertain the possibility that the mechanisms of adult and infantile learning may be so specialized that the study of one does not illuminate the action of the other. The resolution of this issue will reside in specifying the differences between adults and juveniles in learning mechanism, content, and performance rules. These differences can then be studied to see if they are indeed one of the specializations suggested by Shetttleworth or if the apparent differences are the result of incomplete analyses. In this section, we attempt to specify dimensions along which developmental paradigms apparently differ from those studied by learning theorists. These dimensions include differences in (a) conceptualization of experience and its effects on behavior, (b) assumptions about the significance of structural (or maturational) changes, and (c) ideas on the role of the past history of the animal in explanations of current behavior.

Concepts of experience and response

Learning has often been defined as a relatively permanent change in behavior as a result of experience (Bower & Hilgard, 1981). The source of experience that most learning theories refer to is a repetitive change in the external environment. The effect of such experience on discrete behavioral responses is most often the dependent measure. The developmentalist, however, employs a much broader definition of experience along with a broader view of the affected response systems. This view is represented in the writings of Schneirla, as characterized by Lehrman (1970b, p. 20).

Schneirla (1966) has used the concept of 'experience' to mean all kinds of stimulative effects from the environment, ranging from stimulus-involved bio-chemical and biological processes (having effects on the developing nervous system) to what we ordinarily call conditioning and learning.

Here, experience refers to exposure to events or stimuli *per se*, regardless of the frequency of presentation, or the contingency relationships to other stimuli or responses. The response includes behavioral as well as sensory, motor and neural structures altered by environmental events. This difference in the conceptualization of experience and response is reflected in differences in the temporal and qualitative aspects of the units of analysis that are typically employed in each of these traditions. The time frame of independent and dependent variables is frequently orders of magnitude greater in developmental research than in learning research. For example, how does early experience with specific food items and odors (Galef, 1976, 1977), and how does suckling behavior of the newborn (Hall & Williams, 1983) lead to food preferences and eating patterns characteristic of the adult? How do reactions of the newly parturient female to the infant in the postpartum period influence later interactions with and attachments to the child (Klaus & Kennel, 1976; Fleming & Blass, this volume)? As these examples of developmental research questions illustrate, the develop-mentalist is concerned with inputs and behavior that occur over days, weeks, and months.

In contrast, the independent and dependent variables of the learning psychologist are most often measured in seconds and minutes. This raises the possibility that adult and juvenile learning may differ parametrically or even in the underlying mechanism. It is, however, arguable that no new rules of learning are required by this difference in time frame. For example, some have argued that taste aversion learning, where independent and dependent variables are defined in the order of hours, is not outside the domain of a general process learning theory (Logue, 1979; Domjan, 1980). Similarly, it has been argued that Pavlovian conditioning of contextual cues can be conceptualized as learning that occurs across days (Balsam, 1985). In the absence of empirical studies it is unclear whether the difference in time frame requires that different techniques be developed for behavioral analysis, and whether or not different theoretical principles are required to explain the processes of change at these different levels.

Another major difference in the units of analysis used in learning and in developmental research lies in the qualitative nature of the dependent variable. All formal theories of learning and most research in this area employ a quantitative characterization of a discrete response as the dependent variable. Rate and probability of response are the two most common measures. Change in behavior is conceived of as a change in one of these properties. On the other hand, for the developmentalist, significant changes in behavior are defined by qualitative transitions in response form. As we have pointed out, an adequate account of the role of experience in development and learning must include rules about categorical transitions in response form as well as specify the determinants of quantitative response changes.

The significance of structural change

Another difference in the conceptual framework of the learning psychologist and developmentalist is in the assumptions about the role of structural change in producing behavior change. For the developmentalist, behavioral change in the developing organism is guided by epigenetic factors. During development, structural changes are thought to occur as a consequence of experience and as a consequence of maturation (physical change). The central nervous system of the young animal interacts with the environment, and these interactions modify the structural and functional organization of the nervous system. These modifications in turn influence later interactions with the environment.

The changes that occur in the young animal are so salient that the study of learning in development has proceeded as though one could not study developmental changes without taking maturational changes into account. In fact, some of the behavioral and morphological structures of the young animal have been conceived of as 'transient ontogenetic adaptations' (Oppenheim, 1981). The differences between adults and infants of one species, like differences among species, are viewed as specialized adaptations (Domjan, 1983; Galef, 1981; Vogt & Rudy, 1984; Connolly & Prechtl, 1981; Shettleworth, this volume). This view of the young animal suggests that the rules of learning may be determined or constrained by structural changes during ontogeny.

Even if one allows that structural changes accompany learning, it remains to be determined whether the newly emerging structures require different principles of learning at the behavioral level. There are many empirical demonstrations of age-correlated changes in learning (Spear & Campbell, 1979). There is, furthermore, some support for the idea that underlying changes in nervous system structure and connectivity may mediate the age-related changes in learning ability (Nagy, 1979; Rudy, 1992). Whether these structural changes require the postulation of mechanisms other than those already described for adults is not yet known. There have been several suggestions about possible sources of differences between infant and adult learning. First, these might be due to the lack of experimental control over stimuli in young organisms (Rovee-Collier & Lipsitt, 1982). This is a methodological rather than a theoretical problem. Second, the age-related differences in learning ability might be the result of the operation of only subsets of adult learning mechanisms (Hyson & Rudy, 1984; Rudy & Hyson, 1984; Vogt & Rudy, 1984). This alternative formulation requires an understanding of the emergence of new learning but does not require new rules of learning. Finally, changes in performance on learning tasks may in fact reflect changes in the underlying mechanisms (Amsel, 1979). In this third case there is a possibility that an understanding of underlying structure might provide insight into the learning process. Carew (1989), for example, has used a physiological analysis to demonstrate that sensitization and dishabituation are distinct processes: a distinction that could not be achieved strictly at the behavioral level.

In the study of adult organisms, learning theorists do not often refer to structural substrates in seeking explanations of learning. Instead, differences in learning mechanisms are defined paradigmatically in terms of operations and behavioral outcomes. Indeed, there is disagreement among learning theorists as to whether knowledge of the physiological changes underlying learning will ever contribute to understanding learning at the behavioral level (see Holland, 1984b). Hence, while learning theory and developmental psychology differ considerably in their emphasis on structural change, it is unclear that this difference actually has implications for the way in which experientially induced changes in behavior are analyzed.

The place of the past

As Gottlieb (1983, p. 3) points out, one of the basic assumptions of a developmental analysis is that '... certain, if not all, aspects of behavior are determined or influenced by events that occur earlier in ontogenesis (i.e. history of the individual)'. This notion is embodied in many of the important phenomena studied by developmentalists. Critical periods, canalization, developmentalists' description of the facilitating and inducing effects of stimuli (Gottlieb, 1976), and the notion of an 'epigenetic landscape' (Waddington, 1966), all capture the sense that past history determines how current variables affect behavior. More specifically, one must know more than the current state of the organism and the current inputs: one must know the path that led to the current state in order to predict the effects of the current input.

In contrast, in most contemporary learning theories (but see Killeen, 1984; Mackintosh, 1975) the effects of current variables are seen as independent of the history which led an organism to respond in a particular way. For example, in Rescorla & Wagner's (1972) influential theory of Pavlovian conditioning, neither the rate nor the asymptote of learning is in any way dependent on past learning. Even where stimuli interact with one another, that interaction depends only on current associative values, not on the particular treatments or performances used in the past to produce those values.

Nothing is added to our analyses of behavioral change by appealing to an unspecified history of reinforcement (Skinner, 1953) as the cause of the dependency. Specific rules must be worked out for describing the sequential dependencies in behavioral change. Even though traditional learning theories have not considered path dependence, there is an empirical tradition that acknowledges these dependencies (Lindblom & Jenkins, 1981; Miller & Matzel, 1987; Bouton, 1991). For example, the magnitude or delay of reward that will maintain a particular performance is quite different from that required for the acquisition of a performance.

Attempts have been made to articulate the influence of past learning on subsequent learning in the study of transfer. It is accepted that once a particular response is learned, there will be positive transfer to the learning of some new responses, and negative transfer to the learning of some other responses. Negative and positive transfer have been analyzed in the context of verbal learning (Osgood, 1949) and in discrimination learning (see Mackintosh, 1974, 1983). Osgood (1949) attempted to develop rules for specifying the degree and type of transfer based on response and stimulus similarities. The similarity between responses may be structural (e.g. topographically similar responses) or functional (Peterson, 1984; Wickens, 1972). More recently this has been developed in Tulving's (1983) encoding specificity principle, which views performance as a function of the overlap between the information used to guide encoding and the information required by the retrieval task. Positive and negative transfer might, therefore, be predicted on the basis of the similarity between the information encoded during original learning and that required by a transfer task. The nature of this interaction, however, must also be made explicit. Specifically, past history may: (a) change the process of learning, (b) leave process unchanged but alter parameter values, (c) affect neither process nor parameters, but change the way in which learning is expressed in performance. In any given situation, any or all of these possibilities may contribute to the influence of prior experience on current learning.

Conclusions

The major contribution of a learning approach to the study of development may be in the conceptualization of separate stimulus and response transitions during ontogeny. Learning theory, furthermore, can provide prototypical procedures for the analysis of transitions in stimulus control. The phenomena of overshadowing, blocking, and potentiation are probably all present in developmental preparations, and the progress that has been made toward their theoretical understanding in the learning laboratory can provide insights and guides to a developmental analysis.

Similarly, principles of learning can contribute to the analysis of changes in response form during development. Analyzing the role of sensitization, habituation, Pavlovian conditioning, operant conditioning and imitation in producing changes in response form can help to specify how experience produces these developmental changes. As we have pointed out above, identifying rules of response induction is as crucial to the task of understanding adult learning as it is to understanding developmental transitions.

Finally, if learning theories are to encompass developmentally significant phenomena they must be expanded to evaluate/consider the effects of long duration experiences, physiological–structural influences and the path-dependent nature of behavioral change.

Acknowledgements

We thank L. Aber, J. J. Bolhuis, L. Braine, W. Fifer, G. Gottlieb, W. G. Hall, J. A. Hogan, J. Rabinowitz, H. S. Terrace, W. Timberlake and C. L. Williams for comments and criticism on earlier drafts of this manuscript. Work in the authors' laboratories was supported by NSF grant BNS-8919231 to PB and NIMH grant 29380 to RS.

References

Alberts, J. R. (1978). Huddling by rat pups: Multisensory control of contact behavior. *Journal of Comparative and Physiological Psychology*, 92, 220–230.

Alberts, J. R. & Brunjes, P. C. (1978). Ontogeny of olfactory and thermal determinants of huddling in the rat. *Journal of Comparative and Physiological Psychology*, 92, 897–906.

Alessandro, D., Dollinger, J., Gordon, J. D., Mariscal, S. K. & Gould, J. L. (1989). The ontogeny of the pecking response of herring gull chicks. *Animal Behaviour*, 37, 372–382.

Amsel, A. (1979). The ontogeny of appetitive learning and persistence in the rat. In *Ontogeny of Learning and Memory*, ed. N. E. Spear & B. A. Campbell, pp. 189–224. Hillsdale, NJ: Lawrence Erlbaum.

Baer, D. & Wright, J. (1975). Developmental Psychology. *Annual Review of Psychology*, 25, 1–82.

Baerends, G. P. (1976). The functional organization of behaviour. *Animal Behaviour*, 24, 726–738.

Balsam, P. D. (1985). The functions of context in learning and performance. In *Context and Learning*, ed. P. Balsam & A. Tomie, pp. 1–21. Hillsdale, NJ: Lawrence Erlbaum.

Balsam, P. D. (1988). Selection, representation and equivalence of controlling stimuli. In *Steven's Handbook of Experimental Psychology*, ed. R. C. Atkinson, R. J. Herrnstein, G. Lindzey & R. D. Luce, pp. 111–166. New York: John Wiley & Sons.

Balsam, P., Deich, J. & Hirose, R. (1992). The roles of experience in the transition from dependent to independent feeding in ring doves. *Annals of the New York Academy of Sciences*, 662, 16–36.

Balsam, P., Graf, J. S. & Silver, R. (1992). Operant and Pavlovian contributions to the ontogeny of pecking in ring doves. *Developmental Psychobiology*, 25, 389–410.

Biederman, G. B. & Vanayan, M. (1984). Observational learning in pigeons: the function of quality of observed performance in simultaneous discrimination. *Learning and Motivation*, 19, 31–43.

Bijou, S. & Baer, D. (1961). *Child Development*, Vol. 1: *A Systematic and Empirical Theory*. New York: Appleton Century.

Bolhuis, J. J. (1991). Mechanisms of avian imprinting: a review. *Biological Reviews*, 66, 303–345.

Bolhuis, J. J., de Vos, G. J. & Kruijt, J. P. (1990). Filial imprinting and associative learning. *Quarterly Journal of Experimental Psychology*, 42B, 313–329.

Bolhuis, J. J. & Van Kampen, H. S. (1992). An evaluation of auditory learning in filial imprinting. *Behaviour*, 122, 195–230.

Bouton, M. E. (1991). Context and retrieval in extinction and in other examples of interference in simple associative learning. In *Current Topics in Animal Learning: Brain, Emotion, and Cognition*, ed. L. Dachowski & C. F. Flaherty, pp. 27–53. Hillsdale, NJ: Lawrence Erlbaum.

Bower, G. H. & Hilgard, E. R. (1981). *Theories of Learning*. Englewood Cliffs, NJ: Prentice–Hall.

Carew, T. J. (1989). Developmental assembly of learning in Aplysia. *Trends in Neuroscience*, 12, 389–394.

Connolly, K. J. & Prechtl, H. F. R. (1981). *Maturation and Development: Biological and Psychological Perspectives*. Philadelphia: Lippincott

Cramer, C. P., Thiels, E. & Alberts, J. R. (1990). Weaning in rats: I. Maternal behavior. *Developmental Psychobiology*, 23, 479–493.

Davies, N. B. (1976). Parental care and the transition to independent feeding in the young spotted flycatcher (*Muscicapa striata*). *Behaviour*, 59, 280–295.

Davies, N. B. (1978). Parental meanness and offspring independence: an experiment with hand-reared great tits *Parus major*. *Ibis*, 126, 509–514.

Deich, J. D., Allan, R. W. & Zeigler, H. P. (1988). Conjunctive differentiation of gape during food-reinforced keypecking in the pigeon. *Animal Learning & Behavior*, 16, 268–276.

Deich, J. & Balsam, P. (1993). The form of early pecking in the ring dove squab (*Streptopelia risoria*). *Journal of Comparative Psychology*, 107, 261–275.

Deich, J. & Balsam, P. (in press). Development of prehensile feeding in ring doves (*Streptopelia risoria*). In *Perception and Motor Control in Birds*, ed. M. Davies & P. Green, Heidelberg: Springer-Verlag

Domjan, M. (1980). Ingestional aversion learning: unique and general processes. *Advances in the Study of Behavior*, 11, 276–336.

Domjan, M. (1983). Biological constraints on instrumental and classical conditioning: Implications for general process theory. *The Psychology of Learning and Motivation*, 17, pp. 215–277. New York: Academic Press.

Domjan, M. (1992). Adult learning and mate choice: Possibilities and experimental evidence. *American Zoologist*, 32, 48–61.

Durlach, P. & Rescorla, R. A. (1980). Potentiation rather than overshadowing in flavor-aversion learning. *Journal of Experimental Psychology: Animal Behavior Processes*, 6, 175–187.

Epstein, R. (1984). Spontaneous and deferred imitation in the pigeon. *Behavioural Processes*, 9, 347–354.

Fentress, J. C. (1983). Ethological models of species-specific behavior. In *Handbook of Behavioral Neurobiology*, Vol. 6, ed. P. Teitelbaum & E. Satinoff, pp. 185–234. New York: Plenum Press.

Freeman, N. C. G. & Rosenblatt, J. S. (1978). The interrelationship between thermal and olfactory stimulation in the development of home orientation in newborn kittens. *Developmental Psychobiology*, 11, 437–457.

Galef, B. G. Jr. (1976). Social transmission of acquired behavior: a discussion of tradition and social learning in vertebrates. *Advances in the Study of Behavior*, 6, 77–100.

Galef, B. G. Jr. (1977). Mechanisms for the social transmission of acquired food preferences from adult to weanling rats. In *Learning Mechanisms in Food Selection*, ed. L. M. Barker, M. R. Best & M. Domjan, pp. 123–148. Waco, Texas: Baylor University Press.

Galef, B. G. Jr. (1981). The ecology of weaning: Parasitism and the achievement of independence by altricial mammals. In *Parental Care in Mammals*, ed. D. J. Gubernick & P. H. Klopfer, pp. 211–241. New York: Plenum Press.

Galef, B. G. Jr. (1990). The ecology of weaning: Parasitism and the achievement of independence by altricial animals. In *Interpretation and Explanation in the Study of Behavior*, Vol. 1, ed. M. Bekoff & D. Jamieson, pp. 74–95. Boulder: Westview Press.

Galef, B. G. Jr. (1992). Weaning from mother's milk to solid foods: The developmental psychobiology of self-selection of food by rats. *Annals of the New York Academy of Sciences*, 662, 37–52.

Galef, B. G. Jr. & Kennett, D. J. (1987). Different mechanisms for social transmission of diet preferences in rat pups of different ages. *Developmental Psychobiology*, 20, 209–215.

Gallistel, R. (1980). *The Organization of Action: A New Synthesis*. Hillsdale, NJ: Lawrence Erlbaum.

Gauthier, D. & Barrette, C. (1985). Suckling and weaning in captive white-tailed and fallow deer. *Behaviour*, 94, 128–149.

Gewirtz, J. C. & Boyd, E. F. (1977). Experiments on mother infant interaction underlying mutual attachment acquisition: The infant conditions the mother. In *Attachment Behavior*, ed. T. Alloway, P. Pliver & L. Krames, pp. 109–143. New York: Plenum Press.

Goetz, E. M. & Baer, D. M. (1973). Social control of form diversity and the emergence of new forms in children's block building. *Journal of Applied Behavior Analysis*, 6, 209–217.

Golani, I. & Fentress, J. C. (1985). Early ontogeny of face grooming in mice. *Developmental Psychobiology*, 18, 529–544.

Gottlieb, G. (1976). Conceptions of prenatal development: Behavioral embryology. *Psychological Review*, 83, 215–234.

Gottlieb, G. (1983). The psychobiological approach to developmental issues. In *Handbook of Child Psychology*, Vol. 2, ed. P. H. Mussen, pp. 1–26. New York: John Wiley & Sons.

Graf, J. S., Balsam, P.D & Silver, R. (1985). Associative factors and the development of pecking in ring doves. *Developmental Psychobiology*, 18, 447–460.

Hailman, J. P. (1967). The ontogeny of an instinct. The pecking response in chicks of the laughing gull (*Larus atricilla L.*). *Behaviour*, Supplement, 15, 1–159.

Hall, W. G. & Williams, C. L. (1983). Suckling isn't feeding, or is it? A search for developmental continuities. *Advances in the Study of Behavior*, 13, 220–254.

Hearst, E. & Koresko, M. B. (1968). Stimulus generalization and amount of prior training on variable interval reinforcement. *Journal of Comparative and Physiological Psychology*, 66, 133–138.

Heyes, C. M. & Dawson, G. R. (1990). A demonstration of observational learning in rats using a bidirectional control. *Quarterly Journal of Experimental Psychology*, 42B, 59–71.

Higuchi, H. & Momose, H. (1981). Deferred independent and prolonged infantile behaviour in young varied tits, *Parus varius*, of an island population. *Animal Behaviour*, 29, 523–528.

Hirose, R. & Balsam, P. (submitted). Parent–squab interaction during the transition from dependent to independent feeding in the ring dove (*Streptopelia risoria*). *Animal Behaviour*.

Hogan, J. A. (1964). Operant control of preening in pigeons. *Journal of the Experimental Analysis of Behavior*, 7, 351–354.

Hogan, J. A. (1973). How young chicks learn to recognize food. In *Constraints on Learning*, ed. R. A. Hinde & J. Stevenson-Hinde, pp. 119–139. London: Academic Press.

Hogan, J. A. (1974). Responses in Pavlovian conditioning studies. *Science*, 186, 156–157.

Hogan, J. A. (1977). The ontogeny of food preferences in chicks and other animals. In *Learning Mechanisms in Food Selection*, ed. L. M. Barker, M. R. Best & M. Domjan, pp. 71–97. Texas: Baylor University Press.

Hogan, J. A. (1984). Pecking and feeding in chicks. *Learning and Motivation*, 15, 360–376.

Hogan, J. A. (1988). Cause and function in the development of behavior systems. In *Handbook of Behavioral Neurobiology*, Vol. 9, ed. E. M. Blass, pp. 63–106. New York: Plenum Press.

Hogan-Warburg, A. J. & Hogan, J. A. (1981). Feeding strategies in the development of food recognition in young chicks. *Animal Behaviour*, 39, 143–154.

Holland, P. C. (1977). Conditioned stimulus as a determinant of the form of the Pavlovian conditioned response. *Journal of Experimental Psychology: Animal Behavior Processes*, 3, 77–104.

Holland, P. C. (1980). CS–US interval as a determinant of the form of Pavlovian appetitive conditioned responses. *Journal of Experimental Psychology: Animal Behavior Processes*, 6, 155–174.

Holland, P. C. (1984a). Origins of behavior in Pavlovian conditioning. In *The Psychology of Learning and Motivation*, Vol. 18, ed. G. H. Bower, pp. 129–174. Orlando, Florida: Academic Press.

Holland, P. C. (1984b). Biology of learning in non-human mammals. In *The Biology of Learning*, ed. P. Marler & H. S. Terrace, pp. 533–551. New York: Springer-Verlag.

Holman, J., Goetz, E. & Baer, D. M. (1977). The training of creativity as an operant and an examination of its generalization characteristics. In *New Developments in Behavioral Research: Theory, Method and Application*, ed. B. C. Etzel, J. M. LeBlanc & D. M. Baer, pp. 441–471. New York: John Wiley & Sons.

Holt, E. B. (1931). *Animal Drive and the Learning Process*. New York: Henry Holt & Co.

Hyson, R. L. & Rudy, J. W. (1984). Ontogenesis of learning: II. Variation in the rat's reflexive and learned responses to acoustic stimulation. *Developmental Psychobiology*, 17, 263–283.

Innis, N. K., Simmelhag-Grant, V. L. & Staddon, J. E. R. (1983). Behavior induced by periodic food delivery: The effects of interfood interval. *Journal of the Experimental Analysis of Behavior*, 39, 309–322.

Jenkins, H. M. & Harrison, R. H. (1960). Effects of discrimination training on auditory generalization. *Journal of Experimental Psychology*, 59, 246–253.

Jenkins, H. M. & Moore, B. R. (1973). The form of autoshaped reinforcers. *Journal of the Experimental Analysis of Behavior*, 20, 163–181.

Johanson, I. B. & Hall, W. G. (1981). The ontogeny of feeding in rats: V. The influence of texture, home odor, and sibling presence on ingestive behavior. *Journal of Comparative and Physiological Psychology*, 95, 837–847.

Johanson, I. B., Hall, W. G. & Polefrone, J. M. (1984). Appetitive conditioning in neonatal rats: Conditioned ingestive responding to stimuli paired with oral infusion of milk. *Developmental Psychobiology*, 17, 357–381.

Johanson, I. B. & Terry, L. M. (1988). Learning in infancy: A mechanism for behavioral change during development. In *Handbook of Behavioral Neurobiology*, Vol. 9, ed. E. M. Blass, pp. 245–281. New York: Plenum Press.

Kagan, J. (1971). *Change and Continuity in Infancy*. New York: John Wiley & Sons.

Kara, A. & Wahler, R. G. (1977). Organizational features of a young child's behavior. *Journal of Experimental Child Psychology*, 24, 24–39.

Kaufman, C. & Rosenblum, L. A. (1969). The waning of the mother–infant bond in two species of Macaque. In *Determinants of Infant Behavior IV*, ed. B. M. Foss, pp. 41–62. London: Methuen.

Kazdin, A. E. (1982). Symptom substitution, generalization, and response covariation: Implications for psychotherapy outcome. *Psychological Bulletin*, 91, 349–365.

Kehoe, E. J., Gibbs, C. M., Garcia, E. & Gormezano, I. (1979). Associative transfer and stimulus selection in classical conditioning of the rabbit's nictitating membrane response to serial compound CSs. *Journal of Experimental Psychology: Animal Behavior Processes*, 5, 1–18.

Killeen, P. (1984). Incentive Theory III: Adaptive clocks. In *Timing and Time Perception*, ed. J. Gibbon & L. Allen, pp. 515–527. New York: New York Academy of Sciences.

Klaus, M. H. & Kennel, J. H. (1976). *Maternal–Infant Bonding*. St Louis: C. V. Mosby.

Klein, S. B. & Mowrer, R. R. (1989). *Contemporary Learning Theories: Pavlovian Conditioning and the Status of Traditional Learning Theory*. Hillsdale, NJ: Lawrence Erlbaum.

Kuo, Z.-Y. (1967). *The Dynamics of Behavior Development: An Epigenetic View*. New York: Random House.

Kruijt, J. P. (1964). Ontogeny of social behaviour in Burmese red junglefowl (*Gallus gallus spadiceus, Bonaterre*). *Behaviour*, Supplement 12.

LaMon, B. & Zeigler, H. P. (1988). Control of response form in the pigeon: Topography of ingestive behaviors and conditioned keypecks with food and water reinforcers. *Animal Learning & Behavior*, 16, 256–267.

Lehrman, D. S. (1955). The physiological basis of parental feeding behavior in the ring dove (*Streptopelia risoria*). *Behaviour*, 7, 241–286.

Lehrman, D. S. (1962). Interaction of hormonal and experiential factors on the development of behavior. In *Roots of Behavior*, ed. E. S. Bliss, 142–156. New York: Harper.

Lehrman, D. S. (1964). Control of behavior cycles in reproduction. In *Social Behavior and Organization Among Vertebrates*, ed. W. Etkin, pp. 143–166. Chicago: The University of Chicago Press.

Lehrman, D. S. (1970a). Experiential background for the induction of reproductive behavior patterns by hormones. In *Biopsychology of Development*, ed. E. Tobach, pp. 297–302. New York: Academic Press.

Lehrman, D. S. (1970b). Semantic and conceptual issues in the nature–nurture problem. In *Development and Evolution of Behavior*, ed. L. R. Aronson, E. Tobach, D. S. Lehrman & J. S. Rosenblatt, pp. 17–52. San Francisco: W. H. Freeman.

Lett, B. T. (1980). Taste potentiates color–sickness associations in pigeons and quail. *Animal Learning & Behavior*, 8, 193–198.

Lewis, M. & Starr, M. D. (1979). Developmental continuity. In *Handbook of Infant Development*, ed. J. Osofsky, pp. 653–670. New York: John Wiley & Sons.

Lindblom, L. L. & Jenkins, H. M. (1981). Responses eliminated by non-contingent or negatively contingent reinforcement recover in extinction. *Journal of Experimental Psychology: Animal Behavior Processes*, 7, 175–190.

Lipsitt, L. P. & Kaye, H. (1965). Changes in neonatal response to optimizing and non-optimizing suckling stimulation. *Psychonomic Science*, 2, 221–222.

Logue, A. W. (1979). Taste aversion and the generality of the laws of learning. *Psychological Bulletin*, 86, 276–296.

Mackintosh, N. J. (1974). *The Psychology of Animal Learning*. London: Academic Press.

Mackintosh, N. J. (1975). A theory of attention: Variations in the associability of stimuli with reinforcement. *Psychological Review*, 82, 276–298.

Mackintosh, N. J. (1977). Stimulus control: Attentional factors. In *Handbook of*

Operant Behavior, ed. W. K. Honig & J. E. R. Staddon, pp. 481–513. Englewood Cliffs, NJ: Prentice-Hall.

Mackintosh, N. J. (1983). *Conditioning and Associative Learning*. Oxford: Oxford University Press.

Martin, P. (1986). An experimental study of weaning in the domestic cat. *Behaviour*, 99, 221–249.

Miller, R. R. & Matzel, L. D. (1987). Memory for associative history of a conditioned stimulus. *Learning & Motivation*, 18, 118–130.

Moore, B. R. (1973). The role of directed Pavlovian reactions in instrumental learning in the pigeon. In *Constraints on Learning*, ed. R. A. Hinde & J. Stevenson-Hinde, pp. 159–189. New York: Academic Press.

Moreno, J. (1984). Parental care of fledged young, division of labor, and the development of foraging techniques in the Northern Wheateater (*Oenathe oenathe L.*). *Auk*, 101, 741–752.

Mustaca, A. E., Gabelli, F., Papini, M. R. & Balsam, P. D. (1991). The effects of varying the interreinforcement interval on appetitive contextual conditioning in rats and ring doves. *Animal Learning & Behavior*, 19, 125–138.

Nagy, Z. M. (1979). Development of learning and memory processes in infant mice. In *Ontogeny of Learning and Memory*, ed. N. E. Spear & B. A. Campbell, pp. 101–133. Hillsdale, NJ: Lawrence Erlbaum.

Oppenheim, R. W. (1981). Ontogenetic adaptations and retrogressive processes in the development of the nervous system and behavior: A neuroembryological perspective. In *Maturation and Development: Biological and Psychological Perspectives*, ed. K. J. Connolly & H. F. R. Prechtl, pp. 73–109. Philadelphia: Lippincott.

Osgood, C. E. (1949). The similarity paradox in human learning. *Psychological Review*, 56, 132–143.

Palameta, B. & Lefebvre, L. (1985). The social transmission of a food-finding technique in pigeons: what is learned? *Animal Behaviour*, 33, 892–896.

Pavlov, I. P. (1927). *Conditioned Reflexes*. New York: Dover Publications.

Peterson, G. B. (1984). How expectancies guide behavior. In *Animal Cognition*, ed. H. L. Roitblat, T. G. Bever & H. S. Terrace, pp. 135–148. Hillsdale, NJ: Lawrence Erlbaum.

Pryor, K. W., Haag, R. & O'Reilly, J. (1969). The creative porpoise: Training for novel behavior. *Journal of the Experimental Analysis of Behavior*, 12, 653–661.

Reisbick, S., Rosenblatt, J. S. & Mayer, A. D. (1973). Decline of maternal behavior in the virgin and lactating rat. *Journal of Comparative and Physiological Psychology*, 89, 722–732.

Rescorla, R. A. (1968). Probability of shock in the presence and absence of CS in fright conditioning. *Journal of Comparative and Physiological Psychology*, 66, 1–5.

Rescorla, R. A. & Wagner, A. R. (1972). A theory of Pavlovian conditioning: Variations in the effectiveness of reinforcement and non-reinforcement. In *Classical Conditioning II: Current Theory and Research*, ed. A. H. Black & W. F. Prokasy, pp. 64–99. New York: Appleton-Century-Crofts.

Rosenblatt, J. S. (1979). The sensorimotor and motivational bases of early behavioral development of selected altricial mammals. In *Ontogeny of Learning and Memory*, ed. N. E. Spear & B. A. Campbell, pp. 1–38. Hillsdale, NJ: Lawrence Erlbaum.

Rosenblatt, J. S. (1983). Olfaction mediates developmental transition in the altricial newborn of selected species of mammals. *Developmental Psychobiology*, 16, 347–375.

Rovee-Collier, C. K. & Lipsitt, L. P. (1982). Learning, adaptation and memory in the newborn. In *Psychobiology of the Human Newborn*, ed. P. Stratton, pp. 147–190. New York: John Wiley & Sons.

Rudy, J. (1992). Development of learning: from elemental to configural associative networks. In *Advances in Infancy Research*, ed. C. K. Rovee-Collier & L. P. Lipsitt, pp. 247–289. Norwood, NJ: Ablex Publishing.

Rudy, J. W. & Hyson, R. L. (1984). Ontogenesis of learning: III. Variation in the rat's differential reflexive and learned responses to sound frequencies. *Developmental Psychobiology*, 17, 285–300.

Russo, D. C., Cataldo, M. F. & Cushing, P. J. (1981). Compliance training and behavioral covariation in the treatment of multiple behavior problems. *Journal of Applied Behavior Analysis*, 14, 209–222.

Schneirla, T. C. (1966). Behavioral development and comparative psychology. *Quarterly Review of Biology*, 41, 283–303.

Segal, E. F. (1972). Induction and the provenance of operants. In *Reinforcement: Behavioral Analyses*, ed. R. M. Gilbert & J. R. Millenson, pp. 1–34. New York: Academic Press.

Skinner, B. F. (1938). *The Behavior of Organisms*. New York: Appleton-Century-Crofts.

Skinner, B. F. (1953). *Science and Human Behavior*. New York: Free Press.

Slater, P. J. B. (1973). Describing sequences of behavior. In *Perspectives in Ethology*, Vol. 2, ed. P. P. G. Bateson & P. H. Klopfer, pp. 131–153. New York: Plenum Press.

Spear, N. E. (1978). *The Processing of Memories: Forgetting and Retention*. Hillsdale, NJ: Lawrence Erlbaum

Spear, N. E. & Campbell, B. A. (1979). *Ontogeny of Learning and Memory*. Hillsdale, NJ: Lawrence Erlbaum.

Spear, N. E., Kucharski, D. & Miller, J. S. (1989). The CS-effect in simple conditioning snd stimulus selection during development. *Animal Learning & Behavior*, 17, 70–82.

Spear, N. E. & Rudy, J. W. (1991). Tests of the ontogeny of learning and memory: Issues, methods, and results. In *Developmental Psychobiology: New Methods and Changing Concepts*, ed. H. N. Shair, G. A. Barr & M. A. Hofer, pp. 84–113. New York: Oxford University Press.

Staats, A. W. (1975). *Social Behaviorism*. Homewood, Illinois: The Dorsey Press.

Staddon, J. E. R. (1977). Schedule-induced behavior. In *Handbook of Operant Behavior*, ed. W. K. Honig & J. E. R. Staddon, pp. 125–152. Englewood Cliffs, NJ: Prentice Hall.

Staddon, J. E. R. (1983). *Adaptive Behavior and Learning*. Cambridge: Cambridge University Press.

Staddon, J. E. R. & Simmelhag, V. L. (1971). The 'superstition' experiment: A reexamination of its implications for the principles of adaptive behavior. *Psychological Review*, 78, 3–43.

Stern, J. M. & Rogers, L. (1988). Experience with younger siblings facilitates maternal responsiveness in pubertal Norway rats. *Developmental Psychobiology*, 21, 575–589.

Stokes, P. D. & Balsam, P.D (1991). Effects of reinforcing preselected approximations on the topography of the rat's barpress. *Journal of the Experimental Analysis of Behavior*, 55, 213–231.

Suboski, M. D. & Bartashunas, C. (1984). Mechanisms for social transmission of pecking preferences to neonatal chicks. *Journal of Experimental Psychology: Animal Behavioral Processes*, 10, 182–194.

Teitelbaum, P. (1977). Levels of integration of the operant. In *Handbook of Operant Behavior*, ed. W. K. Honig & J. E. R. Staddon, pp. 125–152. Englewood Cliffs, NJ: Prentice-Hall.

Terrace, H. S. (1966). Stimulus control. In *Operant Behavior: Areas of Research and Application*, ed. W. K. Honig, pp. 271–344. New York: Appleton-Century-Crofts.

Timberlake, W. (1983). Rats' response to a moving object related to food or water: a behavior systems analysis. *Animal Learning & Behavior*, 11, 309–320.

Timberlake, W. & Grant, D. (1975). Autoshaping in rats to the presentation of another rat predicting food. *Science*, 175, 690–692.

Timberlake, W. & Lucas, G. A. (1989). Behavior systems and learning: From misbehavior to general principles. In *Contemporary Learning Theories*, ed. S. B. Klein & R. R. Mowrer, pp. 237–274. Hillsdale, NJ: Lawrence Erlbaum.

Tinbergen, N. (1951). *The Study of Instinct*. Oxford: Clarendon Press.

Tinbergen, N. & Perdeck, A. C. (1950). On the stimulus situation releasing the begging response in the newly-hatched herring gull chick (*Larus a. argentatus Pont*). *Behaviour*, 3, 1–38.

Tulving, E. (1983). *Elements of Episodic Memory*. Oxford: Oxford University Press.

Turner, E. R. A. (1964). Social feeding in birds. *Behaviour*, 24, 1–46.

Vogt, M. B. & Rudy, J. L. (1984). Ontogenesis of learning: I. Variation in the rat's reflexive and learned responses to gustatory stimulation. *Developmental Psychobiology*, 17, 11–33.

von Holst, E. & von St. Paul, U. (1963). On the functional organisation of drives. *Animal Behaviour*, 11, 1–20.

Waddington C. H. (1966). *Principles of Development and Differentiation*. New York: Macmillan.

Wahler, R. G. & Fox, J. J. (1981). Response structure in deviant child–parent relationships: Implications for family therapy. *Nebraska Symposium on Motivation: Response Structure and Motivation*, pp. 1–46. Lincoln, Nebraska: University of Nebraska.

Watson, J. B. (1930). *Behaviorism*. New York: W. W. Norton.

Werner, H. (1957). The concept of development from a comparative and organismic point of view. In *The Concept of Development*, ed. D. B. Harris, pp. 107–130. Minneapolis: University of Minnesota Press.

West, M. J., King, A. P. & Arberg, A. A. (1988). The inheritance of niches: The role of ecological legacies in ontogeny. In *Handbook of Behavioral Neurobiology*, Vol. 9, ed. E. Blass, pp. 41–62. New York: Plenum Press.

Wetherington, C. L. (1979). Schedule-induced drinking: Rate of food delivery and Herrnstein's equation. *Journal of the Experimental Analysis of Behavior*, 32, 323–333.

Whiten, A. & Ham, R. (1992). On the nature and evolution of imitation in the animal kingdom: Reappraisal of a century of research. *Advances in the Study of Behavior*, 21, 239–283.

Wickens, D. D. (1972). Characteristics of word encoding. In *Coding Processes in Human Memory*, ed. A. P. Melton & E. Martin, pp. 191–215. Washington, DC: Winston.

Williams, B. A. (1988). Reinforcement, choice, and response strength. In *Steven's Handbook of Experimental Psychology*, 2nd edn, Vol. 1: *Perception and Motivation*, ed. R. A. Atkinson, R. J. Herrnstein, G. Lindzey & R. P. Luce, pp. 167–244. New York: John Wiley & Sons.

Wortis, R. P. (1969). The transition from dependent to independent feeding in the young ring dove. *Animal Behaviour Monographs*, 2, 1–54.

15

The varieties of learning in development: toward a common framework

SARA J. SHETTLEWORTH

During development, factors in the animal combine with its 'experiences' to produce a functioning adult. When it comes to analyzing what those experiences are and how they have their effects, one immediately confronts basic questions about learning. Are any or all of the effects of experience in development fundamentally different in some way from 'learning'? What should we include as learning anyway? Are the effects of discrete, well-defined experiences on adults relevant to understanding the role of experience in development? Can we answer all these questions simply by postulating different kinds of learning?

Historically, learning has been studied from two different perspectives. In psychology, interest in learning stems from the philosophical background of associationism. The enduring appeal of the notion that all behavior and mental life are built up of simple associations is evident in the contemporary interest in connectionist modelling (cf. McLaren, this volume). The biological or more functional approach, in contrast, sees learning as part of the animal's adaptation to its environment, without making assumptions about the form that learning might take. To psychologists, the biological approach seems atheoretical and insufficiently general; to zoologists or developmentalists wanting to understand how particular animals develop in their niches, the psychologists' study of learning in the laboratory seems artificial and irrelevant. Psychologists' attempts to synthesize learning theory and the analysis of phenomena like imprinting that were discovered in a developmental context (e.g. Hoffman & Ratner, 1973) have usually seemed forced and unconvincing.

Such historical fragmentation notwithstanding, it ought to be possible to develop a unified approach to all phenomena involving behavioral plasticity. Balsam & Silver (this volume) discuss one way of doing this. They describe how ideas from learning theory can be used to account for phenomena in development like the transition to pecking at grain in ring dove squabs. At the same time, they suggest, other developmental phenomena, such as the emergence of hierarchical organization in behavior, confront associative learning theory with questions it cannot answer and hence may force it to become broader and more 'ecologically valid'.

This chapter is complementary to Balsam & Silver's in that I suggest a different approach to the same end. Like Balsam & Silver, I base some of what I have to say on general process associative learning theory. What I suggest is that the way to bring together the different perspectives from which behavioral plasticity has been studied is by developing a consistent framework for analyzing and comparing effects of experience on behavior. The way in which associative learning has been studied provides a model for how any effects of experience might be looked at. Rescorla (1988a, 1988b; Rescorla & Holland, 1976) has previously suggested such an approach, but without exploring its implications very far. Using the structure of associative learning theory to suggest a general framework for analyzing effects of experience during development is not the same as explaining any, let alone all, aspects of development as associative learning. What such a framework provides is a specification of the dimensions on which phenomena involving 'learning' can be analyzed and compared. Applying this approach can reveal unanswered, even unasked, questions.

I start with two examples illustrating the need for an explicit framework for comparing effects of experience on behavior. The first is the case of flavor aversion learning, and the second that of imprinting. I then outline a way of thinking about learning that resolves some of the issues raised by discussions of these phenomena, and finally I sketch how this approach can be applied to issues in behavioral development.

The need for an explicit framework: two examples

Flavor aversion learning

In the mid-1960s, the discovery of some anomalous phenomena in the learning laboratory stimulated the interest of psychologists in the possibility that learning was 'constrained', 'prepared', or adapted to the requirements of learning in nature (Rozin & Kalat, 1971; Seligman, 1970; Shettleworth, 1972). Perhaps the best-known and most influential of such phenomena was conditioned taste aversion, or poison-avoidance learning (Garcia & Koelling, 1966). Rats learn to avoid consuming a flavored substance if they become ill after tasting it. This learning has two properties that were exceptional when they were first described. Learning takes place even when several hours intervene between experiencing the flavor and becoming ill, and learning to avoid poisoned substances is specific to flavors. If shock follows immediately upon eating or drinking, aversion develops to exteroceptive stimuli like lights or noises but not to flavors. Conversely, if food or drink is identified by lights or noises, rats learn little or nothing.

Acquisition over long delays and selectivity in what was learned about violated psychologists' prevailing assumptions that learning was possible only with events occurring close together in time, and that one set of learning principles applied equally to all events. Initially, therefore, some investigators attempted to explain away the apparent anomalies as procedural artifacts (cf. Barker, Best & Domjan, 1977; Domjan, 1980). Others, however, accepted (perhaps too readily, cf. Galef, 1991) that conditioned taste aversion was an adaptively specialized kind of learning. Its special properties seemed obviously suited to the requirements of learning about food in nature (Rozin & Kalat, 1971; Seligman, 1970; Shettleworth, 1972). If a food is going to make an animal ill, some delay between eating and its consequences must be bridged, and the best cues for identifying foods should be flavors, at least for a nocturnal feeder like a rat.

The discovery of flavor aversion learning and other apparent 'biological constraints on learning' inspired proclamations that a revolution in the study of learning was on its way. However, in the ensuing years most of these phenomena were absorbed into a liberalized general theory of associative learning (Domjan, 1983; Domjan & Galef, 1983; Rozin & Schull, 1988; but see Bolles, 1985). This was possible because the criteria for comparing conditioned taste aversion to other learning phenomena changed, in part due to rapid developments in associative learning theory itself.

On the face of it, flavor aversion learning is classical, or Pavlovian, conditioning. The animal is exposed to a relatively innocuous stimulus, the flavor or conditioned stimulus (CS). Contingent upon that exposure but maybe several hours later, it experiences a second event, the illness-producing substance or unconditioned stimulus (US). After as little as one pairing of CS and US, a rat shows evidence of learning by reducing its consumption of the CS flavor. The facts about flavor aversion learning were surprising only against certain assumptions about what defined Pavlovian conditioning: it was assumed to depend on the CS and US being only seconds or fractions of seconds apart, and it was assumed not to depend on what the CS and US were. On these criteria, avoiding a flavor but not a light or a tone when it had once signalled illness hours later was clearly anomalous. Some early investigations suggested that this learning also lacked other important features of Pavlovian conditioning such as blocking (Rozin & Kalat, 1971; Seligman, 1970). In blocking, original learning that one CS predicts a given US reduces learning about a second CS that is later added to the first. For example, if a rat learns that a light predicts food, and a tone is then added to the light, the rat learns little or nothing about the tone, although it would have learned if the light and tone had occurred together from the outset. As long as the added CS is redundant with the first CS, it does not support conditioning.

At the same time as conditioned flavor aversion and similar phenomena were being discovered and analyzed, renewed interest in Pavlovian conditioning, which began to be studied in more tractable preparations than Pavlov's fistulated dogs, was leading to a theoretical emphasis on the abstract features of conditioning (Revusky, 1977). What came to be seen as essential for learning was a contingent relation between CS and US, such that the CS predicts an increased rate of occurrence of the US. Blocking, for instance, is understandable in this framework because the second, added, CS does not predict anything different from what is already predicted by the first CS. It became evident that the logical relations necessary for conditioning could obtain without the very constrained temporal relations that had been emphasized in the past. With this sort of abstract approach, the fact that flavor aversions are acquired over long delays ceases to be problematical. Further, contrary to early indications, flavor aversion learning does show blocking, as well as other related effects (Domjan, 1980). The features which were most unusual when flavor aversion learning was first described also turned out not to be unique to flavor aversion learning. Other cases emerged in which learning was possible over long delays and in which the specific combination of CS and US determines the speed of learning (review in Domjan, 1983). Indeed, far from being seen as a special kind of learning, flavor aversion learning came to be used as a preparation for studying the general properties of associative learning (cf. Rescorla, 1988b).

Imprinting

Like flavor aversion learning, imprinting was originally described as a special form of learning. Most modern research on imprinting was stimulated by Konrad Lorenz's (1935) description of the long-lasting attachments some precocial birds form as a result of early experience and his assertion that imprinting was 'not ordinary learning' (i.e. not Pavlovian conditioning; Lorenz, 1970). This assertion was based on four criteria. (1) Imprinting was restricted to a critical period early in life. During this period a single exposure to appropriate stimuli was sufficient to cause imprinting. (2) Imprinting was irreversible after the critical period. (3) Imprinting influences behaviors that are not, and often cannot be, shown at the time of learning, i.e. sexual behaviors. (4) From experience with a particular individual such as the mother, the bird learns the supra-individual characteristics of the species.

A tremendous amount of research was devoted to examining the evidence for one or another of these attributes of imprinting (review in Bolhuis, 1991). Just as in the case of conditioned taste aversion, two things happened to cause views of the 'specialness' of imprinting to change. First, the supposedly special properties of imprinting were seen not to be so clear cut after all. For instance, the term 'critical period', with its implication of all-or-nothing changes in sensitivity, was replaced by 'sensitive period', implying a gradual development and waning of imprintability (Bateson, 1979). Amount of exposure and nature of the imprinting stimuli were seen to be important, just as in other forms of learning; imprinting did appear to be reversible under some conditions; and sexual preferences sometimes appeared to form later in life and through a different process than filial preferences (see Bolhuis, 1991).

Second, at the same time as this research was going on, associative learning theory was evolving in such a way as to be able to encompass some of the apparently unique features of imprinting (cf. Bolhuis, de Vos & Kruijt, 1990). Pavlovian conditioning came to be seen as a process through which a whole behavior system may come under control of new external causal factors (Hogan, 1988; Timberlake & Lucas, 1989). For example, rather than simply becoming the eliciting stimulus for a reflex such as salivation, a CS that signals food excites a variety of behaviors in the feeding system. Which of these behaviors will be observed and described as evidence of learning depends in part on the external stimulus support available for them (Holland, 1984). In this light, the observation that exposure to an imprinting stimulus can influence a whole variety of social behaviors – approaching and snuggling up to the imprinting object, twittering, distress calling in its absence, pressing a pedal to see it, and possibly displaying sexual behaviors toward it – does not make the influence of imprinting on behavior different in kind from that of Pavlovian conditioning. Distinguishing questions about what behaviors are affected by experience from other questions about learning helps to make these commonalities apparent.

A framework for analyzing effects of experience
Operational distinction

A useful starting point for thinking about the diverse forms of behavioral plasticity in response to experience is a simple framework proposed by Rescorla and Holland (1976; Rescorla, 1988a, 1988b). It may be so broad as to embrace most of development, but this characterization of situations involving learning does have the merit of making immediately clear the dimensions along which effects of experience at any time of life can be classified. Most importantly in the present context, Rescorla and Holland stipulate that a test for learning must involve events at two times. At the first time, T1, an animal has some experience of interest and at some later time, T2, the animal's behavior is observed. If behavior at T2 depends on the experience at T1, in that animals without the T1 experience behave differently, then learning is said to have occurred. One could as well say that at T2 the animal exhibits memory for the T1 experience. A theoretically more neutral statement is that the animal's state has changed as a result of the T1 experience (Staddon, 1983). 'Learning' and 'memory' are equivalent labels for such changes in state. They constitute a large subset of developmental changes.

Kinds of learning are often defined operationally, in terms of the kind of experience at T1. For example, behavior may change after repeated experience of a single event, as in habituation. In Pavlovian and instrumental conditioning, presentation of two events is involved, but the two 'kinds of learning' are differentiated by the features of the events presented at T1. In instrumental conditioning, a response made by the animal is followed by a biologically significant event, whereas in Pavlovian conditioning such a biologically significant event (the US, or reinforcer) is signalled by a relatively neutral stimulus. However, the kinds of relationships among events that are critical for bringing about behavioral change seem to be the same whether the paradigm is instrumental or Pavlovian (Mackintosh, 1983).

The importance of the abstract features of experience to accepted classification schemes is evident if one notes that the distinction between operant and Pavlovian conditioning classifies together a pigeon's learning that a light signals food and a rat's learning that a tone signals shock, while it classifies separately the rat's learning to pull a string to get food. The identity of the events matters here, but only up to a point. Thus the rough and ready classification of learning in terms of kinds of experience at T1 is based on an assumption about the appropriate degree of abstraction from the actual events which make up the T1 experience. The same degree of abstraction may not be appropriate when it comes to some phenomena in development. For example, the effects of exposure to a single event during development may depend critically on the event and the species involved. Nevertheless, a more abstract view may help to reveal important commonalities among superficially diverse phenomena.

The identity of the events experienced at T1 figures strongly in defining candidates for specialized kinds of learning. For example, observational learning is classified as such because it results from one animal observing another. However, the process involved could be association of the sight of the demonstrator's actions with its consequences. Imprinting, song learning, entrainment of biological rhythms, and spatial learning are all identifiable as such because they involve special classes of events. The kind of behavioral change observed at T2 is implicitly involved too. For example, Pavlovian and instrumental conditioning were traditionally said to differ, if for no other reason because one affected autonomic behavior and the other skeletal. As another example, habituation and sensitization can both result from repeated presentation of a stimulus, but in one case the initial response decreases and in the other it is enhanced. Similarly, if a bird is exposed to song and later sings this song itself, we say song learning has occurred, but if it merely treats the song as that of a neighbor, the result is attributed to a more general process of individual recognition or habituation. In this last case, however, there is some reason to think that the conditions of learning differ as well as their effects on behavior (McGregor and Avery, 1986).

Theoretical distinctions

Contemporary analyses of associative learning distinguish three theoretical questions about learning processes (Rescorla, 1988a): what are the conditions of learning, what are the contents of learning, and how does learning influence behavior? As I discuss in a later section, these questions have not always been distinguished clearly, nor have all of them been addressed in analyses of developmental phenomena.

Conditions of learning

The conditions of learning are the features of experience at T1 that are critical for the behavioral effect of interest at T2. A study of the conditions for learning consists of varying the events at T1 and observing how behavior changes at T2. For example, varying the species of the tutor bird in a study of song development would be a study of the conditions of learning. The structure of the experience (e.g. whether contiguity or contingency between events is critical), the number, frequency, and duration of events at T1, their nature (e.g. are they sounds or odors; are they similar or dissimilar from each other), and the animal's state (e.g. its age, motivation, and past experience) are among the conditions which often affect learning. Note that, on this analysis, age and past experience are two among possibly many conditions influencing learning, though from the point of view of a developmentalist they may be the most interesting and important.

Contents of learning

Studies of the contents of learning are attempts to characterize the change in the animal resulting from experience under the required conditions. This is the most theory bound of Rescorla's three areas of investigation in that it presupposes the cognitivist assumption that underlying learning can be distinguished from performance based on that learning. In the history of learning theory this question used to be phrased in terms of the conflict between S–R and S–S theories. The issue about contents of learning was whether the animal had connected a response to a stimulus (S–R learning) or whether instead it had associated stimuli and based action on this knowledge in a flexible way (S–S learning). For spatial learning, this became the question whether animals formed cognitive maps. Some studies of bird song development have identified the contents of learning as modifications to an auditory template (cf. Marler, 1976). Theoretical analyses of habituation have distinguished the possibility of simple changes in reflex strength from that of modifications in a neuronal model of the habituating stimulus. In these and other contexts the essential question about the contents of learning is whether learning consists of changes in connection strengths or of changes in a unified representation (e.g. a template, neuronal model, or cognitive map) which cannot be reduced to associations among elements (Gallistel, 1990).

Effects of learning on behavior

Pavlov assumed that what is acquired in conditioning is an excitatory connection between a stimulus–receptor pathway and a response-generating center. In this case explaining how learning influences behavior is trivial: presentation of the stimulus must elicit (i.e. excite) the response. However, if a more complex cognitive structure changes, separate rules are required to specify the behavior it generates, i.e. theorists must specify performance rules in addition to learning rules. For example, the now-classic model of bird-song learning (cf. Marler, 1976) assumes that the progressive refinement of song is attributable to experiential modification of an auditory template (the contents of learning) which the bird then attempts to match with its own vocalizations (effects of learning on behavior).

The relevance of anatomy

Separate kinds of learning or memory are often assumed to have separate neural bases (Tulving, 1985a, 1985b; Bolhuis, this volume). Finding that a capacity like remembering individual episodes, learning song, or imprinting depends on an identifiable area of the brain does seem to give that capacity a kind of reality it otherwise lacks, a reality possessed by the visual and auditory systems with their separable receptors, pathways, and processing areas. On this view, if two kinds of learning or memory can be identified behaviorally, they should also be identifiable physiologically or anatomically. For example, imprinting and song learning, two of the best-studied candidates for special kinds of learning in development, may depend on areas of the brain apparently not involved in other kinds of learning (Horn, 1985; Bolhuis, DeVoogd, this volume). Nevertheless, it is important to recognize that the neural basis of a given learning phenomenon is but one among many criteria by which it may differ from other examples of learning. The behavioral properties of learning and its neural bases are logically distinct areas of investigation.

Distinguishing 'forms of learning'

A widespread assumption among psychologists and zoologists interested in learning and memory is that there are several, perhaps many, different forms of learning (or memory systems, cf. Sherry & Schachter, 1987; Tulving, 1985a) just as there are different sensory systems. For example, Rescorla (1988a, p. 158) writes, (Pavlovian conditioning) 'is, of course, only one of a possibly quite large number of learning processes.' The preceding sections can be summarized, however, by saying that 'the classification problem' in learning and memory (Tulving, 1985b) has no single resolution. Learning, broadly defined as behavioral plasticity in response to experience, has many facets: the identity of the events responsible for behavioral change, their specific effects on behavior, the conditions and outcomes of learning described more abstractly, the contents of learning, its neural basis. The developmental and evolutionary history of a particular learning phenomenon and its function in nature are other features that could be considered. Effects of experience on behavior can be compared and classified with respect to any of these features. The examples of flavor aversion learning and imprinting show that the criteria for distinguishing kinds of effects of experience on behavior are multiple and vary from one case to another, or even from time to time.

When it comes to identifying kinds of learning, what usually matters is evidence of differences along several dimensions. For example, perhaps the most 'special' feature of bird-song learning is that it results in modifying a motor pattern as a result of exposure to auditory stimuli, often long after those stimuli are experienced. As well, however, in many species song learning occurs at a constrained time of life, is sensitive to a species-specific range of stimuli, and is subserved by a special area of the brain (Kroodsma & Konishi, 1991). Furthermore, all these features converge to serve a clear biological function. However, without a strong and perhaps overly restrictive theory, one cannot prescribe a single feature or cluster of features that is always diagnostic of separate 'forms of learning'. It follows from this analysis that rather than separating the effects of experience in development into a variety of separate 'kinds', studied in isolation from one another, it might be more productive to concentrate on investigating specific similarities and differences among examples of plasticity, both within and across species, and understanding the reasons for them. I explore the implications of this point of view in the next section by outlining three applications of the general framework to studies of effects of experience in development.

Three implications

'Avoid acquisition curves'

Rescorla (1988b) so admonishes his readers in a review of Pavlovian conditioning for neuroscientists organized around the framework discussed here. This admonition is equally appropriate for students of development. It follows from specifying that effects of experience can only be assessed when individuals with different experiences at T1 are tested in standard conditions at T2. Simply placing two groups in two different environments and observing that they behave differently in some way during this treatment does not qualify as a proper assessment of learning nor of an enduring effect of experience by any other name.

A classic example of how important this consideration is comes from the analysis of habituation (Rescorla, 1988b). It was long thought that habituation progressed faster with more intense stimuli. This conclusion was based on comparing curves describing the courses of habituation with weak and strong stimuli. However, if animals are tested with a standard intermediate intensity after a fixed number of habituation trials, those previously exposed to the less intense stimuli show greater evidence of habituation (Davis & Wagner, 1968). In this case, then, the conclusion suggested by observing subjects during the different treatments is reversed when they are tested in a standard way.

It is not difficult to think of parallel examples in development. For example, imagine comparing the development of some sort of social behavior in animals raised in small and in large groups. The behavior of interest might appear sooner and occur more often in individuals in the larger groups. However, this observation might simply reflect more frequent elicitation of social activities in the presence of more companions, a fact that perhaps would be of interest but not one that would be evidence of an effect of social experience on development. Only observations of animals with both types of rearing experience in a group of constant size can reveal whether rearing experience influences subsequent behavior.

The importance of Rescorla's admonition can be seen in another way by considering the design of studies of imprinting. Suppose two groups of young birds are exposed to two different imprinting stimuli, say Group A to a red cylinder and Group B to a striped box. Suppose that during the period of exposure, both groups behave identically to their imprinting stimuli, increasing the time they spend near them across trials. If the experiment stops here, nothing can be concluded about the contribution of any aspect of the imprinting experience to changes in behavior with exposure. For example, maybe motivation or ability to stay near an arbitrary object is increasing over time independently of exposure to any object. Maybe imprinting is occurring, but to some common feature of the red cylinder and striped box. And so on. A commonly used design for studies of imprinting (see Bolhuis, 1991) implicitly recognizes this confounding: after exposure to one of two potential imprinting objects, the specificity of changes in following or preference to the experienced object is tested by giving both groups of animals a choice between the two objects. This is the standard test at T2. We are not surprised to see that in such a test the sorts of species that have been studied generally prefer the object they have been exposed to previously. Similar designs are commonly employed in studies of sexual imprinting (e.g. ten Cate, 1987; cf. Clayton, this volume). Again, convincing evidence of imprinting is found in effects on choice in a standard test.

One further point implicit in this example is that the choice of test at T2 is dictated by what the investigator thinks is being learned. For example, testing both groups with a choice between the red cylinder and the striped box is appropriate on the theory that what develops during the exposure to either one is a preference for it. But the experience of being repeatedly placed in some sort of apparatus and seeing or hearing the imprinting stimulus might have other effects such as causing a general improvement in visual–motor coordination, as in 'priming' of young chicks by exposure to unpatterned light (cf. Bolhuis, 1991). Such effects could be revealed only by comparing different groups in a different sort of test at T2. However, the fact that conclusions about specific features of experience must be based on results from specific tests does not invalidate the point that simply observing changes over time in groups with different experiences is not sufficient to infer anything about developmental effects of those experiences. Similar behavior can reflect different learning. Conversely, as in the case of habituation, differences in behavior during different treatments need not indicate that the same sorts of differences will be observed during a standard test.

Of course, establishing that some feature of experience is important in the development of a particular aspect of behavior is just the first step to understanding how that experience works. Among other things, the conditions of learning need to be analyzed in detail to find out just what aspects of the experience at T1 are responsible for the changed behavior at T2 (see ten Cate, 1989, for a step in this direction in the case of song learning and imprinting). What sort of hypothetical internal structure or representation has been changed by experience (i.e. the contents of learning) needs to be investigated. In terms of the framework for analyzing behavior discussed by Hogan in this volume (see also Hogan, 1988), one might ask whether and how perceptual, central, or motor mechanisms have been changed. The aspects of behavior affected can also be characterized within Hogan's (1988) framework. For instance, one might ask whether experience has altered causal factors influencing a whole behavior system or just those relevant to a particular motor pattern.

Know what kinds of questions you are asking

An explicit framework for studying effects of experience makes clear what kinds of questions need to be asked. Seeing proposals for analyses of development (or more narrowly, learning) in terms of such a comprehensive framework can help to make clear which issues are being addressed and which perhaps overlooked. Consider, for example, Gottlieb's (1976) proposal that experience plays three kinds of roles in development: maintenance, facilitation, and induction. As indicated in the following passage, this proposal was put forth as an alternative to viewing early development in terms of conventional categories of learning.

...the notion that experience can play at least three rather different roles (maintenance, facilitation, and induction) in the development of species-typical behavior and the nervous system is applicable to postnatal as well as prenatal development. Traditional forms of learning (habituation, conditioning, and the like) have not proven very useful in explaining the species-typical development of behavior, so it may be helpful to begin to formulate the contribution of experience to behavioral development in terms such as those that have been presented here. (Gottlieb, 1976, p. 232).

The kind of framework for analyzing effects of experience being explored here leads one to ask, 'To which kind of question about effects of experience does Gottlieb's classification provide an answer?' Maintenance, facilitation, and induction are not ways of partitioning the conditions nor the contents of learning. Rather, they are one way of classifying the effects of experience on behavior: does it bring new behaviors into being (induction), keep in the repertoire behaviors that would be there anyway (maintenance), or actually increase their expression (facilitation). This kind of classification is orthogonal to that proposed by Rescorla & Holland (1976) for distinguishing kinds of learning in terms of the conditions of learning, i.e. the characteristics of experience responsible for behavioral change. The modern conception of classical conditioning, for example, emphasizes conditions of learning such as dependence on contingency between events. In principle, conditioning (as well as any other kind of experience) could have inducing, facilitating, or maintaining effects during development. Thus, attempting to see where answers to Gottlieb's question fit in the overall framework for analyzing effects of experience makes clear that there really is no conflict between his descriptive classification and one in terms of conventional forms of learning because they answer different questions.

Another way in which an explicit framework can be helpful is in making apparent important parallels between investigations of different forms of learning. Awareness of such parallels can then help to advance research in one area by means of the insights gained in another. Making use of such insights is not the same as importing an explanation of learning developed in one area to account for phenomena discovered in another. It can mean becoming aware of important theoretical distinctions and effective experimental methods of addressing them. A good example of such a cross-disciplinary interaction arises in recent studies of how early social experience influences later sexual preferences in zebra finches.

Ten Cate (1987) found that some male zebra finches that had been exposed to adults of two species, zebra finch and Bengalese finch, 'dithered' in later tests of sexual preference. Rather than having a clear preference for one species or the other, the 'ditherers' approached and sang more or less equally to females of both species when both were present. Ten Cate asked what kind of representation of appropriate sexual partner might underlie this apparent double imprinting. The birds might have a 'single standard', somewhat like a hybrid, combining features of both species. Alternatively, they might have stored separate representations of both zebra finches and Bengalese finches. To try to discriminate between the two possibilities, ten Cate gave ditherers choices between true zebra finch–Bengalese finch hybrids and either zebra finches or Bengalese finches. What is important here is not the results of these tests (which were interpreted as showing the birds had a single representation), but the similarity of the issues being attacked to issues about how multifeatured events occurring in other learning contexts are represented. Some of these similarities have subsequently been explored by Hollis, ten Cate, & Bateson (1991). They compared ten Cate's (1987) test to one in which animals experience separate Pavlovian conditioning to each of two CSs and are then tested with a compound. A core assumption of modern associative learning theory is that when two stimuli are added in this way, the associative strength of the compound is the sum of that to the individual stimuli. Hollis *et al.* suggest that the same rule for combining the effects of separately trained stimuli holds for imprinting. Whether or not further research supports this hypothesis, it is important because it draws attention to the fact that many forms of learning, including habituation, perceptual learning, Pavlovian conditioning, and song learning, as well as imprinting (cf. Bateson, 1990), involve the acquisition or modification of representations of multifeatured events. A theory of learning in any of these contexts needs to deal with how learning is shared among the separate features.

Distinguish common questions from common answers

In a chapter entitled, 'General principles: The senses considered as physical instruments', Barlow (1982) outlines a framework for analyzing all sensory systems in terms of features such as selectivity, sensitivity, speed, reliability (signal:noise ratio), Weber's law (an example of a principle that holds across a wide variety of specialized systems), selectivity for pattern, and topographic mapping in the brain. Obviously, vision, audition, touch, etc., all have specialized functions served by separate neural systems, but, as Barlow's discussion illustrates, the same questions can be asked about each one. The message of this chapter is that the same kind of approach can profitably be applied to the varieties of behavioral plasticity that occur in response to experience during development. Applying such an approach requires comparing in an abstract way the sorts of questions that are being asked of different developmental phenomena. The comparison of event representation in imprinting and in Pavlovian conditioning outlined in the last section is one example. Hollis *et al.* (1991) report that preliminary analyses are consistent with their suggestion that features learned about separately combine additively in sexual imprinting just as they do in conditioning. The same issue has been tackled for filial imprinting, for example in studies asking whether overshadowing and blocking can be obtained by suitable manipulations of the salient features of a single imprinting object (Bolhuis *et al.*, 1990; de Vos & Bolhuis, 1990; van Kampen & de Vos, 1991).

In imprinting, as in other examples of perceptual learning, the animal appears to 'learn the features of' or come to recognize an object through repeated exposure to it. Independently of asking why exposure to certain kinds of objects results in acquisition of social behavior and why the effects of this exposure are restricted to a certain period of life, one can ask about the principles by which the relevant features are learned and responded to. The issues discussed by Hollis *et al.* (cf. also Bateson, 1990) illustrate one approach to answering this question, namely to compare the conditions of learning across imprinting and Pavlovian conditioning.

Another approach to the same sort of issue, but having more to do with the contents of learning, has been pointed out by Ryan & Lea (1990). One description of imprinting, especially as it occurs naturally, is that the young animal is acquiring a concept of 'family member' or 'conspecific'. Exactly the same issues arise in the analysis of the acquisition of concepts as those discussed in ten Cate's (1987) analysis of double imprinting: does a 'concept' consist of a set of memories of learned examplars (ten Cate's 'double standard' hypothesis) or of a single prototype, a representation extracted from the commonalities among the exemplars but not identical to any of them (ten Cate's 'single standard')? Considerable debate and data from both animals and humans have been applied to this issue, without, as yet, any clear resolution (cf. Herrnstein, 1990; Pearce, 1988). Nevertheless, research like that of ten Cate (1987) and Ryan & Lea (1990) illustrates how one can ask what principles govern how the features are learned in imprinting. Investigation of those principles can be independent of studies of the obviously 'special' aspects of imprinting like why exposure to certain kinds of objects results in acquisition of social behavior and why the effects of this exposure are restricted to a certain period of life.

Questions about how multifeatured objects are represented may or may not turn out to have the same answers for imprinting as for Pavlovian conditioning and/or concept learning. What is important is that rather than assuming that superficially different varieties of learning can be studied in isolation, research of this sort uses modern associative learning theory as a model for the kinds of questions that can be asked about how animals learn the characteristics of complex events and how that learning influences behavior. Using the same concepts to analyze a variety of developmental phenomena should result in the long run in a more comprehensive theory of how animals represent the world and adjust their behavior to it than one based largely on how rats, monkeys, and pigeons represent signals for food and shock. As discussed above, this can mean trying to apply prototype theory (cf. Ryan & Lea, 1990) or connectionist models (cf. McLaren, this volume) to the way chicks or quail learn the characteristics of family members. It can also mean applying notions about perceptual learning that originated in the context of research on imprinting (Chantrey, 1974) to other contexts (cf. Hall, 1991; Bateson, 1990). This is not to deny possible differences among the ways in which experience has its influence during development, but rather to suggest that differences can best be understood with a common set of questions about effects of experience and more cross-disciplinary communication about possible ways of answering them.

Acknowledgements

Preparation of this chapter was supported by the Natural Sciences and Engineering Research Council of Canada. Thanks to Jerry Hogan and Johan Bolhuis for their comments on an earlier version of this chapter. The general framework for analyzing the varieties of learning discussed herein was originally presented at the 14th Symposium on Quantitative Analyses of Behavior at Harvard in June, 1991, and its implications for the issue of 'adaptive specializations versus general processes' in learning are expanded in an article published in the *Journal of Experimental Psychology: Animal Behavior Processes* (Shettleworth, 1993).

References

Barker, L. M., Best, M. R. & Domjan, M. (eds.) (1977). *Learning Mechanisms in Food Selection*. Waco, Texas: Baylor University Press.

Barlow, H. B. (1982) General principles: The senses considered as physical instruments. In *The Senses*, ed. H. B. Barlow & J. D. Mollon, pp. 1–33. Cambridge: Cambridge University Press.

Bateson, P. P. G. (1979). How do sensitive periods arise and what are they for? *Animal Behaviour*, 27, 470–486.

Bateson, P. P. G. (1990). Is imprinting such a special case? *Philosophical Transactions of the Royal Society, B*, 329, 125–131.

Bolhuis, J. J. (1991). Mechanisms of avian imprinting: A review. *Biological Reviews*, 66, 303–345.

Bolhuis, J. J., de Vos, G. J. & Kruijt, J. P. (1990). Filial imprinting and associative learning. *Quarterly Journal of Experimental Psychology*, 42B, 313–329.

Bolles, R. C. (1985). The slaying of Goliath: What happened to reinforcement theory? In *Issues in the Ecological Study of Learning*, ed. T. D. Johnston & A. T. Pietrewicz, pp. 387–400. Hillsdale, NJ: Lawrence Erlbaum.

Chantrey, D. F. (1974). Stimulus pre-exposure and discrimination learning by domestic chicks: Effects of varying interstimulus time. *Journal of Comparative and Physiological Psychology*, 87, 517–525.

Davis, M. & Wagner, A. R. (1968). Startle responsiveness after habituation to different intensities of tone. *Psychonomic Science*, 12, 337–338.

De Vos, G. J. & Bolhuis, J. J. (1990). An investigation into blocking of filial imprinting in the chick during exposure to a compound stimulus. *Quarterly Journal of Experimental Psychology*, 42B, 289–312.

Domjan, M. (1980). Ingestional aversion learning: Unique and general processes. *Advances in the Study of Behavior*, 11, 275–336.

Domjan, M. (1983). Biological constraints on instrumental and classical conditioning: Implications for general process theory. *The Psychology of Learning and Motivation*, 17, 216–277.

Domjan, M. & Galef, B. G. Jr. (1983). Biological constraints on instrumental and classical conditioning: Retrospect and prospect. *Animal Learning and Behavior*, 11, 151–161.

Galef, B. G. Jr. (1991). A contrarian view of the wisdom of the body as it relates to dietary self-selection. *Psychological Review*, 98, 218–223.

Gallistel, C. R. (1990). *The Organization of Learning*. Cambridge, Mass.: MIT Press.

Garcia, J. & Koelling, R. A. (1966). Relation of cue to consequence in avoidance learning. *Psychonomic Science*, 4, 123–124.

Gottlieb, G. (1976). Conceptions of prenatal development: Behavioral embryology. *Psychological Review*, 83, 215–234.

Hall, G. (1991). *Perceptual and Associative Learning*. Oxford: Oxford University Press.

Herrnstein, R. J. (1990). Levels of stimulus control: A functional approach. *Cognition*, 37, 133–166.

Hoffman, H. S. & Ratner, A. M. (1973). A reinforcement model of imprinting: Implications for socialization in monkeys and men. *Psychological Review*, 80, 527–544.

Hogan, J. A. (1988). Cause and function in the development of behavior systems. In *Handbook of Behavioral Neurobiology*, Vol. 9: *Developmental Psychobiology and Behavioral Ecology*, ed. E. M. Blass, pp. 63–106. New York: Plenum Press.

Holland, P. C. (1984). Origins of behavior in Pavlovian conditioning. *The Psychology of Learning and Motivation*, 18, 129–174.

Hollis, K. L., ten Cate, C. & Bateson, P. P. G. (1991). Stimulus representation: A subprocess of imprinting and conditioning. *Journal of Comparative Psychology*, 105, 307–317.

Horn, G. (1985). *Memory, Imprinting, and the Brain*. Oxford: Clarendon Press.

Kroodsma, D. E. & Konishi, M. (1991). A suboscine bird (eastern phoebe, *Sayornis phoebe*) develops normal song without auditory feedback. *Animal Behaviour*, 42, 477–487.

Lorenz, K. (1935). Companions as factors in the bird's environment. In *Studies in Animal and Human Behavior*, Vol. 1, ed. R. Martin, pp. 101–258. London: Methuen.

Lorenz, K. (1970). Notes. In *Studies in Animal and Human Behavior*, Vol. 1, ed. R. Martin, pp. 371–380. London: Methuen.

Mackintosh, N. J. (1983). *Conditioning and Associative Learning*. Oxford: Oxford University Press.

Marler, P. (1976). Sensory templates in species-specific behavior. In *Simpler Networks and Behavior*, ed. J. Fentress, pp. 314–329. Sunderland, Mass.: Sinauer.

McGregor, P. K. & Avery, M. I. (1986). The unsung songs of great tits (*Parus major*): Learning neighbours' songs for discrimination. *Behavioral Ecology and Sociobiology*, 18, 311–316.

Pearce, J. (1988). Stimulus generalization and the acquisition of categories by pigeons. In *Thought without Language*, ed. L. Weiskrantz, pp. 132–161. Oxford: Oxford University Press.

Rescorla, R. A. (1988a). Pavlovian conditioning: It's not what you think it is. *American Psychologist*, 43, 151–160.

Rescorla, R. A. (1988b). Behavioral studies of Pavlovian conditioning. *Annual Review of Neuroscience*, 11, 329–352.

Rescorla, R. A. & Holland, P. C. (1976). Some behavioral approaches to the study of learning. In *Neural Mechanisms of Learning and Memory*, ed. M. R. Rosenzweig & E. L. Bennett, pp. 165–192. Cambridge, Mass.: MIT Press.

Revusky, S. (1977). Learning as a general process with an emphasis on data from feeding experiments. In *Food Aversion Learning*, ed. N. W. Milgram, L. Krames & T. Alloway, pp. 1–51. New York: Plenum Press.

Rozin, P. & Kalat, J. W. (1971). Specific hungers and poison avoidance as adaptive specializations of learning. *Psychological Review*, 78, 459–486.

Rozin, P. & Schull, J. (1988). The adaptive-evolutionary point of view in experimental psychology. In *Stevens's Handbook of Experimental Psychology*, 2nd edn, Vol. 1, ed. R. C. Atkinson, R. J. Herrnstein, G. Lindzey & R. D. Luce, pp. 503–546. New York: Wiley.

Ryan, C. M. E. and Lea, S. E. G. (1990). Pattern recognition, updating, and filial imprinting in the domestic chick (*Gallus gallus*). In *Quantitative Analyses of Behavior*, Vol. VIII: *Behavioral Approaches to Pattern Recognition and Concept Formation*, ed. M. L. Commons, R. J. Herrnstein, S. M. Kosslyn, & D. B. Mumford, pp. 89–110. Hillsdale, NJ: Lawrence Erlbaum.

Seligman, M. E. P. (1970). On the generality of the laws of learning. *Psychological Review*, 77, 406–418.

Sherry, D. F. & Schachter, D. L. (1987). The evolution of multiple memory systems. *Psychological Review*, 94, 439–454.

Shettleworth, S. J. (1972). Constraints on learning. *Advances in the Study of Behavior*, 4, 1–68.

Shettleworth, S. J. (1993). Varieties of learning and memory in animals. *Journal of Experimental Psychology: Animal Behavior Processes*, 19, 5–14.

Staddon, J. E. R. (1983). *Adaptive Behavior and Learning*. New York: Cambridge University Press.

ten Cate, C. (1987). Sexual preferences in zebra finch males raised by two species: II. The internal representation resulting from double imprinting. *Animal Behaviour*, 35, 321–330.

ten Cate, C. (1989). Behavioral development: Toward understanding processes. *Perspectives in Ethology*, 8, 243–269.

Timberlake, W. & Lucas, G. A. (1989). Behavior systems and learning: From behavior to general principles. In *Contemporary Learning Theories*, ed. S. B. Klein & R. B. Mowrer, pp. 237–275. Hillsdale, NJ: Lawrence Erlbaum.

Tulving, E. (1985a). How many memory systems are there? *American Psychologist*, 40, 385–398.

Tulving, E. (1985b). On the classification problem in learning and memory. In *Perspectives in Learning and Memory*, ed. L. G. Nilsson & T. Archer, pp. 67–94. Hillsdale, NJ: Lawrence Erlbaum.

van Kampen, H. S. & De Vos, G. J. (1991). Learning about the shape of an imprinting object varies with its colour. *Animal Behaviour*, 42, 328–329.

16

Representation development in associative systems

I. P. L. McLAREN

This chapter is concerned with mental representations, and their development by learning via association. By representation, I mean to imply
structures that permit recognition and identification of a stimulus without
necessarily entering the conceptual domain. I hope to show that associative
mechanisms can construct and employ representations of this type in such
a way as to permit explanation of phenomena such as latent inhibition,
perceptual learning, and the trade-off between context-specificity and
generalisation. To this end, an outline model of associative learning, that
concerns itself with the formation of associations between elements
representing motivationally neutral stimuli, is developed in the course of
the chapter. The emphasis, however, will be on the representational
assumptions contained within the model and their consequences. The
chapter starts by examining an elemental approach to stimulus representation, which proves surprisingly powerful in an associative context.
Nevertheless, there are drawbacks to an elemental approach, which the
latter half of the chapter attempts to solve by invoking a configural
modification of an elemental account.

Introduction

The modern concept of an association, and hence of an associative learning
system, probably stems from the work of the British empiricists (e.g.
Locke, Hume). At the heart of this approach is the essential idea that one
stimulus can bring about some recollection of another by virtue of the fact
that the two stimuli were associated at some time in the past. Certain laws
of association, e.g. contiguity and similarity, were believed to specify the
conditions under which such associations might be formed. The need for a
theory of representation, i.e. a specification of what internal mental state

stood for the stimulus, was met, if at all, by postulating some relatively basic sense impressions ('simple ideas') that could then be associated by experience to form complex ideas. Representations of external objects in the world would hence be built up over time, by a process of gradual enrichment. Representation was thus, in large part, reduced to association, and the study of associative learning was to become the predominant issue in experimental psychology for the first half of the twentieth century. During this period, however, many stimulus–response (S–R) theorists were to deny the very existence of representations as internal mental states, and focused exclusively on determining the laws governing how stimuli come to be attached to responses. In this scheme it was not 'ideas' that were associated, but observable events.

However, this phase in psychology's development as a science passed, leaving associationism in the hands of those willing to admit mental states as representations of external stimuli. A typically prophetic William James (1890) was their precursor. Tackling the issue of representation formation, he identified two ways in which representations might develop.

First, the terms whose difference comes to be felt contract disparate associates and these help to drag them apart ... The effect of practice in increasing discrimination must then, in part, be due to the reinforcing effect, upon an original slight difference between the terms, of additional differences between the diverse associates which they severally affect.

In other words, different stimuli develop different associates which progressively differentiate them from one another. Note that it is not necessary to read James as restricting himself to stimulus–response associations here. His second suggestion was that experience with stimuli varying along a dimension improved the ability to differentiate stimuli on that dimension, a proposal which receives considerable empirical support (e.g. Lawrence, 1952, transfer along a continuum) but has little to offer by way of mechanism.

Another question concerns the nature of the basic representations utilised by the hypothetical associative learning system. Here, the majority of associative learning theorists were silent, with the notable exception of Estes (1959), who developed the stimulus-sampling approach to learning. The basic assumption adopted by Estes was that stimuli should be conceptualised as sets of elements, the elements themselves being simple primitives (one interpretation of this would have the primitives correspond to what nowadays would be termed the micro-features of a stimulus) that could either be active or not. Stimuli would thus be similar to the extent that the sets of elements representing them overlapped. A further assumption was that the elements that were activated on a given presentation of a stimulus constituted a random sample of elements from the total set of elements that could in principle represent that stimulus. The psychological rationale for this assumption was never completely clear (at least to this author), but the result was a model that was able to simulate many phenomena of interest to learning theorists of the time. In particular, the model coped well with topics such as generalisation from one stimulus to another, and spontaneous recovery of the conditioned response (CR) after extinction. Importantly, stimulus-sampling theory typically only allowed the formation of associations between stimuli and responses, a hangover from the S–R tradition, restricting its ability to develop adequate representations.

This brief review summarises the associationist tradition in representation development. The question now is how a modern connectionist can take this position forward and develop a model of associative learning that tackles these representational issues. I shall begin by considering some of the more general decisions to be taken in constructing such a model, before moving on to specific issues of representation and representation development.

Association

The first question that arises concerns the architecture of the model. It is possible to speak of the architecture of a connectionist system (i.e. how it is organised) by reference to the set of connections between representational elements allowed in the system. Consider the distinction between a connection of zero strength, indicating no association between two elements, and no connection at all, implying that those elements can never be directly associated. By restricting the potential for association in this way, representational domains can be constructed. For example, elements representing motivational properties of stimuli might not be connected to one another or to elements representing motivationally neutral stimuli, whereas the latter elements might be connected amongst themselves and to the elements representing motivationally significant stimulus attributes. Associations could thus be formed from stimulus features such as form and colour to motivational attributes such as hunger and thirst, and between form and colour themselves, but other combinations would be impossible. Such an arrangement slices representation along a motivational axis and Konorski (1967) has argued the case for this architecture in modelling Pavlovian conditioning in infra-humans.

Then there is the question of what associative learning rule should be adopted, i.e. what rule should govern the strengthening and weakening of connections. A wide variety of options are available: Hebbian (Hebb, 1949; Levy & Desmond, 1985); error correcting (Widrow & Hoff, 1960; Rescorla & Wagner, 1972; McClelland & Rumelhart, 1985); or derivative dependent (Konorski, 1948; McLaren & Dickinson, 1990; Sutton & Barto, 1981, 1989). There appears to be a consensus amongst animal learning theorists that a learning rule of an error correcting nature is required; i.e. that learning should be governed by the difference between the output required (in terms of some variable such as associative strength) and the current output (Rescorla & Wagner, 1972; Mackintosh, 1975; Pearce & Hall, 1980). This class of rule is also favoured in modelling humans' cognitive abilities (e.g. McClelland & Rumelhart, 1985), and is recognised as possessing a number of computational advantages over the various Hebbian rules which simply strengthen connections as a function of coactivation of elements (for a discussion see Stone, 1986). One particular error-correcting algorithm, a generalisation of the delta rule known as back propagation (Rumelhart, Hinton & Williams, 1986), is probably the current market leader in connectionist modelling. It should be noted that derivative-dependent rules are, under certain conditions, also error correcting (Sutton & Barto, 1989; McLaren & Dickinson, 1990).

Representation

In general, the transient representation of a stimulus will be taken here to consist of a pattern of activation distributed over a set of elements, rather than a one-to-one correspondence between unitary stimulus and representational element. This adopts one of the central tenets of stimulus-sampling theory (Estes, 1959; Neimark & Estes, 1967), albeit in a modified form. Is there a good case for adopting any other features of the stimulus-sampling approach? The most obvious candidate is the idea of sampling the set of elements potentially representing the stimulus, such that only a randomly selected and partial subset are active at any one moment. This allows some explanation of a number of phenomena in both human and infra-human learning and memory. Gradual improvements in performance with experience, superimposed on substantial variability in the short term, are explained by reference to fluctuations in the elements sampled in a given learning episode. Speed of learning can be related both to the number of elements and to the proportion of elements sampled in any episode, the higher both number and proportion, the greater the speed of learning.

The difficulty with Estes' sampling process as it stands is that it is taken to be random, and as such is not motivated by anything more than a desire to model the data. If, however, the sampling process is assumed to depend on the nature of the stimulus and on the system's ability to inspect or attend to it, as in McLaren, Kaye & Mackintosh (1989), then the account gains both in plausibility and explanatory power. For 'simple' stimuli such as tones and coloured lights, it seems reasonable to suppose that there will be relatively little variability in the sampling from one instant to the next, and that a high proportion of elements will be sampled. For more 'complex' stimuli, whose defining characteristics are multiple and distributed over space, the proportion of elements sampled might be expected to be lower, and sampling variability higher, due to the organism's inability to apprehend simultaneously all the features of the stimulus. An example would be a complex visual stimulus such as a tabby cat, which possesses features at several different scales (in terms of spatial frequency). Some of the features present might be apprehended simultaneously, i.e. a high proportion of elements sampled, at the grosser scales (e.g. it has four legs). A finer grain of detail (e.g. markings on the fur) can only be appreciated over a limited region of the space occupied by the cat because of the visual systems drop-off in acuity with eccentricity, and because of attentional factors .

As already noted, there is a role for the proportion of elements sampled to play in determining the salience of a stimulus representation: one could suppose that the higher the proportion the greater the salience, or, indeed, that the greater the absolute number of active elements representing a stimulus the more salient it will be in that its ability to associate to other stimulus representations will be enhanced. The last is one of the mechanisms envisaged by Konorski (1948) for capturing variations in stimulus salience. The other one is variation in the activity of representational elements, with greater activity for a more salient stimulus.

The core assumption of the elemental approach taken here is that each stimulus will be represented by a set of activated elements, a distributed representation. Variation along a stimulus dimension such as brightness will, for the most part, be represented by different elements corresponding to different values on the dimension, rather than the activation level of an individual element being the primary indicator of value on the dimension (cf. Thompson, 1965). Figure 16.1 represents this schematically. Each element has a 'tuning curve' such that it responds most strongly to a certain value on the dimension and this response drops off fairly rapidly with 'distance' from this optimal value. Note that many elements will be

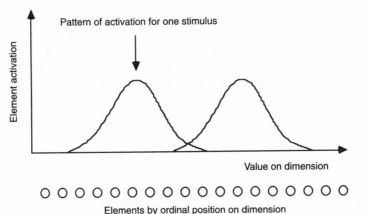

Figure 16.1 Representation of a stimulus dimension in a connectionist system.

active when any stimulus with value on that dimension is present, the coding is via a pattern of activation. The advantage of this scheme is that it permits a degree of independence in coding different values on a dimension, whilst preserving the ability to generalise between neighbouring values.

One way of construing the variability in the sampling of a stimulus is to think of it as noise in the representation of that stimulus. The effect of noise will be to render the representation of some input pattern initially unreliable. This unreliability will attenuate with repeated presentation of the pattern, and the variability in response to the pattern or a portion of the pattern will decline because of the formation of associations between the elements representing the stimulus. Two examples will help to make this clear. In the case of a relatively 'simple' stimulus, such as a tone, it is reasonable to suppose that most of the elements potentially representing that tone will be sampled. The reason for this is that it is not clear what it would be to experience part of the tone, the senses either apprehend the tone or they do not; and there is little structure to the stimulus, no fine-grain detail varying over the region of space occupied by the stimulus. It seems likely that this stimulus does not suffer markedly from the type of noise that results in some elements that should be active being inactive. It will, however, suffer from being accompanied by other active elements that fluctuate from presentation to presentation due to the sampling of the rest of the environment apart from the tone. These elements will associate to the elements representing the tone and will be activated somewhat on subsequent presentations of the tone, acting as unwelcome noise in the representation of the tone. Fortunately, error-correcting rules act as a

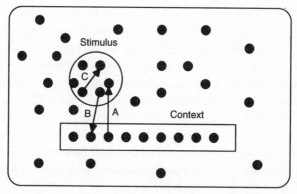

Figure 16.2 The different types of association possible for a stimulus in a context.

'signal-averaging algorithm' in such circumstances (McClelland & Rumelhart, 1985), and will gradually extinguish these unwanted associations. Equally, if an association between the representation of the tone and another stimulus is to be formed, then repeated pairings of the two stimuli will gradually excise contributions from spurious elements that are present as noise on only a few of the trials

The second example concerns a more 'complex' stimulus such as a detailed visual pattern, when the proportion of the total set of elements potentially available to represent that stimulus that are actually activated can be expected to be relatively low. This will be because only part of the stimulus can be apprehended in full detail at once. Of course, the noise due to other stimuli being present by chance on a given presentation of this stimulus will still be a factor, but now the noise due to the sampling process acting on the stimulus representation itself must be contended with. Because over a series of presentations of the stimulus, the elements representing it co-occur more often than the noise elements, they will become associated, and activation of a sample will be able to draw in the rest via these associations. Hence the formation of associations between the representational elements will reduce the variability in stimulus representation.

An example: two components to latent inhibition

As a first example of some of the principles outlined in the preceding sections I shall consider a very simple case of representation development – latent inhibition. Latent inhibition occurs when a stimulus, perhaps a light, is presented to an animal in a Skinner box, say, and the animal is subsequently conditioned to expect some reinforcer contingent on the light. It typically learns more slowly than control animals not pre-exposed to the light (Lubow & Moore, 1959; Hall, 1980). One explanation for this is that the elements representing the light have lost salience, they no longer enter as readily into associations (with the reinforcer representation in this case) as a consequence of pre-exposure (e.g. Wagner, 1978). In Wagner's account, this is taken to be due to the formation of associations between elements representing the context (box) and the stimulus (light). In order to explore this argument further we must consider how the characteristics of latent inhibition due to association formation will depend on what is being associated with what. Following McLaren *et al.* (1989), the representational assumptions in play are illustrated in Figure 16.2. The rectangle contains a large number of elements or nodes which may become associated to one another. The set of nodes depicted within the circle corresponds to those elements representing a particular stimulus.

Given that the laws of association apply equally to all elements, the stimulus elements may develop associations amongst themselves (C) as well as with other elements corresponding to other stimuli (A and B). The upshot of this is that in a stimulus pre-exposure phase of training, there will be two readily identifiable sources of associations to the stimulus elements causing latent inhibition: namely intra-stimulus and context–stimulus associations. Latent inhibition will be determined by the number and strength of the associations in these two classes. The associations from the context to the stimulus elements will be responsible for a context-specific component of latent inhibition. When the subject is shifted to a different context, these associations will no longer be effective, given that the appropriate contextual cues are not present. Hence the activation level and salience of the stimulus elements will be restored. Intra-stimulus associations will be responsible for a contextually non-specific component of latent inhibition. If associations of this type exist, then they will always be expressed when the stimulus is presented, in the sense that they will lower the stimulus elements' salience. The position taken here is that understanding latent inhibition now reduces to specifying to what degree any pre-exposure of a stimulus results in the formation of these two types of association. Some of my own experiments suggest that pre-exposure to the context (which should make its representation less salient) leads to a form of latent inhibition that can transfer across contexts. This is predicted, as a less salient context will not associate as readily to the stimulus representation, and hence latent inhibition will be relatively more dependent on intra-stimulus associations.

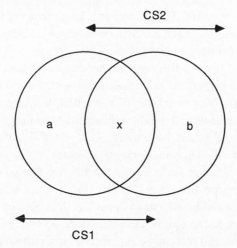

Figure 16.3 Representation of stimuli as sets of elements.

A further example: perceptual learning

Consider the problem of discriminating two cats (one called Drake, the other, Darwin) from one another. Both are tabbies (brothers in fact) and initially very hard to tell apart. Despite clear identification as to which is which, early performance by people unfamiliar with the cats is near chance, this being particularly true when the person tries to identify one of the cats in isolation. After a number of visits to the household containing the cats, however, the situation is quite different. Identification is quick and effortless, the two cats look so different now that it is hard to imagine that they could ever have been confused. We can say that 'perceptual learning' has taken place with experience of the stimuli.

How does this come about?. Again, following McLaren *et al.* (1989), three distinct 'mechanisms' for perceptual learning can be distinguished given the assumptions made so far. I now consider each in turn.

1. Latent inhibition of common elements

Take two stimuli moderately similar to one another. In terms of the representations employed in this model, the state of affairs will be as shown in Figure 16.3.

CS1 has unique elements (the a elements), as does CS2 (b elements). Both, however, share the x elements which are the basis of their similarity and would also be the basis for any generalisation between them. If CS1 is presented for some time before CS2 is paired with a US, then less conditioning should generalise to CS1 compared to a control group that received no pre-exposure. This is because the x elements will be latently inhibited by pre-exposure and hence the b elements in CS2 will overshadow them and acquire most of the associative strength to the US, leaving less strength to accrue to the x elements and hence generalise to CS1. Honey & Hall (1989) have performed this experiment with flavours. This type of study qualifies as a demonstration of perceptual learning because simple stimulus exposure leads to a greater refinement of stimulus representation as measured by the generalisation test. It is not difficult to see that the latent inhibition of common elements in some sense implements the 'tuning out' of common features and the relative increase in salience of distinctive features postulated by Gibson (1969). Even when both stimuli are pre-exposed, a similar effect can occur. As the x elements receive twice as much exposure as the unique elements, their salience will be depressed to a greater extent, and this can be exploited to give more rapid acquisition of a discrimination after pre-exposure.

2. Formation of inhibitory links

There is another potential mechanism for perceptual learning implicit in this associative theory, though it is by no means straightforward to disentangle it from the others considered here. I start by noting that if the process of excitatory association formation is taken to be more rapid than its inhibitory counterpart (as is commonly taken to be the case, cf. Wagner, 1981), then in time the novel stimuli will be in the position of evoking inappropriate unique elements when presented. This is because excitatory associations from the common elements and context to the unique elements will form, and the unique elements of a stimulus will thus be associatively activated even when the other stimulus is presented to the system. Clearly this mechanism acts to increase generalisation between ·the stimuli. Another mechanism can come into play to redress the balance, though if it is to apply then it is critical that the two stimuli should be 'inspected' separately. Given this condition holds, then the following contingency is in force: if the a elements are sampled, then the b ones are not and *vice versa*. This negative contingency will result in the a and b elements forming mutually inhibitory connections, with the result that the active a elements, say, will try to suppress any b elements evoked by the context or x elements. The effect is to refine the system's representation of the stimuli. It might be argued that this will give no advantage over novel stimuli whose elements are not associated to anything, i.e. that there will be no b element activation to suppress when CS1 is presented, and that any inhibitory connections between unique elements are unnecessary. It is the case, however, that novel stimuli will begin forming associations between their elements as soon as they are presented. They will have to pass through a stage that pre-exposed stimuli have already experienced (and in some sense dealt with). Other things being equal, this would put the novel stimuli at a disadvantage compared to the pre-exposed stimuli. Bateson & Chantrey (1972) and Mackintosh, Kaye & Bennett (1991) present evidence supporting this proposal.

3. Unitisation

One further assumption leads to a final mechanism for perceptual learning. Recall that the elements constituting a stimulus are taken to be sampled from moment to moment rather than all continuously present. This is because the representation of a stimulus as a set of activated elements will, in general, be subject to two classes of noise. One, the less interesting class, is that of noise intrinsic to the elements themselves – random fluctuations in activation level, link strength etc. The other class of noise arises from factors extrinsic to the associative learning system, and chief among these will be the variability of the subject's interaction with the environment. An example will clarify what is meant by this. Consider an animal or human examining a relatively complex object. In the case of the animal, an appropriate object might be a black metal equilateral triangle, which is certainly a relatively complex stimulus (in terms of the number of features it possesses) when contrasted to the lights and tones used in conventional conditioning procedures. One aspect of this stimulus's complexity is that its features are distributed over space. The result of this is that an animal's perception of these features will vary according to the point of fixation of gaze on the object. Moreover, it is only those features which correspond to local regions of the object (i.e. the details) that will necessarily be sampled in this way. The gross stimulus features, e.g. 'black', will be apparent wherever the object is fixated. If an equilateral triangle and a circle are compared, then at a low resolution (i.e. a coarse level of detail) both will look like black blobs. It is only at a finer level of detail that the discrimination can be made. The claim is then, that where two stimuli are sufficiently similar and differ only in detail, their unique features will typically be sampled more variably than their common features. Obviously we would not expect this for a black square versus a white square say, as this would not meet the above criteria.

But how does the more variable sampling of unique elements lead to a mechanism for perceptual learning?

Figure 16.4 is an 'exploded' version of Figure 16.3 showing a (unique to

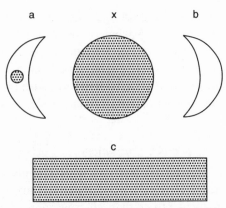

Figure 16.4 An illustration of the more variable sampling of the unique elements of a stimulus.

CS1), b (unique to CS2), and x (common) elements separately; the c elements represent the contextual cues and are included for completeness. The shaded region represents the elements sampled, and hence activated, on a given presentation of CS1. In line with the arguments put forward so far, the a elements are shown as being sampled rather variably (i.e. only a relatively small proportion activated), whilst the x and c elements are shown with a 100% sampling rate. As a result of stimulus presentation, associations form between all the active elements, and as this process is repeated with different samples from a, two important things occur. The x elements will become much less salient than the a elements on a given presentation (as they become latently inhibited more quickly, see above), and the a elements will tend to develop associations between one another. The latter process will enable the currently active subset of a elements to associatively recall other members of that set; and these elements will then be available for learning. The formation of associations among elements of a stimulus set will be referred to here as *unitisation*. Note that associations from x elements to unique elements or from context elements to unique elements will not enhance the discriminability of the stimuli, as these associations will recall those unique elements whether or not the stimulus is presented. What is important is the unitisation of the elements representing the unique features of a stimulus, allowing the sampled and hence active subset of those elements to associatively activate the rest of that set (an idea dating back to Hebb, 1949). If the unique elements are sampled variably and the common elements are not, or at least less so than the unique elements, then this confers an advantage to the unique elements after pre-exposure. This advantage could be absolute if the benefit from unitisation outweighs the harmful effects of latent inhibition. Note that the above argument applies only when the stimuli are 'seen' in isolation; if the stimuli were processed simultaneously, then the unique elements from both would associate, making it harder to discriminate between them. Fanselow (1990) presents some evidence in support of this suggestion.

All three mechanisms might be expected to contribute to perceptual learning in the case of the two cats mentioned earlier. Latent inhibition will act on their grosser features (which will tend to be the same), leaving the more detailed distinctive features relatively more salient. These finer-grained features will unitise, so that sampling a few will bring the rest with them; and separate encounters with both cats will allow the formation of inhibitory links between unique features to form. Thus, we can see that an elemental approach to representation coupled with an associative learning rule can produce a system with considerable potential for representation development. The next section considers some of the problems that remain, and suggests one way of dealing with them.

Extending the elemental account

One of the problems facing the elemental version of an associative learning scheme employing an error-correcting learning rule lies in ensuring that elements do not 'lock themselves out', in the sense that it becomes impossible for associations to them to be formed. This is a general problem for representation in error-correcting systems, as elements that are highly correlated will tend to form strong associations, and hence make it difficult to form associations to them. An example would be if two elements were always to be activated by a red stimulus, then they would associate to one another, lowering their error scores to near zero, and other elements would be unable to associate to them. This is because if there is no error, then there is nothing to drive the formation of associations. The upshot of this is that if this red was now perceived in a novel combination with another feature, e.g. a triangle, the system would be unable to learn that the triangle was red, because the associations from triangle to red could not be formed. Decay (of associative strength) offers a partial solution to this problem. On the one hand, decay will ensure that in the long term there is always some room for association because there will be some need for extra associative strength to replace the decayed component. In the short term, the fact that decay can prevent learning going to asymptote for connections between specific elements also helps ensure that some extra associative input to a node will be welcome, allowing associations to that element to be made. If these mechanisms are coupled with the requirement that no two elements should be correlated too highly, i.e. they should not be representing exactly the same thing, then the problem of 'lockout' can be avoided to a great extent. Nevertheless, a method of avoiding the problem entirely would be preferable.

Another challenge for the elemental position concerns how we trade-off the requirement for some contextual specificity in conditioning with the need to generalise what is learnt in one environment to another. Some degree of contextual specificity is needed to prevent too much interference between learning episodes occurring in different contexts, but involving similar stimuli. Otherwise what is learnt in one context might be inappropriately undone in another. On the other hand, this has to be reconciled with the requirement that some transfer between contexts be allowed for similar stimuli, so that what is learnt in one context can generalise appropriately. Clearly the two requirements are in conflict, and most learning theorists only differ in the nature of the balance struck between the Scylla of specificity and the Charybdis of over-generalisation. I shall argue that this spirit of compromise is misplaced, and that learning can generalise well and yet still be sufficiently context-specific so as to avoid unwarranted interference between learning that has taken place in different contexts. The secret lies in the development of appropriate representations during the course of learning.

I shall begin with the tale of the rat and the bush, which illustrates the desirability of avoiding the compromise referred to earlier. It is a story with several different endings, depending on which learning theory the rat is presumed to instantiate.

Three rats, a bush, and a snake

One day a rat forages near a certain bush, and is startled by the sudden hiss of a snake that was asleep under the bush. It jumps back and scurries away. This was its first experience of both snake and bush. However, the next day whilst foraging elsewhere, it encounters another, similar bush. What does it do? That depends on what kind of rat it is. The Conditional Rat always learns with complete context specificity, and never generalises from one context to another. The Elemental Rat learns about individual cues and responds on the basis of the sum of associative strengths for the cues present; it generalises well and has little context specificity. A Computationally Correct Rat, however, generalises well in the first instance but if proven wrong does not suffer from overmuch interference between learning episodes. Let us see how each rat fares.

As far as the Conditional Rat is concerned, the novel bush is a neutral stimulus, as no learning from the previous context will transfer to this one. It may go up to the bush, it may not. If it does it will discover that it has some very tasty berries on it. The problem it faces now is that as it traverses other regions of its habitat it will fail to make a special effort to investigate these bushes, as it cannot generalise from one encounter to another. This is rather inefficient of it, and other animals capable of generalisation get there first. This rat dies of starvation.

The Elemental Rat has no problem with generalisation. It has learnt that this type of bush is associated with aversive consequences and reacts with fear on encountering a similar bush. Perhaps it never approaches the bush and discovers the berries, more likely it approaches very cautiously over a period of time, discovers no nasty hissing thing and eats the berries. So far so good. The problem here is that the rat will lose all fear of the bush. The next time it forages in the original environment it bounces up to the bush, gets bitten by the snake (which likes to sleep there) and dies.

Two rats down; let us see how the third one does. Our Computationally Correct Rat initially behaves much as the Elemental Rat did. It exhibits fear on its next encounter with the bush, gradually overcomes it, and learns about the berries. This makes it interested in bushes in general (and hence efficient in its foraging behaviour), with one exception. It still avoids the bush in the original context, because the new learning about how good bushes in other contexts are has not interfered with the original learning about the bush and the snake in this context. Our rat leads a happy and contented life, proving that in the world of associative learning, it is not necessarily the case that you are either dead or you are hungry.

Hybrid models

Leaving aside for the moment the question of how the Computationally Correct Rat does it, the time has come to tie our examples a little more closely to theory and practice in animal learning. A number of theorists (e.g. Spear, 1973; Hall & Honey, 1989a, 1989b; Bouton, 1990) have suggested that conditioning be conditional in some way on the context. A popular suggestion is that contextual cues enable retrieval of the relevant information, namely the CS–US association. None of them is so bold as to claim that retrieval is always (or even usually) as strongly dependent on context as portrayed in my example. No doubt there will be a trade-off between context specificity and generalisation. This is exactly what the model espoused by Pearce (1987) is able to do. If a CS, A, is presented in a context, X, then the model requires that a representation of the compound AX be formed and used to associate with any US present. A completely conditional account would now allow only the combination of A and X to access the AX representation and hence any associations it might have to any US. Pearce's model allows that A or X presented alone will activate the AX representation somewhat, permitting some generalisation to occur. For example: if AX supports associative strength of β, then A presented alone might have a generalisation coefficient to AX of k, resulting in a generalised associative strength of kβ. Pearce also offers the suggestion that a typical value for k in this case would be 0.5, as A represents 50% of the AX compound. Note also that if A is presented alone, then a representation of A will be formed and possibly become involved in associations to some US. There will now be some generalisation from A to AX when the compound is presented, and a typical value for the generalisation coefficient in this case would be 1.0, or close to it.

Now let us analyse a slightly simplified version of our example in terms of Pearce's model to get an idea of how context specificity and generalisation might trade-off against one another. Consider the case of a CS, A, reinforced in a context, X, then extinguished in a context, Y, before final testing with A in X. A simple analysis proceeds by assuming that all learning in each phase is taken to asymptote, and that the typical values for our generalisation coefficients apply. Let us say that A and X have coefficients of 0.5 to AX, whereas AX has coefficient 1.0 to either of A or X. Similarly, AY will have a generalisation coefficient of 0.5 to AX as either compound shares 50% of its makeup with the other. The calculation proceeds as follows. Phase 1: reinforce A in X, X alone non-reinforced. AX and X acquire some associative strength. At asymptote the US requires some fixed amount of associative strength, say Ω. Now, clearly the net associative strength for X must be zero at asymptote as it is never reinforced in the absence of A. This leads to the (not unreasonable) conclusion that the context, X, acquires some inhibitory properties to counteract the effects of generalisation from AX.

Phase 2: AY non-reinforced and Y non-reinforced. AY gains inhibitory associative strength to offset generalisation from AX, whereas Y gains excitatory associative strength to compensate for generalisation from AY. We can now form equations which determine exactly how much associative strength each stimulus has. Once we have solved the equations for asymptotic learning during phases 1 and 2, all that remains is to compute the net associative strength for AX on test. This turns out to be zero. Hence our two-phase experiment results in the complete loss of any learning from phase 1. We can also check what the response to AY would be on the first trial of phase 2; this will serve as a measure of generalisation between the contexts. It comes to Ω, indicating perfect generalisation across contexts with the generalisation coefficients chosen for this example. The outcome of this analysis, then, is that if the generalisation coefficients are chosen to give perfect generalisation across contexts, interference between phase 1 and phase 2 learning will be such as to wipe out phase 1 learning. Perhaps this should not surprise us, the real question is whether or not a choice of generalisation coefficients exists that allows both reasonable generalisation and protection from the sort of interference encountered in this example.

Taking the generalisation coefficient from AX to X to be p, from X to AX to be q, from AY to AX to be r, from AY to Y to be s, from Y to AY to be t, and from AX to AY to be u, then we have the most general case possible. In our earlier working, p and s would be 1.0, and q, r, t, and u would be 0.5. It seems reasonable to assume that our generalisation coefficients are all between 0 and 1 (inclusive), and that generalisation from AX to AY should be roughly equal in magnitude to generalisation from AY to AX. This makes u and r equal, and similar cases can be made for p and s, and q and t. The trade-off between generalisation and interference can now be computed. For our earlier values, generalisation is perfect and interference is complete as we calculated earlier. As generalisation is reduced so is interference, and the best trade-off depends on what is defined to be optimal, but note that if generalisation is 50%, then interference is reduced to the extent that 75% of phase 1 learning is spared.

Pearce's model can trade-off context specificity with generalisation across contexts, but it is not alone in being able to do this. Rescorla (1973) provides an alternative approach. It was his suggestion that when A and X are presented in compound, a unique or configural cue for the compound becomes active *as well as* cues representing the individual elements A and X. Only the combination of A and X can activate the unique cue, however, so that any generalisation is dependent on the elemental cues. Applying this model to our example is straightforward, all that is in question are the relative proportions of the final associative strength to the US, Ω, that each cue acquires. This in turn will depend on how salient each cue is. The more salient a cue, the greater the proportion of Ω it will command. For example: if all cues are equally salient, then each of A and AX will come to have an associative strength of $\Omega/2$ during phase 1 (X will asymptote at 0). At the start of phase 2, generalisation to A in Y will be due to the A cue, and hence will be $\Omega/2$. As extinction proceeds in Y, the net associative strength of A, Y, and AY cues must come to be zero. This can be realised in a number of ways, the simplest being that A loses all its associative strength, in which case 50% of the original phase 1 learning would be lost by the end of phase 2. This is unrealistic, however, as it neglects the possibility of inhibitory learning to the Y and AY cues. Following Rescorla & Wagner (1972), we note that Y asymptotes at zero and hence A will asymptote at $\Omega/4$ and AY at $-\Omega/4$. The result is that the elemental approach can achieve the same trade-off as Pearce's model, 50% generalisation whilst retaining 75% of the original learning.

In summary, hybrid versions of the conditional and elemental models can trade-off context specificity and generalisation such that if generalisation between contexts in our example is at 50% (in terms of associative strength), then testing A in X at the end of phase 2 should result in 75% of the phase 1 learning manifesting itself in performance. The key to doing this in the two examples considered appears to be the use of *configural* representations, that allow for the possibility that a compound stimulus might be more than just the sum of its components. This allows a trade-off between context specificity and generalisation which is fixed once the parameters of the model are chosen. If the nature of the function relating performance to associative strength is left open for the moment, we can say that when generalisation is lower, then more of the original learning will be preserved; but if generalisation rises, then interference between phases 1 and 2 increases and more of the original learning will be lost.

Experimental evidence

How do real-world rats do on this task? In at least some circumstances the answer, provided by Bouton & King (1983) (though somewhat in disagreement with Bouton & Peck, 1989; and Lovibond, Preston, & Mackintosh, 1984), is that they do as well as the Computationally Correct Rat discussed earlier. Their experiment used a conditioned suppression procedure, with two groups of rats being conditioned to A in X. One group was then extinguished (non-reinforced presentations of A) in Y, whereas the other was extinguished in X. Both groups extinguished at the same rate and initial responding to A in Y was indistinguishable from that to A in X. By this performance measure, generalisation was essentially perfect. Extinction proceeded until suppression ratios were near asymptote (i.e. 0.5 for no suppression to the CS), which, coupled with the strong generalisation observed, should be a recipe for massive interference with original phase 1 learning when A is tested in X for the group extinguished in Y. In fact, interference was minimal, with near asymptotic suppression to A (i.e. suppression ratio near zero) on test in X. Bouton & King also present evidence against the claim that the contexts X and Y had acquired either excitatory or inhibitory associative strengths due to their procedures. Their conclusion is that you can have your cake (near perfect generalisation) and eat it (near perfect context specificity).

One possible problem with this experiment is that the performance measures taken might not accurately indicate associative strengths. An example of this might be that the suppression ratio cannot assay very high associative strengths because it has a floor of zero. Perhaps the reason why generalisation was so good is that even though there was a reduction in associative strength on shifting contexts from X to Y, this did not show up in performance because of the floor effect just mentioned. Arguments of this general type are difficult to sustain, however, because the group extinguished in X parallels that extinguished in Y so closely. The specific argument given can be countered by noting that if it were true, then the rats extinguished in X should show no drop in responding for the first few extinction trials and should certainly lag behind the rats extinguished in Y, and this was not the case.

The results of this experiment point to a combination of near-perfect context specificity and generalisation in the absence of any evidence for the contexts acquiring any significant excitatory or inhibitory associative strength. The importance of these findings lies in their status as an existence proof for the possibility of just such an outcome. Undoubtedly the trade-off between context-specificity and generalisation can be worse (and was in the other experiments cited), but this experiment indicates that in at least some circumstances there is hardly any trade-off at all. The lack of any evidence for appreciable associative strength to the contexts is important because it suggests that the elemental model will not be able to appeal to some inhibitory strength possessed by Y as a means of bumping up the preservation of phase 1 learning. Figure 16.5 compares the performance of Pearce-type, Rescorla-type, and Computationally Correct Rats when matched for generalisation after phase 1. The Computationally Correct Rat's performance is indistinguishable from Bouton and King's group extinguished in a context different from conditioning.

Both the Pearce and Rescorla-type rats fare similarly, showing poor retention of phase 1 learning after phase 2. Note that the salience of the unique cues has to be low in Rescorla's model to achieve this level of generalisation, which is why so much of the associative strength acquired by A is lost in extinction. If some inhibitory associative strength were allowed to accrue to Y, then the performance of this model could be

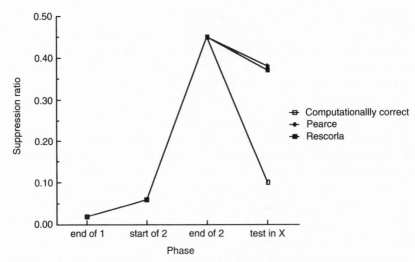

Figure 16.5 Suppression ratios (i.e. response rate in the CS period divided by the sum of the response rates in the CS and pre-CS periods) derived from simulations with APECS (corresponding to the Computationally Correct Rat and Bouton & King's results), Pearce's (1987) model, and Rescorla's (1973) amendment of the Rescorla–Wagner rule to include unique cues. The simulations have been matched for generaliation between contexts.

improved to a best suppression ratio of around 0.22, but this would surely have been detected by Bouton & King in their summation tests. The other option available here is to attempt a best fit of Pearce and Rescorla's models to the data, instead of matching them to it for generalisation and then testing for interference. Even this attempt to rescue them is likely to prove unsuccessful, however, as the best that can be done is to have a suppression ratio of approximately 0.17 on generalisation coupled with one of 0.20 on test. The difference with the computationally and experimentally correct values of 0.06 and 0.10 is still considerable and would probably be experimentally detectable.

Analysis of the problem

Why do the hybrid models have difficulty in achieving a level of performance on this task that is comparable to those found in Bouton & King's experiment? Part of the answer lies in the inflexible nature of the trade-off between contextual specificity and generalisation inherent in these models. As we have seen, once the parameters for generalisation are set, the degree of context specificity follows automatically. Another clue is that both the hybrid models considered could solve the problem posed by this task if alternating sessions in X and Y were given, i.e. reinforcing A in X and extinguishing it in Y. The outcome would be strong responding to A in X and no responding in Y. Given that both models could achieve this solution in these circumstances, there must be something about the sequential nature of the task considered here (phase 1, phase 2, test) that causes difficulty.

A more flexible model could exploit the sequential nature of the task by taking its cue from phase 1 and 2 in such a way as to modify the degree of generalisation between contexts. At the end of phase 1, generalisation could be set high. There is no evidence against the proposition that what has been learnt in this context might apply in others, so this seems a reasonable default as long as a mechanism exists for reducing generalisation if it turns out not to be so. As phase 2 progresses, it becomes clear that what held for X does not hold in Y, and generalisation must be reduced so as to protect phase 1 learning from the effects of phase 2. What is needed is a model that can capture this rather loose characterisation by flexibly developing the representations involved in learning during the various phases. APECS (Adaptively Parametrised Error-Correcting System) is just such a model, and the results for the Computationally Correct Rat in Figure 16.5 showed that it fared very well in the task considered here. I shall now turn to a consideration of the mechanisms underlying the performance of the Computationally Correct Rat running APECS.

Adaptively parametrised error-correcting systems

Figure 16.6 shows the basic architecture of the model, which uses a set of hidden units to mediate between CS and US units via two layers of weights.

In this system, associations between CS and US are more complex than just a simple link allowing one representation to activate another. The mediating hidden unit(s) possess biases (akin to thresholds) which can modulate the effectiveness of the associative link from CS to US units. Figure 16.7 gives an example, in fact the example discussed earlier. It shows the development in APECS of the hidden unit solutions to phases 1 and 2 and on test. Reinforcing A in X during phase 1 leads to the development of positive weights from units corresponding to A and X to a hidden unit that in turn has developed positive weights to the US unit. The weight from A to the hidden unit is somewhat stronger than that from X, and the hidden unit has a small negative bias. This means that X alone cannot activate the hidden unit (and hence the US unit) to any great extent, but that A or A and X are capable of activating this unit and hence activating

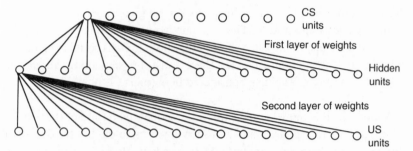

Figure 16.6 The connectivity between CS units and US units via hidden units. Only the connections from one CS unit to the hidden units, and one hidden unit to the US units are shown.

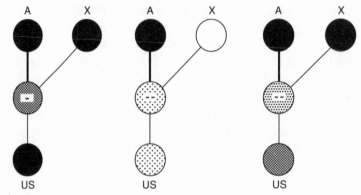

Figure 16.7 The solutions adopted by APECS to the task discussed earlier in this chapter. The left panel corresponds to the end of phase 1, the middle to the end of phase 2, and on the right we see the response to testing A in X. Darker shading indicates higher activity, thicker lines indicate stronger weights, and more − signs indicate a stronger negative bias.

the US unit strongly. There will also be some other hidden unit involvement implementing a weak inhibitory link between X and the US, but this will not concern us here. The figure (leftmost panel) shows A and X units strongly activated (dark shading), activating the hidden unit and hence the US unit as well. Connection strengths are indicated by the thickness of the lines drawn.

When A is presented in Y, the bias on the hidden unit mediating between A, X, and the US becomes increasingly negative, because activation of the US unit is inappropriate and this is the means by which APECS prevents this (rather than weakening the associative links themselves). Now presentation of A on its own has little effect in activating the US unit. Thus at the start of extinction generalisation will be strong (as the bias will not have changed and A can activate the hidden unit), but at the end it will be weak. A feature of APECS is that during this extinction process the weights to and from the hidden unit are frozen, by setting the learning parameter to zero so that only the bias changes, thus preserving the learning instantiated in these weights. Of course, other learning involving other hidden units can proceed at the same time, as long as their biases are not going increasingly negative. The key determinant of whether or not this is so is the error at a hidden unit. If it is negative, then the weights are frozen and the bias becomes more negative. If it is positive, then weights are allowed to change if that unit has won the competition amongst hidden units to be the best match to the conditions, i.e. has the highest positive error. Error is transmitted from the US units in the same fashion as in Back Propagation (Rumelhart *et al.*, 1986), by multiplying the error at a US unit (which is determined by the difference in desired and actual activation of the unit) by the associative strength from a given hidden unit to that US unit, and summing over all US units.

The upshot of all this is that phase 1 learning is protected, as can be clearly seen on test. Now that A and X are present, their combined input to the hidden unit is sufficient to overcome the stronger negative bias and activate it, which in turn activates the US unit. A will no longer generalise its ability to evoke the US representation to Y, yet it has not lost that ability in X. A more formal account of how APECS works will be presented elsewhere. Note, however, that this is not, as it stands, specifically a model of conditioning or animal learning. It is a general approach to the problem identified here, that of reconciling the need for generalisation and the need to avoid interference. A version tailored to the requirements of animal learning theory will be presented at a later date.

A further advantage of this form of configural extension to the elemental account is that it can be applied to the representational problem of 'lockout' referred to earlier. All we have to imagine is that the system is now auto-associating, that is that the same pattern of activation is applied as both input and output during learning. This allows stimuli to associate with themselves, even if they are not represented as sets of elements, though considerations pertaining to generalisation encourage us to keep this feature in the model. Can lockout occur? Will elements associate via some intermediate unit to themselves in such a way that they cannot play a part in further associations? The answer, I think, is no, and for the following reason. No stimulus can occur without a context, and no unit can be active representing that stimulus without others being active as well. Hence associations of units to themselves will always involve other units which will not always be active when our target unit is. Hence, as conditions change, and when a new stimulus is encountered, conditions *must* change, then these associations will become ineffective and new ones involving this unit can form. The reader might question the assumption that when a new stimulus is encountered conditions (i.e. the units active) must change. If a stimulus were to be added to some pre-existing stimulus configuration, would not all the units active before the addition remain active, plus a few more for the new stimulus? If this did happen, then lockout could occur in these circumstances, but it cannot. I shall assert that there is no such thing as pure addition; that adding a stimulus involves deleting others. If this holds true, then lockout cannot occur.

Conclusion

If we adaptively vary the parameters controlling learning, it is possible to escape the straight-jacket imposed on the generalisation/interference trade-off by other models of learning. Remarkably, we do not lose anything that the elemental account offered in doing this, so long as we follow the same elemental approach to the input representations of stimuli. Our analysis of latent inhibition and perceptual learning can proceed as before, though new possibilities are opened by this configural extension of the theory. In particular, the analysis of preparations involving compound stimuli can take some unexpected twists. That is for the future; for the present our computationally correct rat is a clear winner when it comes to eating berries and avoiding snakes.

Acknowledgements

I am grateful to Rob Honey and Johan Bolhuis for their comments on an earlier version of this chapter.

References

Bateson, P. P. G. & Chantrey, D. F. (1972). Retardation of discrimination learning in monkeys and chicks previously exposed to both stimuli. *Nature*, 237, 173–174.

Bouton, M. E. (1990) Context and retrieval in extinction and in other examples of interference in simple associative learning. In *Current Topics in Animal Learning: Brain, Emotion, and Cognition*, ed. L. W. Dachowski & C. F. Flaherty, pp. 25–53. Hillsdale, NJ: Lawrence Erlbaum.

Bouton, M. E. & King, D. A. (1983). Contextual control of the extinction of conditioned fear: test for the associative value of the context. *Journal of Experimental Psychology: Animal Behaviour Processes*, 9, 248–265.

Bouton, M. E. & Peck, C. A. (1989). Context effects on conditioning, extinction, and reinstatement in an appetitive conditioning preparation. *Animal Learning & Behaviour*, 17, 188–198.

Estes, W. K. (1959). Towards a statistical theory of learning. *Psychological Review*, 57, 94–107.

Fanselow, M. S. (1990). Factors governing one-trial contextual conditioning. *Animal Learning & Behavior*, 18, 264–270.

Gibson, E. J. (1969). *Principles of Perceptual Learning and Development*. New York: Appleton-Century-Crofts.

Hall, G. (1980). Exposure learning in animals. *Psychological Bulletin*, 88, 535–550.

Hall, G. & Honey, R. C. (1989a). Perceptual and associative learning. In *Contemporary Learning Theories*, ed. S. B. Klein & R. R. Mowrer, pp. 117–147. Hillsdale, NJ: Lawrence Erlbaum.

Hall, G. & Honey, R. C. (1989b). Contextual effects in conditioning, latent inhibition, and habituation: Associative and retrieval functions of contextual cues. *Journal of Experimental Psychology: Animal Behaviour Processes*, 15, 232–241.

Hebb, D. O. (1949). *The Organisation of Behaviour*. New York: John Wiley & Sons.

Honey, R. C. & Hall, G. (1989). Enhanced discriminability and reduced associability following flavour preexposure. *Learning & Motivation*, 20, 262–277

James, W. (1890). *Principles of Psychology*. New York: Holt.

Konorski, J. (1948). *Conditioned Reflexes and Neuron Organisation*. Cambridge: Cambridge University Press.

Konorski, J. (1967). *Integrative Activity of the Brain*. Chicago: University of Chicago Press.

Lawrence, D. H. (1952). The transfer of a discrimination along a continuum. *Journal of Comparative and Physiological Psychology*, 45, 511–516.

Levy, W. B. & Desmond, N. L. (1985). The rules of elemental synaptic plasticity. In *Synaptic Modification, Neuron Selectivity, and Nervous System Organisation*, ed. W. B. Levy, J. A. Anderson & S. Lehmkuhle, pp. 105–122. Hillsdale, NJ: Lawrence Erlbaum.

Lovibond, P. F., Preston, G. C. & Mackintosh, N. J. (1984). Context specificity of conditioning, extinction and latent inhibition. *Journal of Experimental Psychology: Animal Behaviour Processes*, 10, 360–375.

Lubow, R. E. & Moore, A. U. (1959). Latent inhibition: the effect of nonreinforced preexposure to the conditioned stimulus. *Journal of Comparative and Physiological Psychology*, 52, 415–419.

Mackintosh, N. J. (1975). A theory of attention: variations in the associability of stimuli with reinforcement. *Psychological Review*, 82, 276–298.

Mackintosh, N. J., Kaye, H. & Bennett, C. H. (1991). Perceptual learning in flavour aversion conditioning. *Quarterly Journal of Experimental Psychology*, 43B, 297–322.

McClelland, J. L. & Rumelhart, D. E. (1985). Distributed memory and the representation of general and specific information. *Journal of Experimental Psychology: General*, 114, 159–188.

McLaren, I. P. L. & Dickinson, A. (1990). The conditioning connection. *Philosophical Transactions of the Royal Society of London. B*, 329, 179–186.

McLaren, I. P. L., Kaye, H. & Mackintosh, N. J. (1989). An associative theory of the representation of stimuli: applications to perceptual learning and latent inhibition. In *Parallel Distributed Processing – Implications for Psychology and Neurobiology*, ed. R. G. M. Morris, pp. 102–130. Oxford: Oxford University Press.

Neimark, E. D. & Estes, W. K. (1967). *Stimulus Sampling Theory*. San Francisco: Holden-Day.

Pearce, J. M. (1987). A model for stimulus generalisation in Pavlovian conditioning. *Psychological Review*, 94, 61–73.

Pearce, J. M. & Hall, G. (1980). A model for Pavlovian learning: Variations in the effectiveness of conditioned but not of unconditioned stimuli. *Psychological Review*, 87, 532–552.

Rescorla, R. A. (1973). Evidence for the 'unique stimulus' account of configural conditioning. *Journal of Comparative & Physiological Psychology*, 85, 331–338.

Rescorla, R. A. & Wagner, A. R. (1972). A theory of Pavlovian conditioning: Variations in the effectiveness of reinforcement and nonreinforcement. In *Classical Conditioning II: Current Research and Theory*, ed. A. H. Black & W. F. Prokasy, pp. 64–99. New York: Appleton-Century-Crofts.

Rumelhart, D. E., Hinton, G. E. & Williams, R. J. (1986). Learning internal representations by error propagation. In *Parallel Distributed Processing*, Vol. I, ed. D. E. Rumelhart & J. L. McClelland, pp. 318–362. Cambridge. Mass.: Bradford Books.

Spear, N. E. (1973). Retrieval of memory in animals. *Psychological Review*, 80, 163–175

Stone, G. O. (1986). An analysis of the delta rule and the learning of statistical associations. In *Parallel Distributed Processing*, Vol. I., ed. D. E. Rumelhart & J. L. McClelland, pp. 444–459. Cambridge. Mass.: Bradford Books

Sutton, R. S. & Barto, A. G. (1981). Toward a modern theory of adaptive networks: expectation and prediction. *Psychological Review*, 88, 135–170.

Sutton, R. S. & Barto, A. G. (1989). Time-derivative models of Pavlovian reinforcement in learning and computational neuroscience. In *Learning and Computational Neuroscience*, ed. M. Gabriel & J. W. Moore, pp. 36–49. Cambridge, Mass.: MIT Press.

Thompson, R. F. (1965). The neural basis of stimulus generalisation. In *Stimulus Generalisation*, ed. D. I. Mostofsky, pp. 154–178. Stanford: Stanford University Press.

Wagner, A. R. (1978). Expectancies and the priming of STM. In *Cognitive Processes in Animal Behaviour*, ed. S. H. Hulse, H. Fowler & W. K. Honig, pp. 177–210. Hillsdale, NJ: Lawrence Erlbaum .

Wagner, A. R. (1981). SOP: A model of automatic memory processing in animal behaviour. In *Information Processing in Animals: Memory Mechanisms*, ed. W. E. Spear & R. R. Miller, pp. 5–47. Hillsdale, NJ: Lawrence Erlbaum.

Widrow, G. & Hoff, M. E. (1960). Adaptive switching circuits. *Institute of Radio Engineers, Western Electronic Show and Convention*, Convention Record, Part 4, 96–104.

Author index

Ader, R. 224
Adkins-Regan, E. 59
Adret, P. 104, 106, 107, 108, 135
Agnew, R. L. 131
Akutagawa, E. 58, 59
Albert, P. 293
Alberts, J. R. 225, 227, 228, 230, 292,
  296, 297, 329, 334
Alcock, J. 11
Aldridge, J. W. 163
Allan, R. W. 335
Allen, E. E. 231
Allesandro, D. 330
Al-Majdalawi, A. 171
Altar, C. A. 172, 173
Altman, J. 293
Alvarez-Buylla, A. 27, 58, 59, 62
Ambalavanar, R. 23
Amsel, A. 228, 327, 346
Anderson, V. 219
Andersson, M. 53
Andrew, R. J. 30, 35, 196, 197, 206, 250
Anokhin, K. V. 20
Arai, O. 62
Aram, D. M. 301
Arberg, A. A. 339
Aristotle 3, 4, 6, 8, 9
Arnold, A. P. 31, 53, 56, 58, 59, 60, 62,
  64, 70, 203
Aslin, R. N. 277
Atkinson, J. 273, 275
Avery, M. I. 99, 364

Baer, D. M. 327, 340
Baerends, G. P. xiii, 7, 10, 36, 190, 198,
  249, 341
Baerends-van Roon, J. M. 249
Bagnoli, P. 294
Baker, A. G. 290
Baker, J. R. 8, 9

Baker, M. C. 54
Ball, G. F. 53
Balsam, P. D. 86, 291, 327, 329, 330, 331,
  332, 333, 335, 339, 341, 342, 344, 358
Baptista, L. F. 55, 71, 101, 108, 126, 127
Barclay, D. 306
Barker, L. M. 360
Barlow, G. W. 307
Barlow, H. B. 372
Barnett, A. M. 295
Baron-Cohen, S. 311
Barrette, C. 333
Barry, M. 218
Bartashunas, C. 131, 332
Barto, A. G. 380
Bateson, P. P. G. 21, 27, 29, 35, 36, 82,
  87, 88, 89, 91, 92, 93, 118, 120, 121,
  122, 123, 124, 133, 136, 140, 147, 148,
  255, 371, 372, 387
Bauer, J. A. 281
Bauer, J. H. 216
Beach, F. A. 147, 148
Beauchamp, G. 291
Beaver, P. W. 130
Beck, M. 227, 229, 233
Becker, J. T. 28
Becker, L. E. 270
Beer, C. 12
Behse, J. H. 229
Bekoff, A. 150, 151, 152, 154, 155, 156
Benacerraf, B. R. 301
Bennett, C. H. 387
Berger, J. 317
Bergman, T. 302
Bergmann, H. 188, 189, 191
Berridge, K. C. 10, 36, 147, 163, 168, 169,
  170, 171, 184, 201, 225, 243, 246, 291,
  327, 340
Berryman, J. 124
Best, M. R. 360

Subject index

413